Physics Research and Technology

nova
science publishers
New York

Physics Research and Technology

An Introduction to Charge Carriers
Jai Singh, PhD (Editor)
2022. ISBN: 978-1-68507-456-2 (Hardcover)
2022. ISBN: 978-1-68507-531-6 (eBook)

Handbook of Research on Heat Transfer
Suvanjan Bhattacharyya, PhD
Varun Goel, PhD
(Editors)
2022. ISBN: 978-1-68507-459-3 (Hardcover)
2021. ISBN: 978-1-68507-542-2 (eBook)

Solar PV System: Design, Installation, Operation and Maintenance
L. Ashok Kumar, PhD
K. Mohana Sundaram
2021. ISBN: 978-1-68507-341-1 (Hardcover)
2021. ISBN: 978-1-68507-475-3 (eBook)

An Essential Guide to Thermal Conductivity
S. M. Sohel Murshed, PhD (Editor)
2021. ISBN: 978-1-68507-196-7 (Hardcover)
2021. ISBN: 978-1-68507-289-6978 (eBook)

Neutrino Mass: Past, Present, Future
Nicolai Popov, PhD
Igor Strakovsky, PhD
2021. ISBN: 978-1-68507-130-1 (Hardcover)
2021. ISBN: 978-1-68507-206-3 (eBook)

More information about this series can be found at
https://novapublishers.com/product-category/series/physics-research-and-technology/

Christian Bosch
Editor

A Closer Look at Plasmonics

nova
science publishers
New York

Library of Congress Cataloging-in-Publication Data

ISBN: 978-1-68507-709-9

Published by Nova Science Publishers, Inc. † New York

Contents

Preface

Plasmonics refers to the generation, detection, and manipulation of signals at optical frequencies along metal-dialectric interfaces at the nanometer scale. The four chapters of this book analyze plasmonics and its applications. Chapter One discusses the plasmonic sensor and its use in the detection of brain enzymes. Chapter Two examines the optical constants of plasmonic nanostructures and demonstrates that optical absorption can be used to estimate dielectric constants of noble metal nanostructures. Chapter Three theoretically investigates the magnetoplasmonics of noble metal nanostructures and shows that plasmonic nanostructures have increased magnetic response in visible range at low magnetic fields. Lastly, Chapter Four studies some aspects of classical and quantum optical effects in surface plasmonics.

Chapter 1 - The plasmonic sensor is a versatile technique to recognize biomolecular and toxic chemicals. The plasmonic sensor can display the interaction of biomolecules with biomarker and linker. The biomolecules including DNA, RNA, antibodies, enzyme, and blood sugar can interact with biomarkers, bio-linker, and nanostructures such as gold and silver. The plasmonic sensor can detect the biomolecules and it is an accurate and reliable technique to recognize them. The plasmonic sensor is capable to investigate the interaction of biological cells with antibodies, enzymes, electrolytes, and hormones in the blood and other body fluids, and can detect the IgG (the main type of antibody that can be found in blood and extracellular fluid). Hence, the plasmonic sensors can sense and diagnose the hepatitis antigen, diabetes, leukemia, and cancer. In this study, a brief history of the plasmonic sensor, the analytical method to analyze the plasmonic signals, and the application of plasmonic sensor for displaying DNA, RNA, antibodies, sugar hepatitis, diabetes, leukemia, and cancer have been presented. Recently, the plasmonic sensor is a powerful and accurate method to investigate the brain activity and the neuroplasmonic is an interest intent issue. Hence, the plasmonic senor and the detection of brain enzymes have been considered in this chapter.

Chapter 2 - This research is made to examine the optical constants of plasmonic nanostructures. The noble metals were thoroughly studied for their plasmonic properties in UV and visible range. It is found that optical response of nanostructures is mainly governed by dielectric permittivity of the metal and resonant wavelength depends on size, shape, composition and external matrix. Further, it is observed that the localised surface plasmon resonance (LSPR) dominates optical spectra of metal nanoparticles. During research the optical constants of noble metals (Au and Ag) were investigated theoretically, using Mie's formalism, in quasi static regime. Theoretically, it was evident that resonant wavelength changes with size of nanoparticle and external matrix. It is observed that LSPR is very sensitive tool for analyzing the environment or the matrix. In addition, the Optical absorption studies were performed on Ag and Au nanoparticles, of various sizes and keeping in different environments. It has been demonstrated that the optical absorption can be used to estimate dielectric constants of noble metal nanostructures.

Chapter 3 - Magnetoplasmonics of noble metal nanostructures using Mie's formalism have been investigated theoretically. Optical properties of noble metal structures such as resonant wavelength and extinction coefficients are investigated theoretically, in the presence of static external magnetic field. In this chapter, it is shown that the plasmonic nanostructures have increased magnetic response in visible range at low magnetic fields. Resonant wavelength for these nanostructures is demonstrated to be shifted towards lower wavelength (blue shift) in the presence of magnetic field(B). At low values of applied magnetic field, the spectral shift of resonant wavelength is towards the visible region. This critical value of magnetic field in case of Ag nanostructures is realized as 2 Tesla.

Chapter 4 - In metal optics, including metallic or plasmonic micro/nano structure electromagnetics, electromagnetic or optical surface plasmonic waves (e.g., planar-interface supported surface plasmon polaritons and nanaosphere-sustained localized surface plasmons) have captured a lot of interest and attention of a number of researchers in various research fields, including both fundamental physical principles and application-oriented areas (e.g., materials science and biotechnology). In this chapter, the authors will study some aspects of classical and quantum optical effects in surface plasmonics. For the first, the authors will consider the free-electron Drude model for describing metal optical properties and point out that there are two mysteries in the Drude model (i.e., though the local field correction and plasma oscillation effect have not been taken into account in the Drude model, yet the dielectric constants for metals obtained in such a model still agree well with the experimental observations). The authors resolve this mystery that seems to be puzzling, i.e., it can be found that the local field correction and the plasma oscillation effect cancel each other out exactly, no matter whether the free electron gas is in one-, two- or three-dimensional space. Since the negative permittivity is required in surface plasmonics, the authors suggest some alternative mechanisms for achieving tunable negative permittivity in three- and four-level quantum-coherent atomic vapor (e.g., alkali metal atomic vapor). Then some scenarios of tunable prism coupler (based on such mechanisms relevant to quantum coherence) for exciting surface plasmon polaritons will be reviewed. The field quantization of surface plasmon modes is also considered and some new quantum field theoretical effects (including quantum-vacuum effects) will be suggested. These quantum effects include atom trapping and nanosphere quantum levitation. For example, because of the interaction between nanaosphere-sustained localized surface plasmons and quantized microcavity modes, the metallic nanosphere trapping (or quantum levitation) caused by the nanosphere-sustained localized surface plasmonic modes in a quantum microcavity will be suggested. The concept of metal-surface plasmonic wave modes can be extended to optical parity-time symmetric interface, where the electric permittivities of two electromagnetic media are complex conjugate. Based on this, two new parity time symmetries (i.e., a generalized optical parity-time symmetry and a supersymmetric parity-time invariance) will also be suggested. The authors expect that some of the considerations in this chapter (such as tunable quantum-coherent atomic vapor, quantum surface plasmonic waves, and generalized parity-time symmetries) would open up readers' train of thought and allow them to propose more surface plasmonic effects, providing more theoretical principles and new working mechanisms for future surface plasmonics experiments as well as electromagnetic or photonic plasmonic device design.

Chapter 1

Plasmonic Sensor for Detection of Biomolecular and Medical Diagnostics

A. R. Sadrolhosseini, S. Mehri Hamidi* **and S. Akbari**
Magneto-plasmonic Lab, Laser and Plasma Research Institute,
Shahid Beheshti University, Tehran, Iran

Abstract

The plasmonic sensor is a versatile technique to recognize biomolecular and toxic chemicals. The plasmonic sensor can display the interaction of biomolecules with biomarker and linker. The biomolecules including DNA, RNA, antibodies, enzyme, and blood sugar can interact with biomarkers, bio-linker, and nanostructures such as gold and silver. The plasmonic sensor can detect the biomolecules and it is an accurate and reliable technique to recognize them. The plasmonic sensor is capable to investigate the interaction of biological cells with antibodies, enzymes, electrolytes, and hormones in the blood and other body fluids, and can detect the IgG (the main type of antibody that can be found in blood and extracellular fluid). Hence, the plasmonic sensors can sense and diagnose the hepatitis antigen, diabetes, leukemia, and cancer. In this study, a brief history of the plasmonic sensor, the analytical method to analyze the plasmonic signals, and the application of plasmonic sensor for displaying DNA, RNA, antibodies, sugar hepatitis, diabetes, leukemia, and cancer have been presented. Recently, the plasmonic sensor is a powerful and accurate method to investigate the brain activity and the neuroplasmonic is an interest intent issue. Hence, the plasmonic senor and the detection of brain enzymes have been considered in this chapter.

* Corresponding Author's Email: m_hamidi@sbu.ac.ir.

In: A Closer Look at Plasmonics
Editor: Christian Bosch
ISBN: 978-1-68507-709-9

Keywords: plasmonic, plasmonic nanoparticles, plasmonic sensor, biosensor, surface plasmon resonances sensor, neuroplasmonic

Introduction

The research and development of plasmonic phenomena have been started at the beginning of the last century. The scientists such as Zenneck, Mie, and Sommerfeld considered the theoretical and modeling the phenomena at nano size and they improved and developed the study of nanoscale and plasmonic research areas. The excitation of surface plasmon waves was discovered in 1902 by Wood and it was the starting point of surface plasmon resonance (SPR) sensor (Figure 1(a)). The SPR phenomenon was observed by the p-polarized light beam scattered from a metal-backed diffraction grating (Wood, 1902). Otto (1968) reported the attenuated total reflection (ATR) as a method for excitation of SP wave in 1968 (Figure 1(b)).

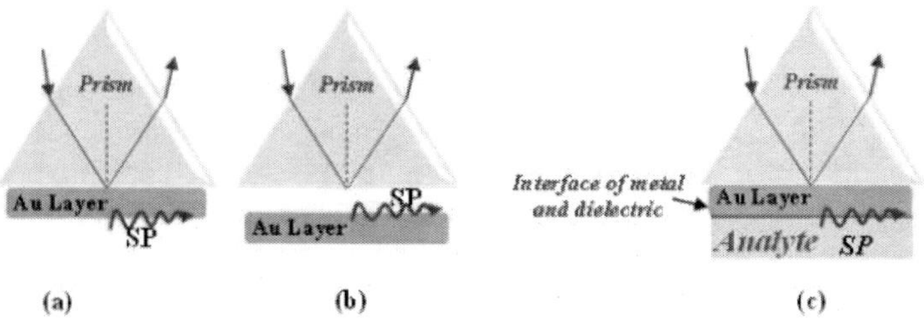

Figure 1. The SPR (a) Kretschmann and (b) Otto configuration, and (c) SPR phenomena in the presence of the analyte.

The concept and the theory of plasmon, plasma frequency, and plasma are significant for an explanation of light interaction with materials in the nanoscale or metal nanoparticles. For example, plasma frequency can be derived from the Drude model to evaluate the transparency of matter and the dielectric function can be achieved in the first approximation. Hence, the behaver of light or electromagnetism wave can be deduced in the materials such as metal and dielectric.

The oscillation of charge density manages the plasma fluctuation which depends on the plasma frequency and the quantum of plasma oscillation is plasmons. The surface plasmon resonance (SPR) is an optical phenomenon that occurs at the interface of a metal and dielectric medium (Figure 1(c)) and it is related to the fluctuation of charge density. The surface plasmon resonance can be investigated using Fresnel reflection theory (Sadrolhosseini, Moksin, Nang, et al., 2011). The SPR and plasmonic properties of metal nanoparticles were observed in the range of visible and the near- infrared wavelength of electromagnetism waves.

A localized surface plasmon resonance (LSPR) appears at the surface of metallic nanoparticles and the surrounding medium (Figure 2) and LSPR has a short length (order) (He, Musick, Nicewarner, et al., 2000). The frequency and the intensity of the plasmon wave depend on the particle size and morphology of the nanoparticles (Jamkhande, Ghule, Bamer, et al., 2019). The Mie theory (Niskanen, Forsberg, Zakrisson, et al., 2019), the Gans theory (Alqahtani, Kavakli-Thorne, and Kumar, 2021), the discrete-dipole approximation method (Yin, Lin, Ruan, et al., 2021) are the models that assume the simulation of the SPR extinction, LSPR, and plasmonic properties of the nanostructure. The finite-difference time-domain method (FDTD) (Amoosoltani, Zarifkar, and Farmani, 2019) and the finite element method (Song, Ooi, and Natarajan, 2018) are the candidates and the most popular methods to solve the differential equation and analyze the SPR, LSPR, and SPRi output. Moreover, Fresnel's formula can be used to explain the SPR signals for sensor and SPR image sensor based on prism configuration.

Plasmonic and surface plasmon resonance have more applications for monitoring, recognize, and detecting of biomolecules (Pacholski, Rosencrantz, Rosencrantz, et al. 2020) and chemical molecules. Plasmonic sensors have the potential to monitor hepatitis (Santos, Alves, Pinto, et al., 2021), diabetes (Pathak, and Viphavakit, 2021), leukemia (Fathi, Rahbarghazi, Movassaghpour, et al., 2019), and cancer (Song, Lee, Kang, et al. 2020), and the plasmonic sensors based on particular a sensing layer were used to detect the enzyme (Miyazaki, Shimizu, Mejía-Salazar, et al., 2017), DNA (Lednický, and Bonyár, 2020), RNA (Xue, Liang, Li, et al., 2019), toxic chemical (Kabashin, and Nikitin, 1998), sugar (Scarano, Pascale, Palladino, 2018), and environmental pollution.

Plasmonic sensors are sensitive, reliable, and flexible sensors for detection and monitoring the chemical and biological molecules in medical applications. Therefore, they were used to detect and diagnose the biomolecules. In this article, plasmonic setup and surface plasmon resonance sensors, the analytical method for analyzing the output of the sensors, and the application of plasmonic sensors for the detection of DNA, RNA, antibodies, and sugar were considered. Recently, the plasmonic sensor was used to investigate the neurological cell and enzyme. In addition, the plasmonic sensor was utilized to monitor the causes of diseases including hepatitis, diabetes, leukemia, and cancer. Hence, in this study, the application of plasmonic sensors in neuroplasmonic (Sohrabi and Hamidi 2016) (Figure 3) and medical diagnostic were presented.

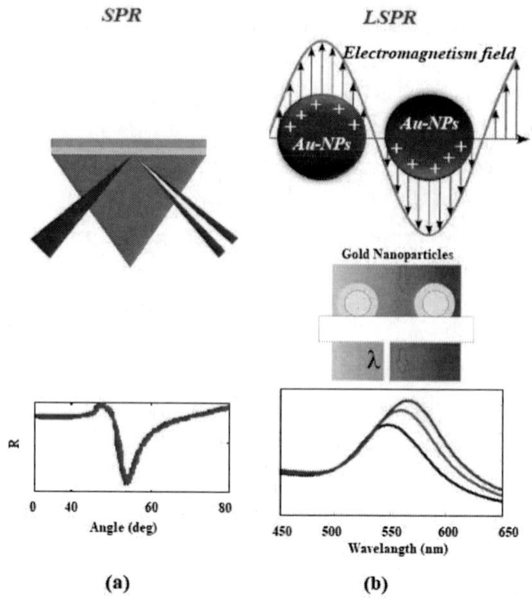

Figure 2. The comparison between SPR and LSPR.

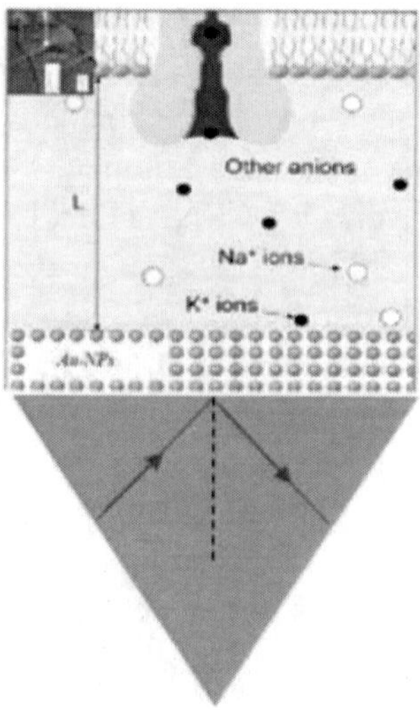

Figure 3. Monitoring of neurons activity using the SPR sensor.

Plasmonic Modelling

Surface Plasmon Resonance Coupling Method

The SPR phenomena is observed with photon-excited charge oscillation at small metals (gold or silver) thin films or particles and it has a wide ranges application in material science, biology, pharmacology, and nanoelectronics. Plasmonic was utilized to enhance the Raman scattering (SERS) (Homola, and Jakub, 2006), fluorescence nanoscopy (Requejo-Isidro, 2013), scanning optical microscopy (plasmonic NSOM) (Yin, Shi, Yang, et al., 2020), and SPR sensor (Omar, Fen, Ramli, et al. 2021; Sadrolhosseini, Shafie, Abdul Rashid, et al., 2020) (Figure 4). Hence, the investigation of coupling methods, resonance condition, the relationship between intensity and resonance angle are the significant parameter to analyze the SPR signal and the surface plasmon resonance image (SPRi) patterns. Prism coupling, grating coupling, and waveguide coupling are the regular and main coupling methods. Kretschmann configuration ATR coupling or prism coupling has the widest application in biosensors and SPR image sensors (Homola, Vaisocherová, Dostálek, et al., 2005).

Prism coupling is the Kretschmann configuration (Figure 5(a)). There is a high index prism and the metal layer such as gold or silver film were coated on the surface of the prism. The monochromatic light beam such as a red laser passes through the high index prism. The light beam is made incident on the metal layer and some part of the light was reflected and part of the light was an inhomogeneous wave propagates in the gold or silver layer. The penetrated part is the evanescent wave and it decays exponentially in the gold layer or silver layer and the penetration deep depends on the metal layer and the energy of the light beam. When the thickness of the gold or silver layer is less than 100 nm, the evanescent wave can penetrate through the gold or silver layer (Singh, Paswan, and Raghuwanshi, 2021). In this case, if the propagation constant of evanescent wave and surface plasmon match together, the evanescent wave can couple with the plasmon wave outer boundary of the gold or silver layer and the resonance occurs (Rothenhäuslar, and Knoll, 1988). The n_p, ε_m, ε_d, and θ_R are the refractive index of the prism, the dielectric constant of gold or silver layer, dielectric constant of the dielectric layer, and the angle of resonance, respectively. The resonance condition is as follows:

$$n_p \sin\theta_R = \mathrm{Re}\left(k\sqrt{\frac{\varepsilon_d \varepsilon_m}{\varepsilon_d + \varepsilon_m}}\right) = n_{sp,ef} \quad (1)$$

The ε_{mr} and ε_{mi} are the real and imaginary parts of the gold or silver layer. Hence, the effective index and the attenuation constant (b) is as follows (Homola, and Jakub, 2006):

$$n_{ef} = \frac{c}{\omega}\mathrm{Re}\left(k\sqrt{\frac{\varepsilon_d \varepsilon_m}{\varepsilon_d + \varepsilon_m}}\right) \quad (2)$$

$$b = 0.087\,\mathrm{Im}\left(k\sqrt{\frac{\varepsilon_d \varepsilon_m}{\varepsilon_d + \varepsilon_m}}\right) \quad (3)$$

Where c and ω are the speed and frequency of light beam and the attenuation b is in dB cm^{-1}.

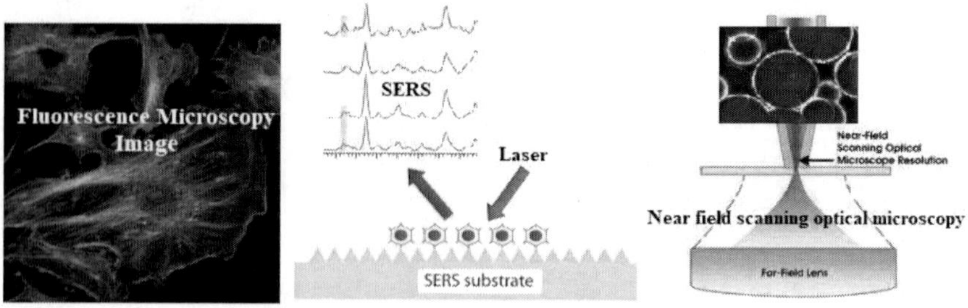

Figure 4. Demonstration of fluorescence microscopy image, surface enhanced Raman scattering, and near field nanoscope.

As mentioned above, the light beam pass through the prism, I_i, I_r are the amplitude of the initial light beam and the amplitude of the reflected light beam after interaction with metal layer, respectively. As Fresnel's theory the amplitude of the reflected light beam is as follows [30]:

$$I_r = I_i \times e^{i\phi} r_{pmd} \quad (4)$$

Where r_{pmd} is the reflection coefficient of light beam from metal-dielectric interface in the prism and it can be expressed as follows (Yariv, and Yeh, 2003):

$$r_{pmd} = \frac{r_{pm} + r_{md} e^{(2ik_{mx}q)}}{r_{pm} + r_{pm} r_{md} e^{(2ik_{mx}q)}} \tag{5}$$

Where r_{pm}, r_{md}, and k_{mx} are amplitude of reflection coefficient from prism-metal, the amplitude of reflection coefficient from metal-dielectric medium which is place after gold or silver layer, and the propagation constant, respectively. The reflection coefficient for TM mode of a light beam and the propagation constant are as follows:

$$r_{pm} = \frac{\varepsilon_m k_{px} - \varepsilon_p k_{mx}}{\varepsilon_m k_{px} + \varepsilon_p k_{mx}}, \tag{6}$$

$$r_{md} = \frac{\varepsilon_d k_{mx} - \varepsilon_m k_{dx}}{\varepsilon_d k_{mx} + \varepsilon_m k_{dx}}, \tag{7}$$

$$k_{mx} = \sqrt{\left(\frac{2\pi}{\lambda}\right)^2 \varepsilon_m - \left(\frac{2\pi}{\lambda}\right) n_p \sin\theta} \tag{8}$$

$$k_{px} = \sqrt{\left(\frac{2\pi}{\lambda}\right)^2 \varepsilon_p - \left(\frac{2\pi}{\lambda}\right) n_p \sin\theta}\ , \tag{9}$$

$$k_{dx} = \sqrt{\left(\frac{2\pi}{\lambda}\right)^2 \varepsilon_d - \left(\frac{2\pi}{\lambda}\right) n_p \sin\theta}\ . \tag{10}$$

Consequently, the reflectance (R) is the square of the reflection coefficient as follows:

$$R = \left| r_{pmd} \right|^2 . \tag{11}$$

The Eq. (11) can be used to simulate the SPR signal. The amplitude of the reflection coefficient (R) is a function of the angle, refractive index. Hence, the refractive index and the resonance angle shift can be calculated from a comparison of the experimental value of the reflectance with the theoretical of the reflectance at the minimum root square.

$$T = \sum \left[R_{th}(\theta, n) - R_{Ex}(\theta, n) \right]. \tag{12}$$

Grating coupling is the other method to excitation to surface plasmon waves (Figure 5(b)). It is based on the diffraction of the light beam on the surface of diffraction grating with high number of groove lines. In this approach, the dielectric layer should be coated on the surface of metallic grating with the grating period p. The light beam passes through the dielectric medium and it interacts with metallic grating. If the propagation constant of diffracted light that is on the surface of the grating plane and perpendicular to the grooves of the grating, becomes equal with the propagating constant of surface plasmon, the resonance occurs and the resonance condition is (Zalevsky, and Abdulhalim, 2010):

$$n_d \sin\theta + \xi \frac{\lambda}{p} = \pm \mathrm{R}\left(k \sqrt{\frac{\varepsilon_d \varepsilon_m}{\varepsilon_d + \varepsilon_m}} \right) \tag{13}$$

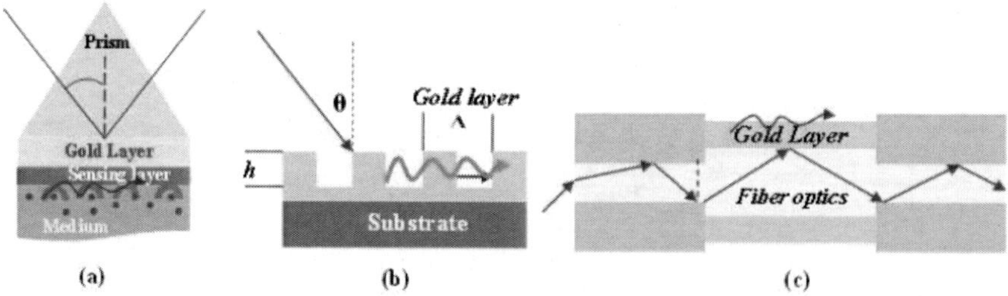

Figure 5. (a) Prism, (b) Grating, and (c) Waveguide coupling.

Where ξ is the order of diffraction. The effective index of the inhomogeneous light wave that is a function of wavelength, is a proportional ratio of ξ / p.

The waveguide coupling is the famous method for excitation of surface plasmon in the waveguide and fiber optics sensor (Figure 5(c)). A mode of fiber optics propagates along with the multimode fiber optics and when it interacts with the gold or silver layer, it can penetrate through the gold or silver layer that is coated on the surface of fiber optics and it can couple with the surface plasmon like the prism coupling. This coupling method is very significant to excite the surface plasmon wave with the broadband light source.

In prism, grating, and waveguide coupling methods, the attenuation of the light beam is the considerable parameter to design the sensor and the attenuation is the proportion with the length (ℓ) of surface plasmon decay as follows (Zalevsky, and Abdulhalim, 2010):

$$\ell = \left[\frac{\varepsilon_{mi}}{(\varepsilon_{mr})^2} \times \frac{\omega}{2c} \times \left(\frac{\varepsilon_{mr}\varepsilon_d}{\varepsilon_{mr} + \varepsilon_d} \right)^{(3/2)} \right]^{-1} \quad (14)$$

Surface Plasmon Resonance Analysis for Multilayer

The SPR sensor signal should be analyzed to find the variation of SPR sensor output such as angle, light intensity, or wavelength. The SPR sensors contain the optical element, metal layer, and sensing layer. The optical properties of the metal layer and sensing layer impress the SPR signals. Therefore, the matrix method based on Fresnel's theory is a significant and capable method to analyze the SPR signals. The reflected incident beam at the interface of the medium should be considered to find the matrix of multilayers. The intensity of the light beam decreases at the coupling angle. So, the condition of resonance can be obtained from considering the energy and momentum conservation laws (Schasfoort, and Tudos, 2008) as follows:

$$k_{glass} \sin\theta_i = \frac{\omega}{c} n_{glass} \sin\theta_i = k_{plasmon} \quad (15)$$

where θ_i is the incident angle at the resonance condition.

The p-polarized light should be useed to generate the surface plasmon wave (SPW) at the interface of medium. Figure 6 shows the electrical (*E*) and magnatic (*B*) fields at the boundary of the medium. The relationship between *E* and *B* are as follows (Kittle 1996; Peyghambarian, Koch, Mysyrowicz 1993):

$$B_a = n_0\sqrt{\varepsilon_0\mu_0}(E_0 + E_{r1}) = n_1\sqrt{\varepsilon_0\mu_0}(E_{t1} + E_{i1}) \quad (16)$$

$$B_b = n_1\sqrt{\varepsilon_0\mu_0}(E_{i2} + E_{r2}) = n_2\sqrt{\varepsilon_0\mu_0}E_{t1} \tag{17}$$

$$E_a = (E_0 - E_{r1})\cos\theta_0 = (E_{t1} - E_{i1})\cos\theta_{t1} \tag{18}$$

$$E_b = (E_{i2} - E_{r2})\cos\theta_{t1} = E_{t2}os\theta_{t2} \tag{19}$$

The entrancing field (E_a, B_a) and exit field (E_b , B_b) are related together and the relationship between them should be derived from phase change due to the light beam passing through the different layers and $E_{i1} = E_{r2}e^{-i\delta}$, $E_{i2} = E_{t1}e^{-i\delta}$, $B_{i1} = B_{r2}e^{-i\delta}$, $B_{i2} = B_{t1}e^{-i\delta}$, $B = \frac{n}{c}E$ and $c = \frac{1}{\sqrt{\varepsilon_0\mu_0}}$ (Kittle 1996; Peyghambarian, Koch, Mysyrowicz 1993), as follows (Sharma 2006):

$$E_a = (e^{-i\delta}\cos\delta)E_b - (i\frac{e^{-i\delta}}{\tau_1}\sin\delta)B_b \tag{20}$$

$$B_a = (-\tau_1 i e^{-i\delta}\sin\delta)E_b + (e^{-i\delta}\cos\delta)B_b \tag{21}$$

as τ_n is

$$\tau_N = \frac{n_N}{\cos\theta_{t_N}}\sqrt{\varepsilon_0\mu_0}\,. \tag{22}$$

Therefore, when the light passes through the dielectric thin metal layer, the phase of light changes and the matrix of layers achieve from the coefficient of equations (21) and (22) as follows:

$$\begin{bmatrix} E_a \\ B_a \end{bmatrix} = M\begin{bmatrix} E_b \\ B_b \end{bmatrix} \tag{23}$$

where *M* is a matrix consist of

$$M = \begin{bmatrix} \cos\delta & -i\dfrac{\sin\delta}{\tau_1} \\ -i\tau_1\sin\delta & \cos\delta \end{bmatrix},\ \delta = \frac{2\pi}{\lambda}dn_1\cos\theta_{t_1} \tag{24}$$

where δ and d are the phase shift and the thickness of the layer. For a system with N-layer, the matrix system can be written as follows (Peyghambarian, Koch, Mysyrowicz 1993; Sharma 2006; Beketov, Shirshov, Shynkarenko, et al., 1998):

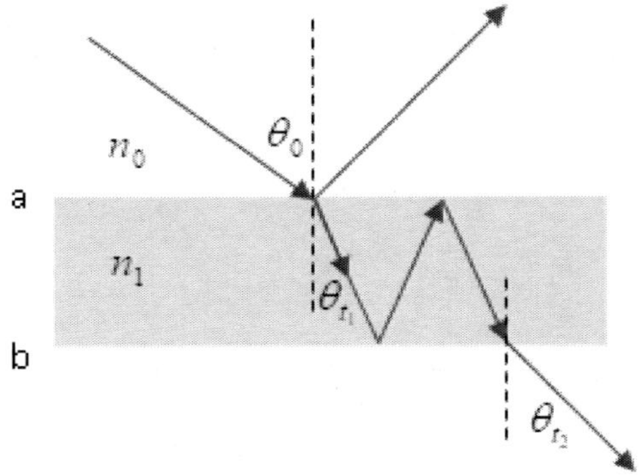

Figure 6. The raytracing of light beam in the thin layer.

$$M = M_1 M_2 M_3 .... M_N \qquad \begin{bmatrix} E_a \\ B_a \end{bmatrix} = M_1 M_2 M_3 .....M_N \begin{bmatrix} E_b \\ B_b \end{bmatrix}. \tag{25}$$

and the of each lyer (Sadrolhosseini, Moksin, Yunus, et al., 2011) is

$$M = \begin{bmatrix} m_{11} & m_{12} \\ m_{21} & m_{22} \end{bmatrix} \tag{26}$$

If one layer is deposited on the prism, the reflection and transmission coefficients are defined respectively as

$$r = \frac{E_{r_1}}{E_0}, \; \Gamma = \frac{E_{t_2}}{E_0} \; . \tag{27}$$

By simplifying the mentioned formula, the reflection coefficient is as follows:

$$r = \frac{m_{21} + m_{22}\tau_2 - m_{11}\tau_0 - m_{12}\tau_2\tau_0}{m_{21} + m_{22}\tau_2 + m_{11}\tau_0 + m_{12}\tau_2\tau_0}, \tag{28}$$

and the reflectivity or reflectance is

$$R = rr^* \tag{29}$$

If there are three media of prism, n_0, metal layer, n_1 and air ($n_2 = 1$); and the thickness of the metal layer, t , is less than the wavelength ($\lambda = 632.8nm$) then the characteristic matrix is

$$M = \begin{pmatrix} 1 & -i(2\pi/\lambda)d\cos^2\theta_{t1} \\ -i(2\pi/\lambda)dn_1^2 & 1 \end{pmatrix} \tag{30}$$

According to Eqs. (26, 28, and 30) the reflection coefficient is

$$r = \frac{\overbrace{\left(\tau_0 n_2\sqrt{\varepsilon_0\mu_0} - \cos\theta_{t_2}\right)}^{A} - i\,\alpha\overbrace{\left(n_1^2\cos\theta_{t_2}\tau_0 - \cos^2\theta_{t_1}\sqrt{\varepsilon_0\mu_0}\right)}^{B}}{\underbrace{\left(\tau_0 n_2\sqrt{\varepsilon_0\mu_0} + \cos\theta_{t_2}\right)}_{C} - i\,\alpha\underbrace{\left(n_1^2\cos\theta_{t_2}\tau_0 + \cos^2\theta_{t_1}\sqrt{\varepsilon_0\mu_0}\right)}_{D}} \tag{31}$$

Substituting α for $2\pi t/\lambda$, and after multiplying the numerator and denominator of Eq. (31) with the complex conjugate of the denominator, the reflection coefficient is derived as follows:

$$r = \frac{AC + \alpha^2 BD - i\alpha(BC - DA)}{C^2 + \alpha^2 D^2} \tag{32}$$

and the phase change (Yuan, Ong, Tan, et al., 2006) is

$$\tan\phi = \frac{\alpha(BC - DA)}{AC + \alpha^2 BD}\,. \tag{33}$$

Plasmonic Sensor Parameter

The sensitivity, limit of detection, linearity, resolution, selectivity, accuracy, and plasmonic quality factor are the main parameters for characterizing the performance of the SPR sensors. The SPR sensors use for the detection of biomolecules or chemical materials in low concentrations. When the biomolecules and toxic chemical bind to the sensing layer, the refractive index or thickness of the sensing layer changes and the sensor output such as intensity, angle and wavelength are modulated.

The sensitivity of the SPR sensor depends on the variation of the refractive index and the thickness of sensing layer. Therefore, it is the ratio of output variation to measurable parameters variation including refractive index or thickness of sensing layer. When the SPR sensors are used for affinity biosensors, the selectivity of the sensor is significant parameter and it depends on the affinity of biomolecules to bind the sensing layer.

The limit of detection (LOD) is the ability of the SPR sensor to detect the biomolecules or chemical analyte. The LOD assigns the minimum sensitivity of the SPR output such as variation in reflectance, angle shift, and intensity of a particular wavelength. In accordance with the literature (Armbruster, and Pry, 2008), LOD is three times greater than the ratio of standard deviation (σ) to sensitivity (S) as follows:

$$LOD = 3 * (\sigma / S) \tag{34}$$

The linearity describes the relationship between the output of the SPR sensor with the thickness and refractive index of the sensing layer and it depends on kinetic models which describe the binding of molecules with sensing layer.

The resolution of the SPR sensor is the smallest change in the refractive index or thickness of the sensing layer which generates during the association process due to binding the molecules on the surface of the sensing layer and it impresses the output of the SPR sensor.

The SPR sensor uses to measure the refractive index of the layer, the concentration of analyte, and thickness of the thin layer which is coated on the surface of the gold layer. The accuracy of the SPR sensor is the closeness of real value to the measured value.

The plasmonic quality factor is the parameter to characterize the quality of the sensing layer system. As a literature, the plasmonic quality factor (P_{Qf})

depends on real (ε_r) and imaginary (ε_i) parts of the dielectric constant of the layer as follows (Sadrolhosseini, Krishnan, Safie, et al., 2020):

$$P_{Qf} = -\frac{\varepsilon_r}{\varepsilon_i} \tag{35}$$

The SPR signals generate when the light interacts with a metal layer. Sometimes, the sensing layer has conductivity and it can improve the SPR signals. Hence, the plasmonic properties of the sensing layer system can evaluate using the plasmonic quality factor.

Plasmonic Properties of Metal Nanoparticles

Plasmonic nanomaterials have been considered for their excellent properties such as localized surface plasmon resonance (LSPR), magneto-optical (MO) effects, and surface-enhanced Raman scattering (SERS). Metal and conductive nanoparticles have free electron or charge density around the molecular cluster. Therefore, the electromagnetic can interact with nanoparticles via charge density located around the metal cluster. Figure 7 shows metal nanoparticles in the electromagnetism field. The oscillation of charge density can couple with the electromagnetic wave when the wavelength of the electromagnetic wave is larger than the particle size and the surface plasmon wave (SPw) generate at the interface of metal cluster nanoparticle and the dielectric that is surrounded the metal cluster. The SPw length is short and it depends on the nanoparticle size. Consequently, the interaction of light with nanoparticles is significant in the photonic devices including plasmonic solar cells, photonic integrated circuits, sensor chips, and optical switches. The plasmonic quality factor (P_{Qf}) depends on imaginary parts of the dielectric function. Hence, gold, silver, and copper nanoparticles are the common metal nanoparticles.

Gold Nanoparticles

The gold nanoparticles (Figure 8) usually form in a spherical shape and they can be synthesized using laser ablation (Sadrolhosseini, Abdul Rashid, Shafie, 2019; Sadrolhosseini, Abdul Rashid, Zakaria, 2017, Haddawi, Humud,

Hamidi, 2020), microwave (Das, Babu, Gogoi, et al., 2012), and sonochemical method (Okitsu, Ashokkumar, and Grieser, 2005). Moreover, the composite and the core shell of gold nanoparticles can be fabricated beside other metals

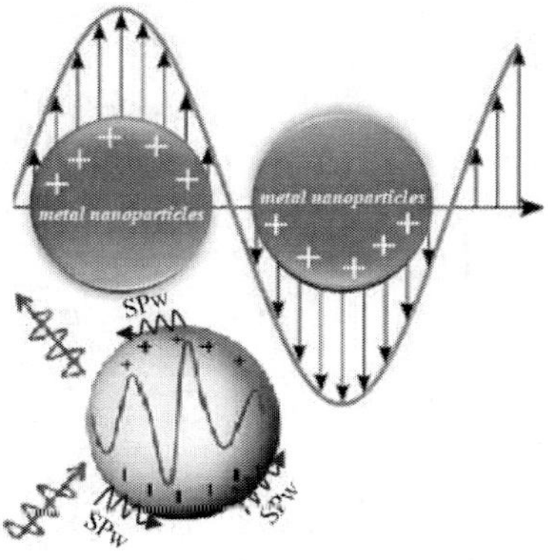

Figure 7. Metal nanoparticles in the electromagnetic field.

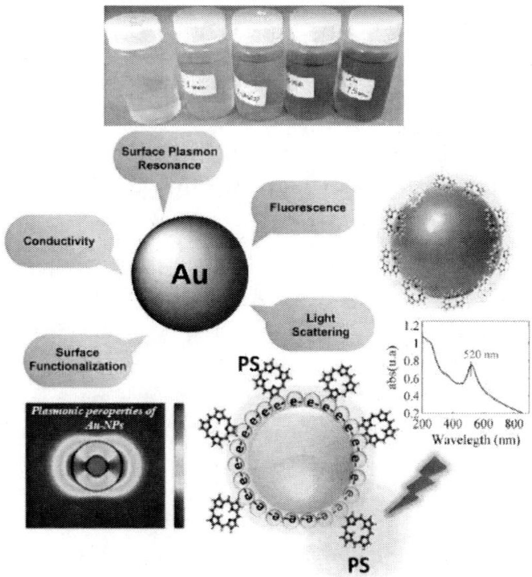

Figure 8. Gold nanoparticles in a liquid solution.

and ceramic such as silver and titanium oxide (Kodeary, Gatea, Haddawi, et al., 2020; Hamidi, Oskuei, 2013) using laser ablation. The gold nanoparticles have a sharp peak around 520 nm in the UV-vis spectrum. The abortion peak is related to localized surface plasmon resonance. The gold nanoparticles have a wide range of applications in medicine, electronic device sensors, and biosensors. The important application of the gold nanoparticles can be considered in photothermal therapy, gene delivery, computed tomography (CT), and photodynamic therapy (Hu, Zhang, Ding, et al., 2020) The gold nanoparticles (Draz, and Shafiee, 2018) and the composite of gold nanoparticles (Kim, Kim, Lee, et al., 2020) were used to detect the various virus and RNA. Consequently, the plasmonic properties of gold nanoparticles are considerable for medical and biological applications.

Silver Nanoparticles

The silver nanoparticles (Figure 9) have plasmonic properties in the visible range of the electromagnetic spectrum. Normally, the silver nanoparticles have a spherical shape and have a sharp peak around 405 nm. The particle size is less than 100 nm and the silver nanoparticles are antimicrobial and anticancer (Xu, Wang, Huang, et al., 2020). The silver nanoparticles are used for cutting-edge medical applications (Xu, Wang, Huang, et al., 2020), such as antidiabetic agents, wound repair, dental material filling, bone healing, and bioimaging (Xu, Wang, Huang, et al., 2020). The silver nanoparticles can be fabricated in physical, chemical, and biological methods such as laser ablation, photochemical (Zheng, Peng, Lombardi, et al., 2016), electrochemical (Kuntyi, Kytsya, Mertsalo, et al. 2019), microwave-assisted (Chung, Kim, Ko, et al., 2018), and sonochemical methods (Elsupikhe, Shameli, Ahmad, 2015). The silver nanoparticles are used to improve the sensitivity and selectivity of biosensors for the detection of microRNA (Elhakim, Azab, Fekry, 2018).

Copper Nanoparticles

Copper nanoparticles have attracted great interest in metal nanoparticles. The copper nanoparticles (Figure 10) have unique physical and chemical properties including high thermal and electrical conductivities. Some polymer and plastic material can cap the copper nanoparticles and the copper nanoparticles operate as antifungal, antimicrobial, and antibiotic (Betancourt-Galindo, Reyes-Rodriguez, Puente-Urbina, et al., 2014). The copper nanoparticles can be synthesized using the pulsed laser ablation (Yeh, Yang, Lee, et al., 1999), vacuum vapor deposition (Liu, Bando, 2003), pulsed wire discharge (Yatsui, Grigoriu, Kubo, et al.,) and mechanical milling (Oleszak,

Shingu, 1996) microemulsion techniques (Solanki, Sengupta, Murthy, 2010), sonochemical reduction (Rajeshkumar, Rinitha, 2018), electrochemical (Miao, Zhang, Xu, Tang, 2018), microwave-assisted (Tahseen, Ikram, Sher, 2019), and hydrothermal (Ashok, Hariram, Siengchin, 2020) methods. The copper nanoparticles have a sharp UV-vis peak around 580 nm and the copper nanoparticles form in a spherical shape. The particle size decreases when the blue shift appears in the UV-vis spectrum (Figure 10).

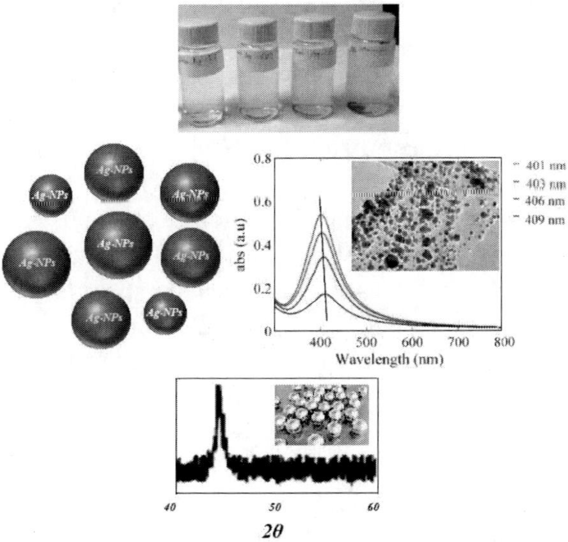

Figure 9. The prepared silver nanoparticles in CQD solutions.

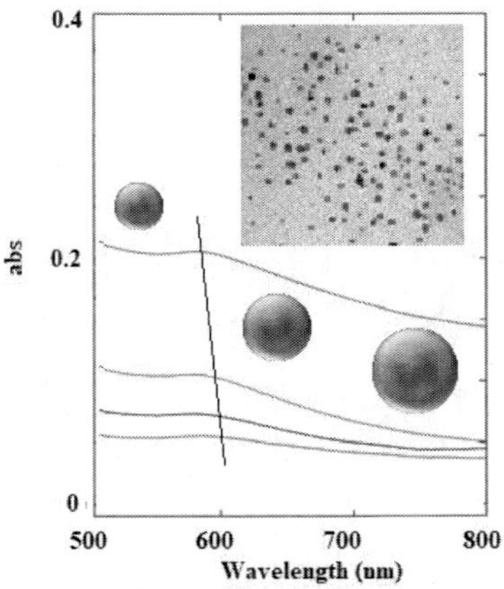

Figure 10. Copper nanoparticles were dispersed in liquid solution using laser ablation.

Application of Plasmonic

Plasmonic has wide range application in industrial and medicine including sensor, biosensor, solar cell, photodetector, enhancement Raman surface spectroscopy, and antenna. *Plasmonic sensors have extraordinary sensitivity and they are reliable and flexible sensors for biological and medical applications. Therefore, in this study, the application of plasmonic sensors* have been considered for *monitoring the DNA, and RNA interaction with the biomarkers, virus detection, and cancer display*

Plasmonic Sensor to Recognize DNA, RNA, and Antibody

The plasmonic sensors have more applications to recognize and detect DNA and RNA. The SPR based on the prism, LSPR, SPRi, and fiber optics surface plasmon resonance sensors are the famous and popular plasmonic methods to detect the DNA and micro-RNA in biological and medicine and the portable plasmonic sensor is based on fiber optics. Usually, the sensing layer system contains a glass, gold, or silver layers, and a linker layer. The gold, silver nanoparticles, and carbon nanostructures (graphene, cabon nanotube, and carbon dots) were used to generate and enhance the plasmonic response. The DNA ligand was immobilized on the surface of the linker layer and it can interact with the analyte after immobilization of DNA or RNA (Figure 11). The plasmonic sensor is a versatile method to investigate the interaction of light with RNA and DNA.

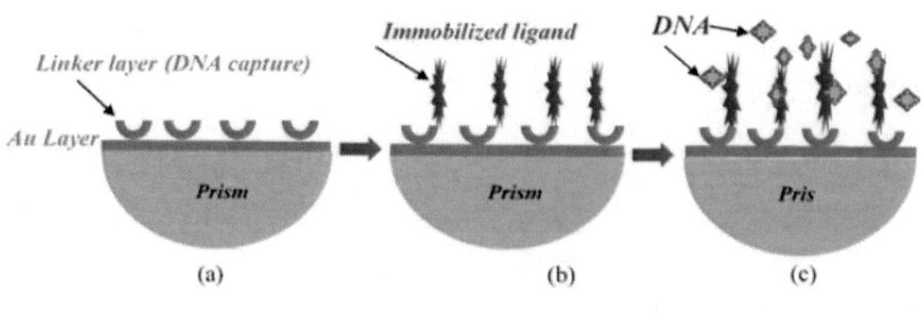

Figure 11. (a) The gold layer and linker layer, (b) immobilized ligand were coated on the substrate. (c) the DNA interacts with the sensing layer system.

The nanoscale part of DNA is a DNA origami and it can create arbitrary two-and three-dimensional shapes at the nanometer size. The DNA can use for the assembly of nanostructures (Green, Subramanian, Mardanlou, 2019) to detect the biomolecule in biosensors (Zeng, Zhou, Sun, et al., 2021) and drug delivery. The DNA origami technique has emerged as a powerful method for the fabrication of complex nanostructures that can serve as templates for the assembly of various functional components with nanometer precision (Huang, Nguyen, Natarajan, et al., 2018). DNA origami can use to enhance the sensitivity and selectivity of the plasmonic sensors to recognize the aptamers (Huang, Nguyen, Natarajan, et al., 2018). Figure 12 shows the gold nanorod that was combined with DNA origami. The sensing system was based on biorecognition with the DNA aptamer, DNA origami and gold nanorod. The DNA origami molecules locked between two set the gold nanorod which have formed in the chiral plasmonic form. The optical signals depend on the angle between two cross gold nanorod. Therefore, the selectivity and the sensitivity of plasmonic sensor can be improved using this sensing layer system.

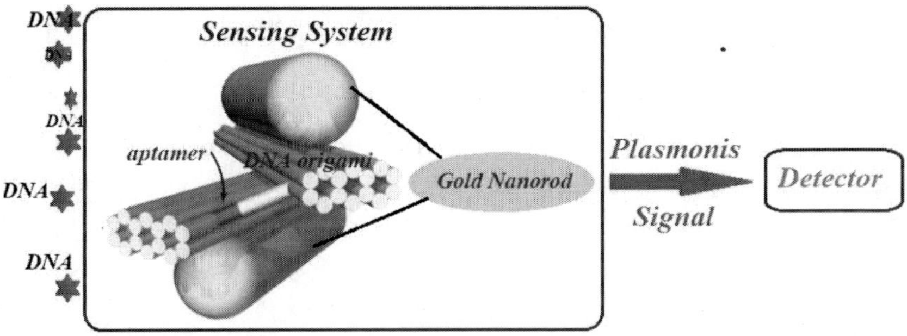

Figure 12. Sensing layer based on DNA origami.

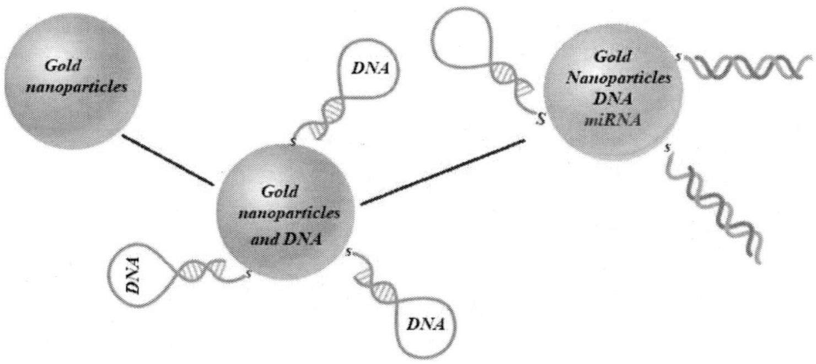

Figure 13. The demonstration of sensing system for plasmonic DNA sensor using hybridization on the single gold nanoparticles surface.

Hu et al., (2015) used the plasmonic sensor based on hairpin DNA for the detection of oligonucleotides to recognize cancer. They used gold nanoparticles and single strand DNA with SH-labeled at its 3'end (Figure 12). They used this composite to detect and recognize the oligonucleotides with the limit of detection about 3 nM (Hu, Zhang, Zhang, 2015). The LSPR technique was used to investigate the DNA hybridization (Figure 12) process in binary solution. Rapisarda et al., 2017 used the LSPR for the detection of DNA hybridized (Figure 13) using gold nanoparticles with the limit of detection in the range of 10nmol/L (Rapisarda, Giamblanco, Marletta, 2017).

DNA and RNA are the main biological macromolecules and they are nucleic acid. RNA is a ribonucleic acid and it is a polymeric molecule essential in various biological roles. They have the main role in coding, decoding, regulation, and expression of genes. Along with lipids, proteins, and carbohydrates, nucleic acids constitute one of the four major macromolecules essential for all known forms of life. MicroRNAs (miRNAs) are a class of small, endogenous RNAs of 21–25 nucleotides (nts) in length. They play an important regulatory role in animals and plants by targeting specific mRNAs for degradation or translation repression. The mRNA is one of RNA and it has not sub-type of other RNA. mRNA contains more nucleotides, and it has a role to produce a protein but RNA is a nucleic acid and it can generate the ribonucleotide. RNA can transfer the amino acid to ribosomes. Hence, the detection, recognition, and control the RNA and mRNA are significant activities in biotechnology. The plasmonic technique is a considerable method for the detection of mRNA and RNA. The LSPR sensor was used to detect the particular miRNAs using the immobilization of nanoparticles in the presence of thiolate oligonucleotide probes (Drozd, Pietrzak, Malinowska, 2018)] (Figure 14). Miti et al., (2020) used the mentioned probe for the detection of miRNA as a short nucleic acid.

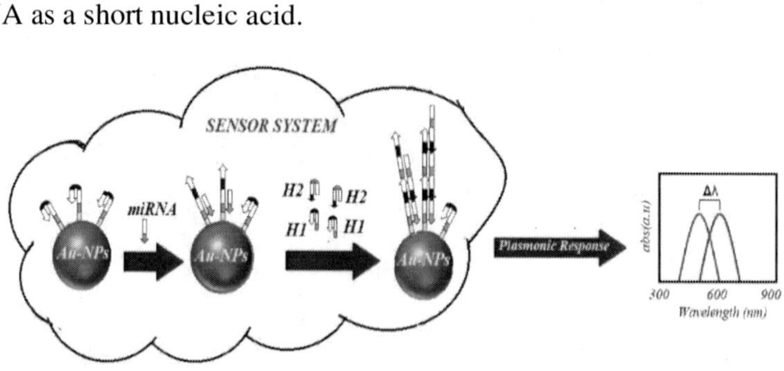

Figure 14. The demonstration of LSPR sensor and Hybridization of chain reaction for of RNA.

Li et al., (2017) improved the sensitivity of the SPR sensor using two layers of graphene oxide-gold nanoparticles (GO-AuNPs) nanocomposites to analyze and detect the miRNA and small molecule adenosine (Khoa, Kim, Yoo, et al., 2014). The binary layer provided an effective functionalized substrate for immobilization of mRNA and DNA in a biosensor chip. Hence, the sensitivity and selectivity were improved and the measurement of miRNA-141 with a detection limit of 0.1 fM was obtained. Moreover, the SPR sensor based on this nanochip was developed for the investigation of the miRNA-200 family members (Li, Wang, Yang, 2017). The SPR sensor based on the gold nano-slit array (Figure 15) was used to detect and investigate the urinary mRNA biomarker (Yeung, Li, Chung, et al., 2020). The low concentrations of biomarkers are about 1 mL which are datable using the SPR sensor (Mousavi, Chen, Lee, 2015). The SPR sensor was improved for label-free detection of miRNA (Khoa, Kim, Yoo, et al., 2014). The SPR biosensor was designed base on streptavidin and the signal was amplified using connecting the biotin to the long DNA. The limit of detection was increased down to 9 pM (Ding, Yan, Li, 2015). Vaisocherova et al., 2015 applied the SPRi biosensor based on streptavidin-functionalized gold nanoparticles for detection of multiple miRNAs in erythrocyte lysate (EL) (Vaisocherová, Šípová, Víšová, et al., J., 2015) and the sensitivity was about 0.5 pM (Vaisocherová, Šípová, Víšová, et al., 2015).

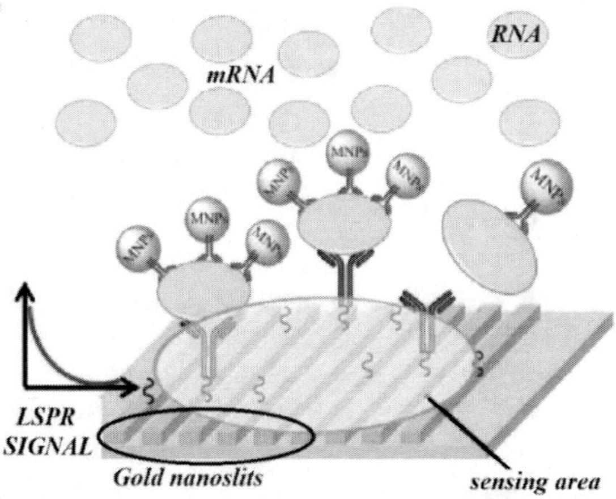

Figure 15. Scheme of gold nano-slits array for display of mRNA or RNA using LSPR.

Plasmonic Sensor for Detection of Sugar

Sugar is a soluble carbohydrate and any of numerous sweets which are present in food and beverage. The colorless and water solubility of sugar are the main properties. Glucose, fructose, and galactose are monosaccharides and simple sugar. Sucrose, maltose, and lactose are disaccharides, and their molecules are derived from a combination of two monosaccharides. They were achieved them by replacing a hydroxy group from monosaccharide and a hydrogen nucleus from the other simple sugar.

Table 1. The list of main sugar and components

Monosaccharide ($C_6H_{12}O_6$)	Disaccharide ($C_{12}H_{22}O_{11}$)	Components
Fructose	Sucrose	fructose + glucose
Glucose	Maltose	glucose + glucose
Galactose	Lactose	glucose + galactose

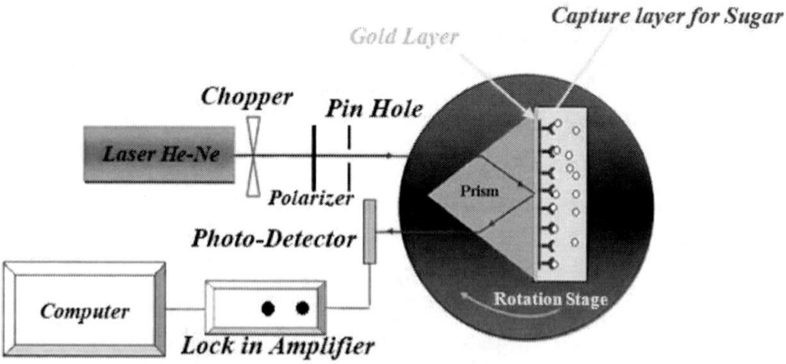

Figure 16. The SPR setup contains a He-Ne laser, a chopper, a pinhole, a prism, a gold layer, a capture layer, a rotation stage, a photodetector, and a lock-in amplifier.

Generally, there is sugar with hemiacetal and without hemiacetals or acetals. Sugar such as α-D-glucose and D-fructose are hemiacetal and they can react with silver (Ag) or copper (Cu) ions. Therefore, the plasmonic nanoparticles can easily absorb sugars. The sugar without hemiacetal cannot reduce with any agent. Normally, glyceroxide was used to measure and recognize glucose. So, gluconic acid and hydrogen peroxide are achieved from the oxidation of glucose in the presence of glyceroxide and enzymes.

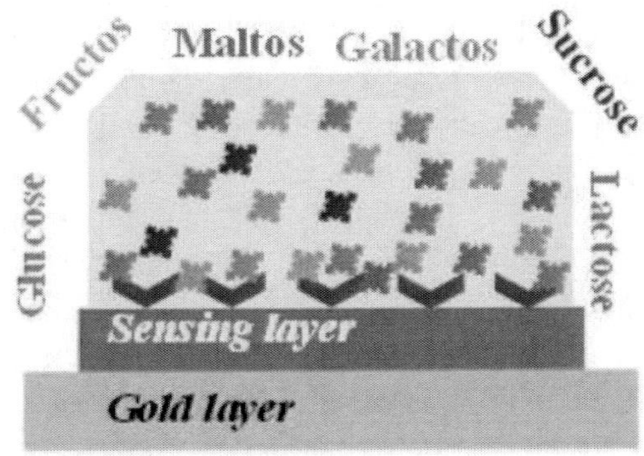

Figure 17. Scheme of the SPR sensor for monitoring the sugar level using binding the sugar on the surface of a sensing layer such as PPy-magnatic nanoparticles, or enzymatic capture system.

The measurement of glucose concentration can be carried out through indirect and direct methods. In indirect methods, the concentration of hydrogen peroxide must be measured using electrochemical signals. The SPR sensor is a direct sensor and the surface of the gold layer can be improved with the capture layer to interact and absorb glucose, fructose, and sucrose. Figure 16 shows the SPR setup to measure a sugar with the capture layer that improved the gold layer. The glucose/glucose/galactose-binding (GGB) protein is an immobilizer for the detection of glucose.

Li et al. (2017) detected the glucose using the GGB protein in the range of 0.1-10 mg/dL of concentrations. Wong Wang, Banerji, Menegazzo (2011) used the imprinting of glucose phosphate barium salt into a poly (allylamine hydrochloride) network layer to improve the gold layer for detection of glucose (Figure 17). This sensor was sensitive to measure the low concentration of glucose about 0.02 mg/mL and it was used to detect the glucose in human urine. Stephenson-Brown, Wang, Iqbal, Preece (2013) applied the bis-boronic acid bearing a thioctic acid moiety complex on the surface of the gold layer. They fabricated the sensitive sensor for recognizing the gluconic acid and detection of glucose. The response of the sensor was registered in the presence of D-glucose, D-galactose, and D-fructose and the sensor can detect the glucose better than others. The limit of sensors for detection of glucose, fructose, and galactose was about 0.3 mM, 0.6 mM, and 1.25 mM, respectively. The carbon-based nanostructure (Li, Su, Yang, 2017), and magnetic nanoparticles are the other nanostructures that can be used for the detection of sugars.

Magnetic nanoparticles including Fe_2O_3 and Fe_3O_4 nanoparticles have some potential to interact with sugars (Figure 17). The advantages of magnetic nanoparticles are high catalytic efficiency, good stability, and mono-dispersion (Wang, Banerji, Menegazzo, 2011). Hence, many researchers used $ZnFe_2O_4$ (Wang, Banerji, Menegazzo, 2011), Fe_3O_4, and polypyrrole–chitosan/Fe_2O_3 nanocomposite to detect the sugars. The role of magnetic nanoparticles is the improve the sensitivity, limit of detection, selectivity, and response time. For example, the limit of detection is in the range of 1 to 3 μM, and the ratio of signal to noise is about 3. The polypyrrole–chitosan–$BaFe_2O_4$ nanocomposite (Sadrolhosseini, Moozarm Nia, Naseri, et al., 2020) was used to improve the gold layer. The nanocomposite layer was to detect glucose, fructose, and sucrose. The limit of detection was 0.005 ppm and the sensor was sensitive for recognition of glucose.

Neuroplasmonic

Recently, the functions of brain have considerably attracted the worldwide attention. Certainly, the precise and detailed studies would help to create a dynamic understanding of brain function as well as facilitate medical research related to healing and brain development. In addition, they would improve the engineering of computer chips which could subsequently play an important role in neurorobotics, neuromorphic computing, etc. The destruction of brain tissue is due to a tumor, trauma, infection, or various degenerative conditions (Alzheimer's, macular degeneration, and osteoporosis). The investigation and check of the functions of the brain attract research topics in biology, medicine, and biotechnology. A dynamic understanding of brain function should be considered in detail using medical instruments, medical probes, and medical images analysis. The improvement of biomolecular simulation and bio-computer chips have a great impact on the development of neurorobotics, neuromorphic computing, and neuroimage (Figure 18). The neuroimaging and recording of brain signals are important activities to find unique information about the functions of the brain and disturbing the brain tissue (Ahn, Kim, Kim, et al., 2021). Therefore, the record of the high-quality image, the activity of large neuron populations, and the high-quality signals can be effective in examining the brain tissue. Hence, the improvement of resolution, analysis algorithm, and short response time are the challengeable and considerable parts of the research in the medical and biotechnology instruments (Beletskiy, 2013).

The investigation of neuronal signals is the best topic for research in the scale of biomolecular, cellular, and large neuronal networks. The spectrum of scales for researching and observing neuronal signals in large neuronal networks is the main challenge. Recently, the physio-pathological function of the brain and the activity of the brain can investigate using neuronal circuits.

The plasmonic sensor, plasmonic electrode, and plasmonic antenna have been provided to find information about neuro-interface, neuro-activity, brain enzyme, neuron, and brain-related chemicals. The neuro-transmitter and the transportation of chemical molecular to inside the biomolecule are the main activity between neurons (axons and neurites). The investigation of biochemical processes and transportation at the membrane of biomolecules are possible using the application of nanoparticles and plasmonic properties of metal nanoparticles. Therefore, the main topics of neuroscience are the structure, functions, and electrophysiological properties of neurons and the study of high-performance tools for investigation the neurobiology at the biomolecular and bio-cellular. Hence, the application of plasmonic and the surface plasmon resonance sensor provide the high-sensitivity optical biosensors and high-resolution imaging to investigate the neurons and plasmonic nanoprobes, plasmonic antenna, and plasmonic sensor can be used for modeling and investigation of a particular area of neurons.

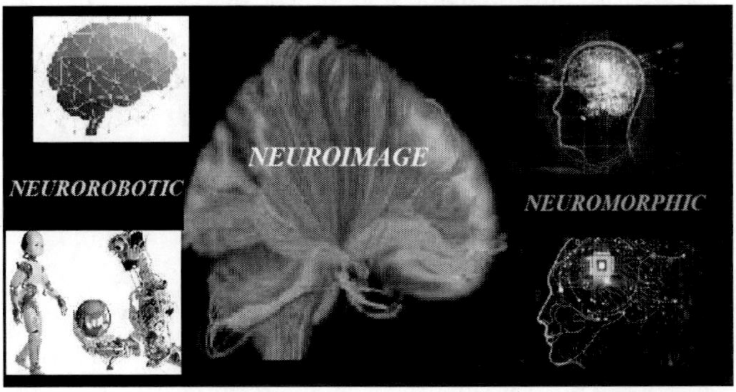

Figure 18. Demonstration of neurorobotics, neuroimage, and neuromorphic.

Three plasmonic methods have been considered for the investigation of neurons including plasmonic sensors for neuronal activities and neuron-related chemicals, high-resolution optical images based on plasmonic enhancements, and plasmonic neuromodulations (Ahn, Kim, Kim, et al., 2021). Moreover, prism coupler configurations, the combination of SPR and fluorescence microscopy plasmonic properties of nanorods, and plasmonic crystals are the experimental approach in neuroplastic for deriving the neuron information (Sohrabi, Hamidi, 2016).

Many researchers were reported to use prism coupling and fiber optics coupling for the investigation of neural activity. Baac, Jun, Turner, (2005) carried out an experiment for the investigation of neural action potential using SPR configuration. The SPR setup based on the prism that was used on this test, contained He-Ne laser (632 nm), glass slide, a gold layer, spacer, Ti adhesion layer, Si photodiode, index matching liquid, and a prism. Two amplifiers, filters, and analog oscilloscope were used to register the SPR signals. In the first step, there was no biological sample (i.e., neurons) and the space between two gold-coated glass slides was fulfilled the phosphate buffer solution (PBS) and external electric field would be applied and the SPR signals have been modulated electrochemically at the interface of metal-dielectric. The reflectance was obtained in the different incident angles and SPR signals for four applied voltages of 50mV, 30mV, and 10mV with the frequency of 1 kHz were achieved. In this case, the SPR sensor can detect the angular shift about the 10^{-4}–10^{-5} in neural action potential recording. The SPR signals can be registered for investigation of the neuron-neuron communication in-network level.

In 2008, Kim, Byun, Lee, et al., (2008) applied the low-noise the SPR sensor as a label- and artifact-free method for the extracellular recording of neural activity (Figure18). The results were reported for a real nervous tissue by recording both optical and electrical signals simultaneously. The sciatic nerves from the knee (distal) were a real biological sample to the proximal of the male SD rat. Lidocaine (2%) as a nerve blocker was applied for confirmation the neural SPR signals. The only limitation of this system was the required close proximity of the metal surface to the nerve.

Choi, Kim, Im, (2011) used the remarkable work with the aim of "theoretically" exploring the mechanism of neural activity using the SPR phenomenon. The Kretschmann ATR system, artificial cerebrospinal fluid, and nerve membrane of rat sciatic nerve were illuminated with the light and the dynamic interactions between the rat sciatic nerve and the SPR were depicted. The simulation was carried out using Poisson-Boltzmann equation, numerically analyzed by finite element method (FEM), and Drude-Lorentz electron model and that was numerically analyzed by Green's method and a Fresnel's 3-layered model have been utilized for numerical analysis. The dielectric constant, resonance angle shift, and the variation of refractive index were achieved. Consequently, the parameters for the evaluation of neural signals were experimentally measured using the SPR signals.

The LSPR was used to detect the mammalian brain cell activity. Zhang, Atay, Nurmikko, (2009) measured the cellular potential induced shift in the surface plasmon wave in the presence of gold nanoparticles. Hippocampal neural cells were grown on the surface of the planer gold nanoparticles array and the cellular level individual transient signals were detected optically when the chemically triggered neurons switched their potential.

Flexible patterned plasmonic substrate is used as a sensing surface for phase-sensitive neural recording of cerebellum tissue slice under electrical and chemical stimulations (Sohrabi, Etezadi, Perin, et al., 2020). While the traditional reflection spectrum cannot represent the changes in neural activity with high-precision, phase-sensitive neuroplasmonics not only can reveal the neural activity level but also distinguish different electrical and chemical stimulation types with considerable phase splitting factor.

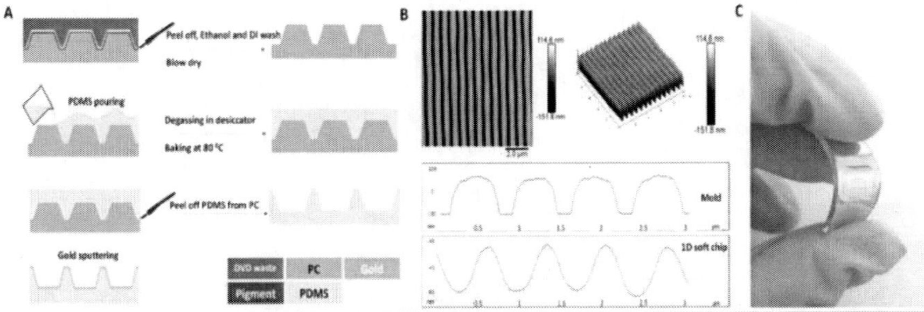

Figure 19. (A) Process flow of the chip fabrication starting from extracting the grating from a DVD using a scalpel, removing the pigments using Ethanol following DI rinsing and drying with N2. Prepared PDMS is poured on the extracted grating mold and let it degassed and cured at 80°. After 1h40, the top PDMS layer is peeled from the mold. Thin gold film is deposited on fabricated PDMS grating using sputtering machine. (B) AFM image of the grating mold. The topography of the polycarbonate mold and fabricated 1D soft chip. The periodicity and height in polycarbonate mold and fabricated chip are respectively [741 nm, 162 nm] and [724 nm, 67 nm]. (C) Real photo of the fabricated flexible chip (Sohrabi, Etezadi, Perin, et al., 2020).

In addition, cell activity detection in three different types of cultured cells (primary hippocampal neurons, neuroblastoma SH-SY5Y cells, and human embryonic kidney 293 (HEK293) cells) on the grating surface has been investigated. By introducing KCl solution as a chemical stimulus, one can differentiate the neural activity of distinct cell types and observe the signaling event in a label-free, optical recording platform. This method has potential applications for recording neural signal activity without labeling and stimulation artifacts (Zimmerman, Tian, 2018).

The plasmonic sensor can use to measure dopamine as a brain chemical. Dopamine is a type of neurotransmitter and it has a role in human metabolism and it is a significant biochemical molecule. Kumbhat et al. prepared the conjugate of DA using bovine serum albumin protein to improve the gold layer. The prepared layer system was used to interact with dopamine and the surface plasmon resonance signal was achieved for 85 pg/ml concentration of dopamine (Kumbhat, Shankaran, Kim, et al., 2007) The dopamine can be detected using a surface plasmon resonance sensor. For example, chitosan and graphene quantum dots have a tendency to interact with the graphene and can detect dopamine with the limit of detection about 1.0 fM (Eddin, Fen, Omar, et al., 2021).

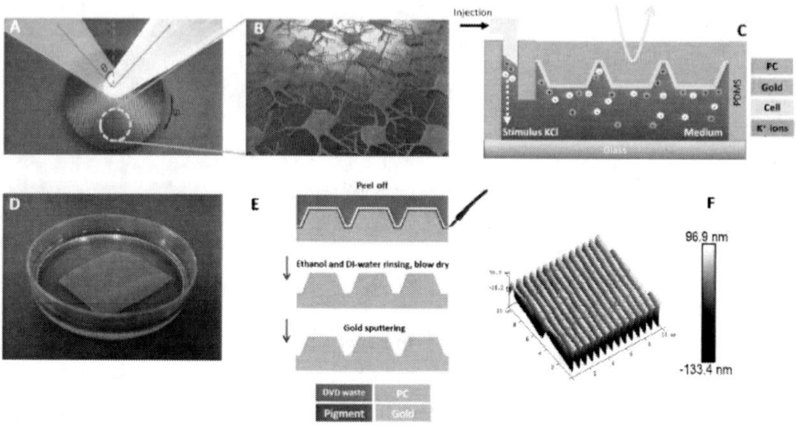

Figure 20. (A) Schematic of the impinged light on the grating chip with the incident and azimuthal angles denoted by θ and φ. (B) Schematic of the cultured neurons on the chip. (C) Schematic of the chamber with embedded chip and stimulus KCl medium. (D) Real photo of the grating chip with the size of 25 mm × 30 mm in the cell culture medium. (E) Process flow of chip fabrication starting from extracting the grating from a DVD and then depositing gold on it after removing the pigments. (F) AFM image of the grating chip (Sohrabi, F, Jahani, Y., Sanchez-Mut, et al., 2020.)

Plasmonic Sensor for Monitoring Hepatitis, Diabetes, Leukemia, and Cancer

The plasmonic biosensor is an accurate and user-friendly tool to investigate biomolecular activity. The output of plasmonic biosensors depends on the interaction between biomolecules and the sensing layer. It causes a change in the optical properties of the sensing layer based on nanomaterials. On the other hand, when the interaction occurs fluorescence intensity, plasmonic absorbance peck, and angle of resonance will be shifted to other values to monitor the concentration and presentation of biomolecules. Therefore, the plasmonic sensor or surface plasmon resonance sensor are versatile optical tools to monitor the biomolecule. The analytical methods can be used to discover the antigen, antibody, enzyme, and some lipid. Moreover, the effect of viruses and bacteria on human cells can investigate using biological methods. The plasmonic sensors can monitor the interaction of biomolecule (antigen, antibody, enzyme, etc.) with biomarker and immobilized ligand (Sohrabi, F., Saeidifard, S., Ghasemi, e al, 2021). Hence, the plasmonic sensor can discover the virus, hepatitis antigen, diabetes, leukemia, and cancer (Figure 21).

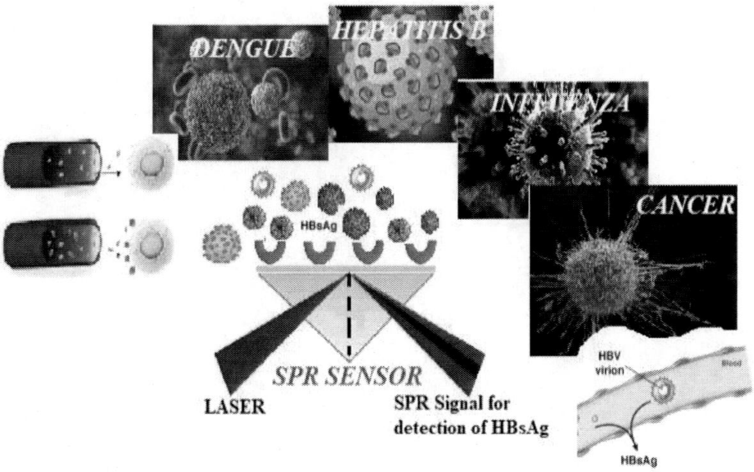

Figure 21. The SPR sensor can monitor the virus, hepatitis, diabetes, and cancer.

To detect a virus, a biomolecule, an antibody, the SPR signal should be analyzed based on kinetic theory. The practical analysis for biosensors based on the SPR includes the kinetic, equilibrium, and concentration analysis using the association and dissociation processes (Figure 22). The interaction of virus, biomolecules such as an antibody, DNA, ligand–analyte binding, antibody-antigen, nucleic acid, and enzyme-linked immunosorbent assay (ELISA) with the sensing layer and immobilizer were explained using kinetic and equilibrium analyses. The quantity of analyte can measure using the SPR technique from the analysis of the sensorgram plot and it has application in clinical point-of-care (POC) diagnostics, evaluation of pollution in environmental, healthcare, measurement of food components, bio-pharmaceutical production (Singh, Paswan, and Raghuwanshi, 2021).

The virus detection and viral diagnosis can be carried out using direct and indirect techniques. In the direct techniques, the interaction of viruses and viral nucleic acids, or proteins have been considered with immobilizer. In this technique, sensing layer or bioreceptor surfaces are viz antibody-, glycan-, and aptamer-based. The bioreceptor based on aptamer is usually used to prepared the sensing layer for the SPR sensor. The indirect techniques involve culturing of clinical samples followed by infection of cells, eggs, or animals to isolate viruses. The SPR technique can use for viral diagnosis and detection of viruses in the environmental and invitro sample in the laboratory.

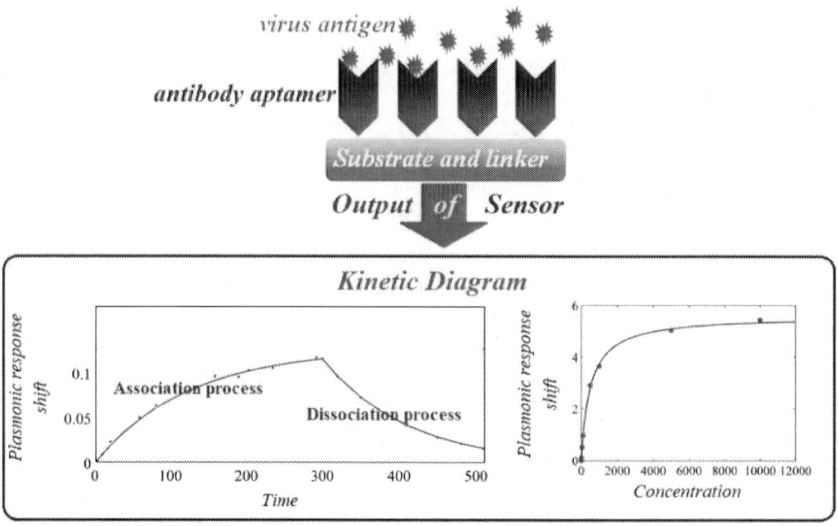

Figure 22. Model for monitoring the virus using kinetic model.

For the first time, Schofield and Dimmock have used the SPR sensor for the detection of the influenza A virus (Schofield and Dimmock, 1996). The analysis of sensogram that was derived from the analysis of the SPR signal, was used to diagnose different stages of Epstein–Barr virus through detection of antibodies against. Therefore, the different antigens, in the virus (Riedel, Rodriguez-Emmenegger, Santos, 2014) antibiotics in milk samples, and mycotoxin patulin (Hunt and Armani, 2010; Damborský, Švitel, Katrlík, 2016) can be displayed using analysis the sensogram. SPR sensor can be used to detect the dengue virus (Omar, Fen, Ramli, et al. 2021) and investigate the binding kinetics of dengue antibody with dengue antigen (Jahanshahi, Wei, Jie, 2016). The SPR sensor can be used to detect Enterovirus 7. The mechanism of detection is the based-on interaction of the sensing layer with major capsid protein VP1 of EV71 (Prabowo, Wang, Secario, 2017) and the virous was displayed using surface plasmon resonance technique. The limit of detection was reported about 1 (vp/ml). Influenza Viruses such as hemagglutinin and neuraminidase were detected with the SPR probe using DNA aptamer (Gopinath and Kumar, 2013). The DNA aptamer was used to recognize the avian flu viruses, and influenza virus H1N1 as a sensing layer for the SPR sensor. Ebola VP40 is another aptamer for the detection of viruses. The Ebola virus was examined using the SPR sensor via binding with immobilized VP40. Consequently, the various virus can recognize using the SPR sensor by designing the suitable aptamer layer on the surface of the gold thin film.

When the level of blood glucose increases abnormally, diabetes appears in type I or type II. The plasmonic sensor also can use to diagnose diabetes type I and II. For discovering type I diabetes (T1D), the surface of the gold layer was improved with a proteoliposome-based full-length ZnT8 self-antigen (PLR-ZnT8). The PLR-ZnT8 was immobilized on the surface of the gold layer. The interaction of PLRZnT8 with ZnT8A has been monitored using a plasmonic-based reflectance shift (Tokel, Inci, Demirci, 2014). To diagnose diabetes in type II, Glycosylated Hemoglobin (HbA1c) must be traced in the human blood and the amount of HbA1c gives convincing accurate results. when the blood glucose is bound to the theβ-chain of hemoglobin with N-terminal valine, the level of HbA1c increased and it appears in the blood. The gold thin layer was modified with 4-vinyl phenylboronic acid to prepare the SPR sensor chip. The measurement was carried out to find the concentration of HbA1c using the interaction between glucose and boronic acid and the limit of detection was about 2.86μg/mL (Chen, Zhao, Zhang, et al., 2015).

Hepatitis is inflammation of the liver tissue and the reason is the high use of alcohol, the virus hepatitis A, and B infection. The Surface plasmon resonance has the potential to discover hepatitis (Figure 23). In 2005, Hwang, Yoon, Jeon, et al., (2005) carried out the SPR sensor experiment to recognize hepatitis B. The hepatitis B surface antigen polyclonal antibody was immobilized to the dextran layer on a CM5 chip surface. The dextran layer on a CM5 chip had been activated by N-hydroxysuccinimide previous to the preparation biochip. The reflectance was measured to investigate the binding of hepatitis B. The SPR sensor was sensitive for changing the HBsAg concentration, and the variation of antigen was detectable about 40 μg/mL. Another plasmonic sensor for recognizing hepatitis B has been designed based on the Pichia pastoris-derived recombinant hepatitis B surface antigen (HBsAg) and it was used as a bioreceptor for detection of hepatitis B. The plasmonic sensor detected hepatitis B in the dynamic range of ~0.00098-0.25 mg/L (Tam, Zeenathul, Rezaei, et al., 2017). The detection of hepatitis B was investigated from saliva using immobilization of hepatitis antigen in poly[(N-(2-hydroxypropyl) methacrylamide)-co-(carboxybetaine methacrylamide)] brush as a sensing layer. The surface plasmon coupled emission setup was used to detect the enhancement fluorescence signals. The sensor layer was attached to the prism and the fluorescence enhanced when hepatitis B interact with the bioreceptor and the limit of detection was about 1 IU/mL (Riedel, Hageneder, Surman, et al., 2017).

Blood cancer or leukemia starts with the leukemic proliferation of B-cell tumors. They contain plasma cell neoplasms, Waldenström's disease, chronic lymphocytic leukemia (CLL), and lymphomas. Recently, the standard screening methods need for recognition of leukemia instant of morphological and phenotypic examinations. The current techniques for diagnosis the leukemia are based on the testing of blood cell morphology, flow cytometry, or bone marrow biopsy. These methods are very expensive, need a long analysis time, and require highly qualified professionals. Therefore, the user-friendly, affordable, reliable, and accurate method should be replaced by the mentioned methods. The latest method is based on testing the serum components, such as immunoglobulin (Bradwell, Harding, Fourrier, et al., 2013) and B-cell neoplastic perturbation to recognize the biomarkers for leukemia diagnosis. The light chain biomarker (Campbell et al., 2013) was used to discover leukemia. For example, IgGκ, IgGλ, Ig MkandIg Mλ are the immunoglobulin classes and the light chains which can be recognized by particular antibodies (Figure 23).

To discover leukemia, the fast and accurate method is the measurement of immunoglobulins concentrations in the human serum. The SPR sensor was designed for monoclonal immunoglobulin as a leukemia tumor marker using evaluation of the immunoglobulin level in the human serum. The ratio of kappa and lambda light chains concentrations have been determined and this ratio is assigned to the concentration of monoclonal immunoglobulin. The resolution of this SPR platform is about 5×10-7 in the variation of refractive index unit (RIU) of the immunoglobulin solution (Maisonneuve, Valsecchi, Wang, et al., 2015).

The plasmonic sensor based on Cytop (a fluoropolymer with a low refractive index that matches biologically compatible fluids) was used for the detection of B-cell leukemia. The coupling method is based on a waveguide (Figure 24) which was embedded in Cytop and the top of Cytop was etched into a gold layer to fluidic contact. The fiber optic was used to excite the long-range SPR when the light was coupled to the waveguide. This biosensor detects the leukemia markers in the human serum using a functionalization strategy based on Protein G. The distribution of G kappa (IgGκ) and lambda (IgGλ) were evaluated to detect leukemia using the ratio of them and the normal ratio is in the range of 1.4 to 2.0. The ratio can be determined in the limit of 0.01. Moreover, the real-time and label-free biosensor was designed for detection of leukemia using cell surface biomarker CD133 antibody which was immobilized on the surface of the gold thin film. The EDC/NHS coupling method was utilized to prepare the sensing layer and develop the gold layer to improve the selectivity (Fathi, Rahbarghazi, Movassaghpour, et al., 2019). The sensor has the potential to detect acute myeloid leukemia via the detection of cell surface biomarkers including CD133 antibody (Fathi, Rahbarghazi, Movassaghpour, et al., 2019).

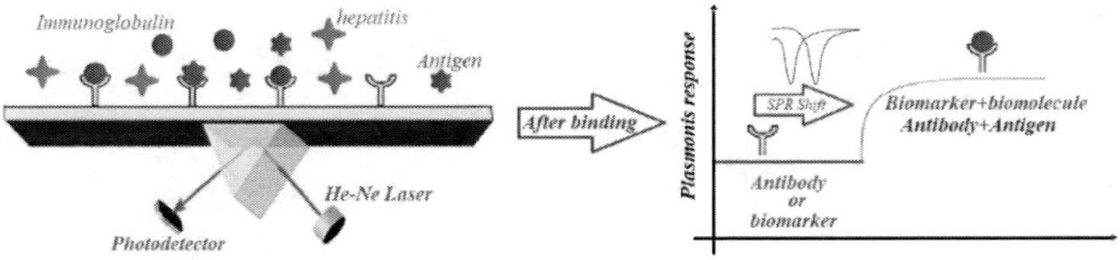

Figure 23. Antigen, hepatitis virus, and immunoglobulin were monitoring using SPR shift and binding diagram.

The SPR biosensor has potential to discover the other cancer such as lung cancer, breast cancer, gastric cancer and oral cancer. To discover the lung cancer, the simple sensing layer was prepared using carboxyl-functionalized graphene oxide (GO-COOH). The cytolerayin 19 (CK19) protein biomarker can discover in spiked human plasma using the interactions between GO-COOH and anti-CK19. Therefore, the CK19 protein can monitor as low concentration as 1 fg/mL (Chiu, Lin, Kuo, 2018).

To evaluate and discover lung cancer in a short time, the minoritized the SPR chip was prepared based on detection of exosomal epidermal growth factor receptor (EGFR) using programmed death-ligand 1 (PD-L1) as a biomarker. The sensor chip was less than 100 nm in diameter and the detection sensitivity of refractive index is about 0.009 RIU and the resolution of the sensor is about 8.311×10-6 RIU Liu, Zeng, An, et al., 2018.

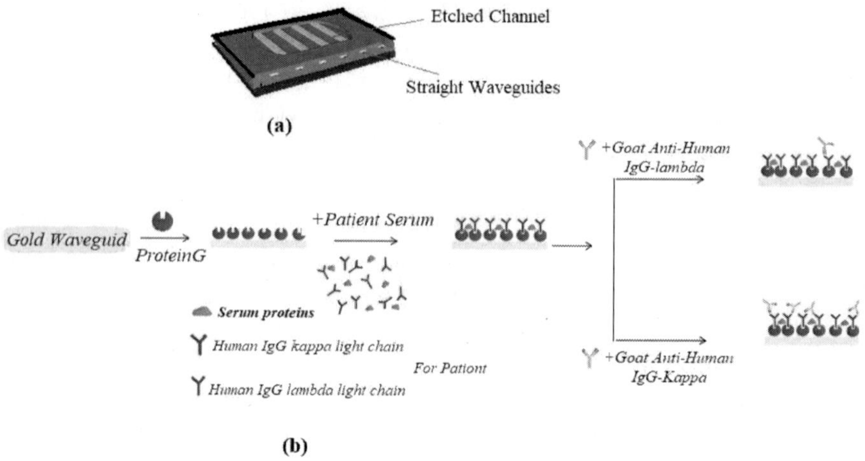

Figure 24. The demonstration of long-range SPR sensor based on waveguide coupling for detection of leukemia.

Breast cancer also can monitor using magnetic nanoparticles. The MUC1 aptamer can interact with breast cancer cells (MCF-7) and to improve the selectivity, folic acid conjugated monodisperse magnetic nanoparticles were used as a second binding reagent to prepare the sandwich SPR chip. The limit of detection is 500 cells m/L for discovery the MCF-7 (Chen, Hou, Ye, et al., 2014). The SPR immunosensor was suggested to detect breast cancer, was based on ErbB2 or HER2 a tyrosine kinase receptor (biomarker), and the SPR signals were enhanced using gold nanoparticles. The sensor assay has been evaluated with clinical sample and the limit of detection was about 180 pg m/L (Eletxigerra, Martinez-Perdiguero, Barderas, et al., 2016).

The plasmonic sensor has application to monitor the gastric cancer. Gastric cancer can monitor using monoclonal antibodies MG7 as a bio-probes to detect antigen MG7-Ag. The gold film should be modified with bio-probes. When the human sera contacts with the sensing layer that contains bio-probes, the SPR signals change and the abnormal situation can be detected (Fang, Tie, Xie, et al., 2010) and the antibodies related to gastric cancer have been monitored.

The gold nanoparticles can enhance the Raman shift. So, the complex of the gold nanoparticles and anti-epidermal growth factor receptor (EGFR) were used to discover oral cancer using surface-enhanced Raman scattering (SERS) spectrum. The spectra of saliva have a shift in the presence of oral cancer (Kah, Kho, Lee, 2007) and the growth of oral cells was monitored. Therefore, oral cancer diagnosis using SERS.

Conclusion

The surface plasmon resonance and plasmonic sensor based on the interaction of monochromatic and broadband light sources have been assigned to display, recognize, and detect the chemical and biochemical moleculars. The sugars, enzymes, antigen, and antibodies can intract with biomarkers and it is the base of the plasmonic sensor. The plasmonic sensor can display the variation of biomolecule concentration less than 10 ppm. The selectivity and sensitivity of the plasmonic sensors can control with biomarkers and linkers. Consequently, plasmonic sensors have the potential to monitor hepatitis, cancer, and virus.

References

Ahn, H., Kim, S., Kim, Y., Kim, S., Choi, J. & Kim, K. (2021). "Plasmonic sensing, imaging, and stimulation techniques for neuron studies." *Biosensors and Bioelectronics*, *182*, 113150.

Alqahtani, H., Kavakli-Thorne, M. & Kumar, G. (2021). "Applications of Generative Adversarial Networks (GANs): An Updated Review." *Archives of Computational Methods in Engineering*, *28*, 525–552.

Amoosoltani, N., Zarifkar, A. & Farmani, A. (2019). "Particle swarm optimization and finite-difference time-domain (PSO/FDTD) algorithms for a surface plasmon resonance-based gas sensor." *Journal of Computational Electronics*, *18*, 1354–1364.

Armbruster, D. A. & Pry, T. (2008). "Limit of Blank, Limit of Detection and Limit of Quantitation." *Clinical Biochemist Reviews*, *29*, S49-S52.

Ashok, B., Hariram, N., Siengchin, S. & Rajulu, A. V. (2020). "Modification of tamarind fruit shell powder with *in situ* generated copper nanoparticles by single step hydrothermal method." *Journal of Bioresources and Bioproducts*, *5*, 180-185.

Baac, H., Jun, S. B., Turner, J. N., Shain, W., Smith, K. L., Shuler, M. L. & Kim, S. J. (2005). "Extracellular Optical Recording Configuration for Neuronal Action Potential Detection by using Surface Plasmon Resonance: Preliminary Experiment"., *Conference Proceedings. 2nd International IEEE EMBS Conference on Neural Engineering*, 332-335.

Beketov, G. V., Shirshov, Y. M., Shynkarenko, O. V. & Chegel, V. I. (1998). "Surface plasmon resonance spectroscopy: prospects of superstrate refractive index variation for separate extraction of molecular layer parameters." *Sensors and Actuators B*, *48*, 432.

Beletskiy, R. V. (2013). "Imaging electrical activity of neurons with metamaterial nanosensors", in *Research Proposal Supplementary* https://arxiv.org/abs/1308.6636.

Betancourt-Galindo, R., Reyes-Rodriguez, P. Y., Puente-Urbina, B. A., Avila-Orta, C. A., Rodríguez-Fernández, O. S., Cadenas-Pliego, G., Lira-Saldivar, R. H. & García-Cerda, L. A. (2014). "Synthesis of Copper Nanoparticles by Thermal Decomposition and Their Antimicrobial Properties." *Journal of Nanomaterials*, *10*, 10.

Bradwell, A., Harding, S., Fourrier, N., Mathiot, C., Attal, M., Moreau, P., Harousseau, J. L. & Avet-Loiseau, H. (2013). "Prognostic utility of intact immunoglobulin Ig′κ/Ig′λ ratios in multiple myeloma patients." *Leukemia*, *27*, 202–207.

Burke, J. J., Stegeman, G. I. & Tamir, T. (1986). "Surface-polariton-like waves guided by thin, lossy metal films." *Physical Review B*, *33*, 5186. Accessed April 15, 1986.

Campbell, J. P., Wang, Y., Cobbold, M., Goodall, M., Bonney, Sarah L., Chamba, A., Birtwistle, J., Plant, T., Afzal, Z., Jefferis, R. & Drayson Mark, T. (2013). "Development of a highly-sensitive multi-plex assay using monoclonal antibodies for the simultaneous measurement of kappa and lambda immunoglobulin free light chains in serum and urine." *Journal of Immunological Methods*, *391*, 1-13. Accessed May 31, 2013.

Chung, D. S., Kim, H., Ko, J., Lee, J., Hwang, B., Chang, S., Kim, B. & Chung, S. J. (2018). "Microwave Synthesis of Silver Nanoparticles Using Different Pentose Carbohydrates as Reducing Agents.". *Journal of Chemistry and Chemical Engineering*, *12*, 1-10.

Choi, S. H., Kim, S. J., Im, C. H., Kim, S. A. & Kim, D. (2011). "Quantitative model for the change of optical resonance in neural activity detection systems based on surface plasmon resonance." *Optics & Laser Technology*, *43*, 938-948.

Chen, S., Zhao, Q., Zhang, L., Wang, L., Zeng, Y. & Huang, H. (2015). "Combined detection of breast cancer biomarkers based on plasmonic sensor of gold nanorods." *Sensors and Actuators B: Chemical*, *221*, 1391-1397.

Chiu, N. F., Lin, T. L. & Kuo, C. T. (2018). "Highly sensitive carboxyl-graphene oxide-based surface plasmon resonance immunosensor for the detection of lung cancer for cytokeratin 19 biomarker in human plasma." *Sensors and Actuators B: Chemical*, *265*, 264-272.

Chen, H., Hou, Y., Ye, Z., Wang, H., Koh, K., Shen, Z. & Shu, Y. (2014). "Label-free surface plasmon resonance cytosensor for breast cancer cell detection based on nano-conjugation of monodisperse magnetic nanoparticle and folic acid." *Sensors and Actuators B: Chemical*, *201*, 433-438.

Das, R., Babu, P. J., Gogoi, N., Sharma, P. & Bora, U. (2012). "Microwave-mediated rapid synthesis of gold nanoparticles Using Calotropis procera latex and study of optical properties." *International Scholarly Research Notices*, 2012, Article ID650759.

Damborský, P., Švitel, J. & Katrlík, J. (2016). "Optical biosensors." *Essays in Biochemistry*, *60*, 91–100.

Ding, X., Yan, Y., Li, Sh., Zhang, Y., Cheng, W., Cheng, Q. & Ding, S. (2015). "Surface plasmon resonance biosensor for highly sensitive detection of microRNA based on DNA super-sandwich assemblies and streptavidin signal amplification." *Analytica Chimica Acta*, *874*, 59-65.

Draz, Mohamed S. & Shafiee, H. (2018). "Applications of gold nanoparticles in virus detection." *Theranostics*, *8*, 1985–2017.

Drozd, M., Pietrzak, M. D. & Malinowska, E. (2018). "SPRi-Based Biosensing Platforms for Detection of Specific DNA Sequences Using Thiolate and Dithiocarbamate Assemblies"., *Frontiers. Chemistry*, *6*, 173. Accessed May 2018.

Eddin, F. B. K., Fen, Y. W., Omar, N. A. S., Liew, J. Y. C. & Daniyal, M. (2021). "Femtomolar detection of dopamine using surface plasmon resonance sensor based on chitosan/graphene quantum dots thin film." *Spectrochimica Acta Part A: Molecular and Biomolecular Spectroscopy*, *263*, 120202.

Elsupikhe, R. F., Shameli, K. & Ahmad, M. B. (2015). "Sonochemical method for the synthesis of silver nanoparticles in κ-carrageenan from silver salt at different concentrations." *Research on Chemical Intermediates*, *41*, 8515–8525.

Elhakim, H. K. A., Azab, S. M. & Fekry, A. M. (2018). "A novel simple biosensor containing silver nanoparticles/propolis (bee glue) for microRNA let-7a determination." *Materials Science and Engineering*, *92*, 489-495.

El-Kady, M. & Kaner, R. (2013). "Scalable fabrication of high-power graphene micro-supercapacitors for flexible and on-chip energy storage." *Nature Communications*, *4*, 1475.

Eletxigerra, U., Martinez-Perdiguero, J., Barderas, R., Pingarrón, J. M., Campuzano, S. & Merino, S. (2016). "Surface plasmon resonance immunosensor for ErbB2 breast cancer biomarker determination in human serum and raw cancer cell lysates." *Analytica Chimica Acta*, *905*, 156-162.

Fathi, F., Rahbarghazi, R., Movassaghpour, Ali A. & Rashidi, Mohammad-R. (2019). "Detection of CD133-marked cancer stem cells by surface plasmon resonance: Its application in leukemia patients." *Biochimica et Biophysica Acta (BBA) - General Subjects*, *1863*, 1575-1582.

Fang, X., Tie, J., Xie, Y., Li, Q., Zhao, Q. & Fan, D. (2010). "Detection of gastric carcinoma-associated antigen MG7-Ag in human sera using surface plasmon resonance sensor." *Cancer Epidemiology*, *34*, 648-651.

Green, L. N., Subramanian, H. K. K. & Mardanlou, V. (2019). "Autonomous dynamic control of DNA nanostructure self-assembly." *Nature Chemistry*, *11*, 510–520.

Gopinath, Subash C. B. & Kumar Penmetcha K. R. (2013). "Aptamers that bind to the hemagglutinin of the recent pandemic influenza virus H1N1 and efficiently inhibit agglutination." *Acta Biomaterialia*, *9*, 8932-8941.

Haddawi, S. F., Humud, H. R. & Hamidi, S. M. (2020). "Signature of plasmonic nanoparticles in multi-wavelength low power random lasing." *Optics & Laser Technology*, *121*, 105770.

Hamidi, S. M. & Oskuei, M. A. (2013). "Magneto-plasmonic effect in cobalt thin film incorporating core–shell Ag@ Au nanoparticles." *Journal of Superconductivity and Novel Magnetism*, *27*, 1469-1472.

He, L., Musick, M. D., Nicewarner, S. R., Salinas, F. G., Benkovic, S. J., Natan, M. J. & Keating, C. D. (2000). "Colloidal Au-enhanced surface Plasmon resonance for ultrasensitive detection of DNA hybridization." *Journal of the American Chemical Society*, *122*, 9071–9077.

Homola, J. & Jakub, D. (2006). "*Surface plasmon resonance based sensors*." Berlin, Springer.

Homola, J., Vaisocherová, H., Dostálek, J. & Piliarik, M. (2005). "Multianalyte surface plasmon resonance biosensing." *Methods*, *37*, 26–36.

Hu, X., Zhang, Y., Ding, T., Liu, J. & Zhao, H. (2020). "Multifunctional Gold Nanoparticles: A Novel Nanomaterial for Various Medical Applications and Biological Activities." *Frontiers in Bioengineering and Biotechnology*, *8*, 990. Accessed August, 13, 2020.

Hu, Y., Zhang, L., Zhang, Y., Wang, B., Wang, Y., Fan, Q., Huang, W. & Wang, L. (2015). "Plasmonic Nanobiosensor Based on Hairpin DNA for Detection of Trace Oligonucleotides Biomarker in Cancers." *ACS Applied Materials*, *7*, 2459–2466.

Huang, Y., Nguyen, M. K., Natarajan, A. K., Nguyen, V. H. & Kuzyk, A. (2018). "A DNA Origami-Based Chiral Plasmonic Sensing Device." *ACS Applied Materials*, *10*, 44221–44225.

Huang, Y., Nguyen, M. K., Natarajan, A. K., Nguyen, V. H. & Kuzyk, A. (2018). "A DNA Origami-Based Chiral Plasmonic Sensing Device." *ACS Applied Materials & Interfaces*, *10*, 51. Accessed December 7, 2018.

Hunt, H. K. & Armani, A. M. (2010). "Label-free biological and chemical sensors." *Nanoscale*, *2*, 1544–1559.

Hwang, S. Y., Yoon, C. H., Jeon, J. Y., Choi, Sung C. & Lee, Eun K. (2005). "Quantitative assay of hepatitis B surface antigen by using surface plasmon resonance biosensor." *Biotechnology and Bioprocess Engineering*, *10*, 309. Accessed Augest 2005.

Jamkhande, Prasad G., Ghule, Namrata W., Bamer, Abdul H. & Kalaskar, Mohan G. (2019). "Metal nanoparticles synthesis: An overview on methods of preparation, advantages and disadvantages, and applications." *Journal of Drug Delivery Science and Technology*, *53*, 101174.

Jahanshahi, P. Wei, Q., Jie, Z., Ghomeishi, M., Sekaran, Shamala D. & Adikan, Faisal R. M. (2017). "Kinetic analysis of IgM monoclonal antibodies for determination of dengue sample concentration using SPR technique." *Bioengineered.*, *4*, 239-247. Accessed May 2017.

Kabashin, A. & Nikitin, P. (1998). "Surface plasmon resonance interferometer for bio-and chemical-sensors." *Optics Communication*, *150*, 5–8.

Kah, J. C. Y., Kho, K. W., Lee, C. G. L., Richard, C. J., Shen, Z. X., Soo, K. C. & Olivo, M. C. (2007). "Early diagnosis of oral cancer based on the surface plasmon resonance of gold nanoparticles." *International Journal of Nanomedicine*, *2*(4), 785–798.

Khoa, N. T., Kim, S. W., Yoo, D. H., Kim, E. J. & Hahn, S. H. (2014). "Size-dependent work function and catalytic performance of gold nanoparticles decorated graphene oxide sheets." *Applied Catalysis*, *469*, 159-164.

Kittle, C. (1996). "*Solid State Physics*." New York, Wiley.

Kim, J. W., Kim, M., Lee, Kyung K., Chung, Kwang H. & Lee, C. S. (2020). "Effects of Graphene Oxide-Gold Nanoparticles Nanocomposite on Highly Sensitive Foot-and-Mouth Disease Virus Detection." *Nanomaterials*, *10*, 1921. Accessed September 25, 2020.

Kim, S. A., Byun, K. M., Lee, J., Kim, J. H., Kim, D. G. A., Baac, H., Shuler, M. L. & Kim, S. J. (2008). "Optical measurement of neural activity using surface plasmon resonance." *Optics Letters*, *33*, 914-916.

Kodeary, A. K., Gatea, M. A., Haddawi, S. F. & Hamidi, S. M. (2020). "Tunable thermo-piezo-plasmonic effect on core/shell nanoparticles under laser irradiation and external electric field." *Optical and Quantum Electronics*, *52*, 1-19.

Kuntyi, O. I., Kytsya, A. R., Mertsalo, I. P. Mazur, A. S., Zozula, G. I., Bazylyak, L. I. & Topchak, R. V. (2019). "Electrochemical synthesis of silver nanoparticles by reversible current in solutions of sodium polyacrylate". *Colloid and Polymer Science.*, *297*, 689–695.

Kumbhat, S., Shankaran, D. R., Kim, S. J., Gobi, K. V., Joshi, V. & Miura, N. (2007). "Surface plasmon resonance biosensor for dopamine using D3 dopamine receptor as a biorecognition molecule." *Biosensors Bioelectronics*, *23*(3), 421-427.

Lednický Tomáš, L. T. & Bonyár, A. (2020). "Large Scale Fabrication of Ordered Gold Nanoparticle–Epoxy Surface Nanocomposites and Their Application as Label-Free Plasmonic DNA Biosensors." *ACS Applied Materials Interfaces*, *12*, 4804–4814.

Liu, Z. & Bando, Y. (2003). "A Novel Method for Preparing Copper Nanorods and Nanowires." *Advanced Materials*, *15*, 303-305.

Li, Q., Wang, Q., Yang, X., Wang, K., Zhang, H. & Nie, W. (2017). "High sensitivity surface plasmon resonance biosensor for detection of microRNA and small molecule based on graphene oxide-gold nanoparticle composites." *Talanta*, *174*, 521-526.

Li, D., Su, J., Yang, J., Yu, S., Zhang, J., Xu, K. & Yu, H. (2017). "Optical surface plasmon resonance sensor modified by mutant glucose/galactose-binding protein for affinity detection of glucose molecules." *Biomedical Optics Express*, *1*, 5206–5217.

Liu, C., Zeng, X., An, Z., Yang, Y., Eisenbaum, M., Gu, X., Jornet, J. M., Dy, G. K., Reid, M. E., Gan, Q. & Wu, Y. (2018). "Sensitive Detection of Exosomal Proteins via a Compact Surface Plasmon Resonance Biosensor for Cancer Diagnosis." *ACS Sensors*, *3*, 1471–1479.

Logo, Z. D., Qian, J., Li, Y., Lin, R., Wang, J., Cheng, P. & Xu, Z. (2021). "Reducing molecular simulation time for AFM images based on super-resolution methods." *Beilstein Journal of Nanotechnology*, *12*, 775–785.

Maisonneuve, M., Valsecchi, C., Wang, C., Brolo, A. G. & Meunier, M. (2015). "Leukemic marker detection using a spectro-polarimetric surface plasmon resonance platform." *Biosensors and Bioelectronics*, *63*, 80-85.

Miyazaki, Celina M., Shimizu, Flávio M., Mejía-Salazar, J. R., Oliveira, Jr. Osvaldo N. & Ferreira, M. (2017). "Surface plasmon resonance biosensor for enzymatic detection of small analytes." IOP Publishing Ltd, *Nanotechnology*, *28*, 145501.

Miao, P., Zhang, T., Xu, J. & Tang, Y. (2018). "Electrochemical Detection of miRNA Combining T7 Exonuclease-Assisted Cascade Signal Amplification and DNA-Templated Copper Nanoparticles." *Analytical Chemistry*, *90*, 11154–11160.

Miti, A., Thamm, S., Müller, P., Csáki, A., Fritzsche, W. & Zuccheri, G. (2020). "A miRNA biosensor based on localized surface plasmon resonance enhanced by surface-bound hybridization chain reaction." *Biosensors and Bioelectronics*, *167*, 112465.

Mousavi, M. Z., Chen, H. Y., Lee, K. L., Lin, H., Chen, H. H., Lin, Y. F., Wong, Ch. Sh., Li, H. F., Wei, P. K. & Cheng, J. Y. (2015). "Urinary micro-RNA biomarker detection using capped gold nanoslit SPR in a microfluidic chip." *Analyst*, *140*, 4097-4104.

Niskanen, I., Forsberg, V., Zakrisson, D., Reza, S., Hummelgård, M., Andres, B., Fedorov, I., Suopajärvi, T. Liimatainen, H. & Thungström, G. (2019). "Determination of nanoparticle size using Rayleigh approximation and Mie theory." *Chemical Engineering Science*, *201*, 222-229.

Okitsu, K., Ashokkumar, M. & Grieser, F. (2005). "Sonochemical synthesis of gold nanoparticles: effects of ultrasound frequency." *Journal of Physical Chemistry B*, *109*, 20673– 20675.

Oleszak, O. & Shingu, P. H. (1996). "Nanocrystalline metals prepared by low energy ball milling." *Journal of Applied Physics*, *79*, 2975.

Omar, Nur A. S., Fen, Y. W., Ramli, I., Sadrolhosseini, Amir R., Abdullah, J., Yusof, Nor A., Kamil, Yasmin M. & Mahdi, Mohd A. (2021). "An Optical Sensor for Dengue Envelope Proteins Using Polyamidoamine Dendrimer Biopolymer-Based Nano-composite Thin Film: Enhanced Sensitivity, Selectivity, and Recovery Studies." *Polymers*, *13*, 762.

Otto, A. (1968). "Excitation of nonradiative surface plasma waves in silver by the method of frustrated total reflection." *Zeitschrift für Physik A Hadrons and Nuclei*, *216*, 398–410.

Pacholski, C., Rosencrantz, S., Rosencrantz, R. R. & Balderas-Valadez, Ruth F. (2020). "Plasmonic biosensors fabricated by galvanic displacement reactions for monitoring biomolecular interactions in real time." *Analytical and Bioanalytical Chemistry*, *412*, 3433–3445.

Pathak, A. K. & Viphavakit, C. (2021). "VOC biomarker monitoring for diabetes through exhaled breath using Ag/P-TiO2 composite plasmonic sensor." *IEEE Sensors Journal*, 1-1. Accessed August 13, 2021..

Peyghambarian, N., Koch, S. W. & Mysyrowicz, A. (1993). "*Introduction to Semiconductor Optics*." New Jersey, Prentice Hall, 1993.

Prabowo, Briliant A., Wang, Robert Y. L., Secario, Muhammad K., Ou, Po-T., Alom, A., Liu, J. J. & Liu, K. C. (2017). "Rapid detection and quantification of Enterovirus 71 by a portable surface plasmon resonance biosensor." *Biosensors and Bioelectronics*, *92*, 186-191. Accessed Jun 15, 2017.

Rajeshkumar, S. & Rinitha, G. (2018). "Nanostructural characterization of antimicrobial and antioxidant copper nanoparticles synthesized using novel Persea americana seeds." *OpenNano*, *3*, 18-27.

Rapisarda, A., Giamblanco, N. & Marletta, G. (2017). "Kinetic discrimination of DNA single-base mutations by localized surface plasmon resonance." *Journal of Colloid and Interface Science*, *487*, 141-148.

Requejo-Isidro, J. (2013). "Fluorescence nanoscopy. Methods and applications." *Journal of Chemical Biology*, *6*, 97–120.

Riedel, T., Rodriguez-Emmenegger, C. & de los Santos Pereira, A. (2014). "Diagnosis of Epstein–Barr virus infection in clinical serum samples by an SPR biosensor assay." *Biosensors and Bioelectronics*, *55*, 278–284.

Riedel, T., Hageneder, S., Surman, F., Pop-Georgievski, O., Noehammer, C., Hofner, M., Brynda, E., Rodriguez-Emmenegger, C. & Dostálek, J. (2017). "Plasmonic Hepatitis B Biosensor for the Analysis of Clinical Saliva." *Analytical Chemistry*, *89*, 2972–2977.

Rothenhäuslar, B. & Knoll, W. (1988). "Surface-plasmon microscopy." *Nature*, *332*, 615–617.

Ruffato, G., Zacco, G. & Romanato, F. (2012). "Innovative Exploitation of Grating-Coupled Surface Plasmon Resonance for Sensing." In: *Plasmonics - Principles and Applications*, edited by Ki Y. Kim, London, *IntechOpen*, Chapter 17, 419-444. Accessed October 24, 2012.

Sadrolhosseini, Amir R. Moksin, M. M., Nang, H. L. L., Norozi, M., Yunus, W. M. M. & Zakaria, A. (2011). "Physical Properties of Normal Grade Biodiesel and Winter Grade Biodiesel." *International Journal of Molecular Sciences*, *11*, 2100-2111.

Santos, Gabriel M. C. d. S., Alves, Carlos R., Pinto, Marcelo A., Leon, Luciane, A. A. & Souza-Silva, F. (2021). "Detection of Antibodies against Hepatitis A Virus (HAV) by a Surface Plasmon Resonance (SPR) Biosensor: A New Diagnosis Tool Based on the Major HAV Capsid Protein VP1 (SPR-HAVP1)." *Sensors*, *21*(9), 3167.

Sadrolhosseini, Amir R., Shafie, S., Abdul Rashid, S. & Mahdi, Mohd A. (2020). "Surface plasmon resonance measurement of arsenic in low concentration using polypyrrole-graphene quantum dots layer." *Measurement*, 108546.

Sadrolhosseini, Amir R., Moksin, M. M., Yunus, W. M. M., Talib, Z. A. & Abdi, M. M. (2011). "Surface Plasmon Resonance Detection of Copper Corrosion of Biodiesel Using Polypyrrole-Chitosan Layer Sensor." *Optical review*, *18*, 331-337.

Sadrolhosseini, Amir R., Krishnan, G., Safie, S., Beygisangchin, M., Abdul Rashid, S. & Harun, Sulaiman W. (2020). "Enhancement of fluorescence property of carbon quantum dots based on laser ablated gold nanoparticles to evaluate pyrene." *Optical Material Express*, *10*, 8. Accessed Sep 23, 2020.

Sadrolhosseini, Amir R., Abdul Rashid, S., Shafie, S. & Nezakati, H. (2019). "Laser ablation synthesis of gold nanoparticle to enhance the fluorescence properties of graphene quantum dots." *Journal of Laser Applications*, *31*, 022006. Accessed February 1, 2019.

Sadrolhosseini, Amir R., Abdul Rashid, S. & Zakaria, A. (2017). "Synthesis of Gold Nanoparticles Dispersed in Palm Oil Using Laser Ablation Technique." *Journal of Nanomaterials*, v *2017*, Article ID 6496390. Accessed May 5, 2017.

Sadrolhosseini, Amir R., Moozarm Nia, P., Naseri, M., Mohammadi, A., Fen, Y. W., Shafie, S. & Mohamed Kamari, H. (2020). "Surface Plasmon Resonance Sensor Based on Polypyrrole–Chitosan–BaFe2O4 Nanocomposite Layer to Detect the Sugar." *Applied Sciences*, *10*, 2855.

Scarano, S., Pascale, E., Palladino, P., Fratini, E. & Minunni, M. (2018). "Determination of fermentable sugars in beer wort by gold nanoparticles@polydopamine: A layer-by-layer approach for Localized Surface Plasmon Resonance measurements at fixed wavelength." *Talanta*, *183*, 24-32.

Schasfoort, R. & Tudos, A. (2008). "*Handbook of Surface Plasmon Resonance*." Cambridge, U.K., RSC Publishing.

Schofield, D. J. & Dimmock, N. J. (1996). "Determination of affinities of a panel of IgGs and Fabs for whole enveloped (influenza A) virions using surface plasmon resonance." *Journal of Virological Methods*, *62*, 33–42. Accessed October, 1996.

Sharma, K. (2006). *Optics*. Academic Press, California, 301-316.

Singh, Y., Paswan, M. K. & Raghuwanshi, S. K. (2021). "Sensitivity Enhancement of SPR Sensor with the Black Phosphorus and Graphene with Bi-layer of Gold for Chemical Sensing." *Plasmonics* Accessed April 9, 2021.

Song, C., Ooi, Ean T. & Natarajan, S. (2018). "A review of the scaled boundary finite element method for two-dimensional linear elastic fracture mechanics." *Engineering Fracture Mechanics*, *187*, 45-73.

Sohrabi, F. & Hamidi, Seyedeh M. (2016). "Neuroplasmonics: From Kretschmann configuration to plasmonic crystals." *The European Physical Journal Plus*, *131*, 221. Accessed July 1, 2016.

Song, S., Lee, Jong U., Kang, J., Park, Kyong H. P. & Sim, Sang J. (2020). "Real-time monitoring of distinct binding kinetics of hot-spot mutant p53 protein in human cancer cells using an individual nanorod-based plasmonic biosensor" *Sensors and Actuators B: Chemical*, *322*, 128584. Accessed November 1, 2020.

Solanki, J. N., Sengupta, R. & Murthy, Z. V. P. (2010). "Synthesis of copper sulphide and copper nanoparticles with microemulsion method." *Solid State Sciences*, *12*, 1560-1566.

Sohrabi, F., Etezadi, D., Perin, R., Jahani, Y., Mohammadi, A. & Hamidi, S. M. (2020). "Phase-sensitive optical neural recording of cerebellum tissue on a flexible interface." *Journal of Applied Physics*, *127*, 113101.

Sohrabi, F., Jahani, Y., Sanchez-Mut, J. V. E., Mohammadi, E., Barzegar, Z., Li, X. & Hamidi, S. M. (2020). "Membrane activity detection in cultured cells using phase-sensitive plasmonics." *Optics Express*, *28* (24), 36643.

Sohrabi, F., Saeidifard, S., Ghasemi, M., Asadishad, T., Hamidi, S. M. & Hosseini, S. (2021). "Role of plasmonics in detection of deadliest viruses: a review." *The European Physical Journal Plus*, *136* (6), 1-71.

Stephenson-Brown, A., Wang, H. C., Iqbal, P., Preece, J. A., Long, Y., Fossey, J. S., James, T. D. & Mendes, P. M. (2013). "Glucose selective Surface Plasmon Resonance-based bis-boronic acid sensor." *Analyst*, *138*, 7140-7145.

Tahseen, K., Ikram, A., Sher, B. Kh., Ul-Islam, M. & Asiri, A. M. (2019). "Microwave Assisted Synthesis and Carboxymethyl Cellulose Stabilized Copper Nanoparticles on Bacterial Cellulose Nanofibers Support for Pollutants Degradation." *Journal of Polymers and the Environment*, *27*, 2867–2877.

Tam, Y. J., Zeenathul, N. A., Rezaei, M. A., Mustafa, N. H., Azmi, M. L. M., Bahaman, A. R., Cen Lo, S., Tan, J. S., Hani, H. & Rasedee, A. (2017). "Wide dynamic range of surface-plasmon-resonance-based assay for hepatitis B surface antigen antibody optimal detection in comparison with ELISA." *Biotechnology and Applied Biochemistry*, *64*, 735-744.

Tokel, O., Inci, F. & Demirci, U. (2014). "Advances in Plasmonic Technologies for Point of Care Applications." *Chemical Review*, *114*, 5728–5752.

Vaisocherová, H., Šípová, H., Víšová, I., Bocková, M., Špringer, T., Ermini, M. L., Song, X., Krejčík, Z., Chrastinová, L., Pastva, O, Pimková, K., Merkerová, D., Dyr, J. E. & Homola, J. (2015). "Rapid and sensitive detection of multiple microRNAs in cell lysate by low-fouling surface plasmon resonance biosensor." *Biosensors and Bioelectronics*, *70*, 226-231.

Wang, J., Banerji, S., Menegazzo, N., Peng, W., Zou, Q. & Booksh, K. S. (2011). "Glucose detection with surface plasmon resonance spectroscopy and molecularly imprinted hydrogel coatings." *Talanta*, *86*, 133-141.

Wood, R. W. (1902). "On a remarkable case of uneven distribution of light in a diffraction grating spectrum." *The London, Edinburgh, and Dublin Philosophical Magazine and Journal of Science*, *4*, 396–402. Accessed April 15, 2009.

Xue, T., Liang, W., Li, Y., Sun, Y., Xiang, Y., Zhang, Y., Dai, Z., Duo, Y., Wu, L., Qi, K., Shivananju, Bannur N., Zhang, L., Cui, X., Zhang, H. & Bao, Q. (2019). "Ultrasensitive detection of miRNA with an antimonene-based surface plasmon resonance sensor." *Nature Communications*, *10*, 28. Accessed January 3, 2019.

Xu, L., Wang, Y. Y., Huang, J., Chen, C. Y., Wang, Z. X. & Xie, H. (2020). "Silver nanoparticles: Synthesis, medical applications and biosafety." *Theranostics*, *10*, 8996–9031.

Yariv, A. & Yeh, P. (2003). "*Optical waves in crystals: propagation and control of laser radiation*." New Jersey, Hoboken, Wiley.

Yatsui, K., Grigoriu, C., Kubo, H., Masugata, K. & Shimotori, Y. (1995). "Synthesis of nanosize powders of alumina by ablation plasma produced by intense pulsed light-ion beam." *Applied Physics Letters*, *67*, 1214.

Yeh, M. S., Yang, Y. S., Lee, Y. P., Lee, H. F., Yeh, Y. H. & Yeh, C. S. (1999). "Formation and Characteristics of Cu Colloids from CuO Powder by Laser Irradiation in 2-Propanol." *Journal of Physical Chemistry B*, *103*, 6851–6857.

Yeung, W. K., Li, H. F., Chung, C. L., Lee, K. L., Wei, P. K., Lin, H., Chen, H. H. & Cheng, J. Y. (2020). "Promising urinary miRNA biomarkers t-SPR profiling for urothelial cell carcinoma." *Sensors and Actuators B*, *322*, 2020.

Yin, P., Lin, Q., Ruan, Y. & Chen, J. J. (2021). "Investigation of multiple metal nanoparticles near-field coupling on the surface by discrete dipole approximation method." *Optoelectronics Letters*, *17*, 257–261.

Yin, X., Shi, P., Yang, A., Du, L. & Yuan, X. (2020). "Surface plasmon coupled nano-probe for near field scanning optical microscopy." *Optics Express*, *28*, 14831-14838.

Yuan, X. C., Ong, B., Tan, Y., Zhang, D., Irawan, R. & Tjin, S. (2006). "Sensitivity-stability-optimized surface plasmon resonance sensing with double metal layers." *Journal of optics*, *959*, 963-968.

Zalevsky, Z. & Abdulhalim, I. (2010). "Integrated Nanophotonic Devices." London, UK, Elsevier, 175-205.

Zeng, Z., Zhou, R., Sun, R., Zhang, X., Cheng, Z., Chen, C. & Zhu, Q. (2021). "Nonlinear hybridization chain reaction-based functional DNA nanostructure assembly for biosensing, bioimaging applications." *Biosensors and Bioelectronics*, *173*, 112814.

Zhang, J., Atay, T. & Nurmikko, A. V. (2009). "Optical Detection of Brain Cell Activity Using Plasmonic Gold Nanoparticles." *Nano Letters*, *9*, 519–524.

Zheng, X., Peng, Y., Lombardi, J. R., Cui, X. & Zheng, W. (2016). "Photochemical growth of silver nanoparticles with mixed-light irradiation." *Colloid and Polymer Science*, *294*, 911–916.

Zimmerman, J. F. & Tian, B. (2018). "Nongenetic Optical Methods for Measuring and Modulating Neuronal Response." *ACS Nano*, *12*, 4086–4095.

Biographical Sketches

Amir Reza Sadrolhosseini

Affiliation: Postdoctoral Researcher Fellow in Plasmonic

Education: PhD in Applied Optics

Business Address: Magneto-plasmonic Lab, Laser and Plasma Research Institute, Shahid Beheshti University, Daneshju Boulevard, 1983969411, Tehran, Iran

Research and Professional Experience: Optics, Plasmonic, Optical Sensor, and Laser Ablation

Honors:

Silver Medal in PRPi14

Project: Measurement of Metal Nanoparticles Concentration in Aqueous Solution Using Surface Plasmon Resonance Sensor.

Silver Medal in PRPi14

Project: Green Fabrication of Metal Nanoparticles in Graphene Oxide

Gold Medal in ITEX 2013

Project: Heavy Metal Detection Based on Chitosan

Silver Medal in PRPi12

Project: Measurement of Very Low Water Concentration in Palm Oil Biodiesel and Blend

Silver Medal in PRPi12

Project: Surface Plasmon Resonance Corrosion Sensor

Gold Medal in PRPi11

Project title: Fabrication of Silver Nanoparticle Dispersed in palm oil using Laser Ablation

University Putra Malaysia, Graduate Research Fellowship (GRF) 2007-2009

Publications from the Last 3 Years:

2021:

Alif Syafiq Kamarol Zaman, Tong Ling Tan, Yamuna A/P Chowmasundaram, Norhanisah Jamaludin, Amir Reza Sadrolhosseini, Umer Rashid, Suraya Abdul Rashid. Properties and molecular structure of carbon quantum dots derived from empty fruit bunch biochar using a facile microwave-assisted method for the detection of Cu^{2+} ions. *Optical Materials*, 112, 110801 (1/Feb/2021). https://doi.org/10.1016/j.optmat.2021.110801 IF:2.779 Q2.

Amir Reza Sadrolhosseini, Mina Habibiasr, Hassan Soleimani, Mohd Nizar Hamidon, Yap Wing Fen, Lim, HN. Surface Plasmon Resonance Sensor to Detect n-Hexane in Palm Kernel Oil Using Polypyrrole Nanoparticles Reduced Graphene Oxide Layer, *Journal of Sensors*, V2021, *Article ID 8813801* (4/Jan/2021) IF:1.595 Q2-Q3. https://doi.org/10.1155/2021/8813801.

Amir Reza Sadrolhosseini, Suhaidi Shafie, Hassan Soleimani, Mohd. Adzir Mahdi. Polypyrrole-$BaFe_2O_4$ Sensing Layer for Detection of Strontium Ion in Aqueous Solution using Surface Plasmon Resonance Curves Fitting. *Optics and Laser Technology*, 140, 106970 (9/March/2021) IF:3.232 Q1-Q2. https://doi.org/10.1016/j.optlastec.2021.106970

Amir Reza Sadrolhossini, Mahnoush Beygisangchin, Suhaidi Shafie, Suraya Abdul Rashid, Hossein Nezakati, Laser ablated Titanium Oxide Nanoparticles in Carbon Quantum Dots Solution for Detection of Sugar Using Fluorescence Spectroscopy. *Materials Research Express*, Accepted, (22/Jan/2021) IF:1.929. Q2-Q3. https://iopscience.iop.org/article/10.1088/2053-1591/abdf0f.

Nur Alia Sheh Omar, Yap Wing Fen, Irmawati Ramli, Amir Reza Sadrolhosseini, Jaafar Abdullah, Nor Azah Yusof, Yasmin Mustapha Kamil, Mohd Adzir Mahdi. An Optical Sensor for Dengue Envelope Proteins Using Polyamidoamine Dendrimer Biopolymer-Based Nanocomposite Thin Film: Enhanced Sensitivity, Selectivity, and Recovery Studies. *Polymers*, 13, 762(12/Feb2021). https://doi.org/10.3390/polym13050762 IF:3.426 Q2.

2020:

Amir Reza Sadrolhosseini, Ganesan Krishnan, Suhaidi Safie, Mahnoush Beygisangchin, Suraya Abdul Rashid, Sulaiman Wadi Harun. Enhancement of fluorescence property of carbon quantum dots based on laser ablated gold nanoparticles to evaluate pyrene. *Optical Material Express*, 10, 8 (23/Sep/2020) IF:2.673, Q2 https://doi.org/10.1364/OME.396914.

Amir Reza Sadrolhosseini, Ganesan Krishnan, Suhaidi Shafie, Suraya Abdul Rashid and Sulaiman Wadi Harun. Optical and Photoacoustic Properties of Laser-Ablated Silver Nanoparticles in a Carbon Dots Solution. *Molecules 25*(24), 5798 (2/Dec/2020) IF:3.267 Q1-Q2 https://doi.org/10.3390/molecules25245798.

Amir Reza Sadrolhosseini, Mohr. Ikmal Hafizi Bin Azaman, Amir Masoud Arabi, Suhaidi Shafie, Nur Hazahsha Shamsudin, Mohd. Adzir Mahdi, Suraya Abdul Rashid, Photoluminescence property of laser-ablated zinc oxide-carbon quantum dots nanocomposites for detection of Hg and Pb ions. *Journal of Nanophotonics*, *14*(4), 046004-1 (30/Oct/2020) IF: 1.415 Q3. https://doi.org/10.1117/1.JNP.14.046004.

Amir Reza Sadrolhosseini, Pooria Moozarm Nia, Mahmoud Naseri, Ahmad Mohammadi, Yap Wing Fen, Suhidi Shafie, Halimah Mohamed Kamari. Surface Plasmon Resonance Sensor Based on Polypyrrole–Chitosan–$BaFe_2O_4$ Nanocomposite Layer to Detect the Sugar. *Appl. Sci. 10*, 8, 2855 (20/Apri/2020) IF:2.17, Q2. https://doi.org/10.3390/app10082855.

Amir Reza Sadrolhosseini, Pooria Moozarm Nia, Suhaidi Shafie, Kamyar Shameli, Mohd Adzir Mahdi. Polypyrrole-Chitosan-$CaFe_2O_4$ Layer Sensor for Detection of Anionic and Cationic Dye Using Surface Plasmon Resonance, *International Journal of Polymer Science*, v2020, Article ID 3489509 (16/Jan/2020). IF: 1.892, Q2 https://doi.org/10.1155/2020/3489509.

Amir Reza Sadrolhosseini, Suhaidi Shafie, Suraya Abdul Rashid, Mohd Adzir Mahdi. Surface plasmon resonance measurement of arsenic in low concentration using polypyrrole-graphene quantum dots layer, *Measurement*, 108546 (1/March/2020) IF:3.364 Q1-Q2. https://doi.org/10.1016/j.measurement.2020.108546.

Fahad Usman, John Ojur Dennis, Abdelaziz Yousif Ahmed, Khe Cheng Seong, Yap Wing Fen, Amir Reza Sadrolhosseini, Fabrice Meriaudeau, Pradeep Kumar, Olumide Bolarinwa Ayodele, Structural characterization and optical constants of p-toluene sulfonic acid doped polyaniline and its composites of chitosan and reduced graphene-oxide. *Journal of Material Research and Technology.*, *9*, 2, 1468-1476 (March/2020) IF:3.327, Q1-Q2. https://doi.org/10.1016/j.jmrt.2019.11.072.

Fahad Usman, John Ojur Dennis, Mkawi, EM; Yas Al-Hadeethi, Fabrice Meriaudeau, Yap Wing Fen, Amir Reza Sadrolhosseini, Thomas L. Ferrell, Ahmed Alsadig, Abdelmoneim Sulieman, Acetone Vapor-Sensing Properties of Chitosan-Polyethylene Glycol Using Surface Plasmon Resonance Technique, *Polymers*, 2020, 12(11), 2586 IF: Q1-Q2 https://doi.org/10.3390/polym12112586.

Noor Zirwatul Ahlam Naharuddin, Amir Reza Sadrolhosseini, Muhammad Hafiz Abu Bakar, Nizam Tamchek, Mohd Adzir Mahdi. Laser Ablation Synthesis of Gold Nanoparticles in Tetrahydrofuran. *Optical Material*

Express, 10, 2, 323-331 (1/Feb/2020) IF:2.673, Q2 https://doi.org/10.1364/OME.381427.

Norhanisah Jamaludin, Tongling Tan, Alif Syafiq Kamarol Zaman, Amir Reza Sadrolhosseini, Suraya Abdul Rashid. Acid-Free Hydrothermal-Extraction and Molecular Structure of Carbon Quantum Dots Derived from Empty Fruit Bunch Biochar Materials. *Materials*, *13*(15), 3356 (29/Jul/2020) IF:3.057 Q2 https://doi.org/10.3390/ma13153356.

Nur Alia Sheh Omar, Yap Wing Fen, Jaafar Abdullah, Amir Reza Sadrolhosseini, Yasmin Mustapha Kamil, Nurul'Illya Muhamad Fauzi, Hazwani Suhaila Hashim, Mohd Adzir Mahdi. Quantitative and selective surface plasmon resonance response based on a reduced graphene oxide–polyamidoamine nanocomposite for detection of dengue virus e-proteins. *Nanomaterials*, v10, 3, 569(21/March/2020) IF:4.034 Q2. https://doi.org/10.3390/nano10030569.

Nur Alia Sheh Omar, Yap Wing Fen, Jaafar Abdullah, Yasmin Mustapha Kamil, Wan Mohd Ebtisyam Mustaqim Mohd Daniyal, Amir Reza Sadrolhosseini, Mohd Adzir Mahdi, Sensitive Detection of Dengue Virus Type 2 E-Proteins Signals Using Self-Assembled Monolayers/Reduced Graphene Oxide-PAMAM Dendrimer Thin Film-SPR Optical Sensor. *Scientific Reports*, v10, 1 pp 1-15. (11/Feb/2020) IF: 3.998 Q1. https://doi.org/10.1038/s41598-020-59388-3.

2019:

Amir Reza Sadrolhosseini, Mahmoud Naseri, Investigation of Corrosiveness Biodiesel Blends Using Polypyrrole Chitosan-Cobalt/Ferrite Nano-composite. *Protection of Metals and Physical Chemistry of Surfaces*, 55, 1, 72-79 (13/May/2018) IF:0.985 Q3 https://doi.org/10.1134/S2070205119010179.

Amir Reza Sadrolhosseini, Mina Habibiasr, Suhaidi Shafie, Hassan Solaimani and Hong Ngee Lim. Optical and Thermal Properties of a Laser-Ablated Platinum Nanoparticles Graphene Oxide Composite, International Journal of molecular Sciences. *Int. J. Mol. Sci.*, 20, 6153 (6/Dec/2019) IF:4.18, Q2. https://doi.org/10.3390/ijms20246153.

Amir Reza Sadrolhosseini, Suhaidi Shafie, Yap Wing Fen. Nanoplasmonic Sensor Based on Surface Plasmon-Coupled Emission: Review. *Appl. Sci.*, 9, 1497 (10/Apri/2019) IF:2.17, Q2. https://doi.org/10.3390/app9071497.

Amir Reza Sadrolhosseini, Suraya Abdul Rashid, Norhanisah Jamaludin, Arun Mohan Isloor. Experimental and molecular modeling of interaction of carbon quantum dots with glucose. *Applied Physics A*, 125, 529, (19/Jul/2019) IF: 1.784, Q3. https://doi.org/10.1007/s00339-019-2753-z.

Amir Reza Sadrolhosseini, Suraya Abdul Rashid, Norhanisah Jamaludin, ASM Noor and Arun MohanIsloor, Surface plasmon resonance sensor using polypyrrole-chitosan/ graphene quantum dots layer for detection of sugar. *Mater. Res. Express*, 6, 075028 (10/Apri/2019) IF:1.929, Q3 https://doi.org/10.1088/2053-1591/ab0b7a.

Amir Reza Sadrolhosseini, Suraya Abdul Rashid, Suhaidi Shafie, and Hossein Nezakati, Laser ablation synthesis of gold nanoparticle to enhance the fluorescence properties of graphene quantum dots. *Journal of Laser Applications*, 31, 022006, (1/Feb/2019), IF:1.44, Q3 https://doi.org/10.2351/1.5046062.

Amir Reza Sadrolhosseini, Suraya Abdul Rashid, Suhaidi Shafie, Hassan Soleimani. Laser ablation synthesis of Ag nanoparticles in graphene quantum dots aqueous solution and optical properties of nanocomposite. *Applied Physics A*, 125, 82 (8/Jan/2019), IF: 1.784, Q3. https://doi.org/10.1007/s00339-018-2233-x

Fahad Usman, John Ojur Dennis, Khe Cheng Seong, Abdelaziz Yousif Ahmed, Thomas L. Ferrell, Yap Wing Fen, Amir Reza Sadrolhosseini, Olumide Bolarinwa Ayodele, Fabrice Meriaudeau, and Aminu Saidu. Enhanced Sensitivity of Surface Plasmon Resonance Biosensor Functionalized with Doped Polyaniline Composites for the Detection of Low-Concentration Acetone Vapour. *Journal of Sensors*, v 2019, Article ID 5786105, 13 pages (20/Nov/2019) IF: 2.024, Q2 https://doi.org/10.1155/2019/5786105.

Nur Hazahsha Shamsudin, Suhaidi Shafie, Mohd Zainal Abidin Ab Kadir, Fauzan Ahmad, Amir Reza Sadrolhosseini, Yusran Sulaiman, Siti Amaniah Mohd Chachuli, Power conversion efficiency (PCE) performance of back-illuminated DSSCs with different Pt catalyst contents at the optimized TiO2 thickness. *Optik*, 203, 163567 (31/Oct/2019) IF:1.914, Q3 https://doi.org/10.1016/j.ijleo.2019.163567.

Silvan Saleviter, Yap Wing Fen, Wan Mohd Ebtisyam Mustaqim Mohd Daniyal, Jaafar Abdullah, Amir Reza Sadrolhosseini, and Nur Alia Sheh Omar, Design and analysis of surface plasmon resonance optical sensor for determining cobalt ion based on chitosan-graphene oxide decorated quantum dots-modified gold active layer. *Optics Express*, 27, 32294, (5/Nov/2019), IF:3.561, Q1 https://doi.org/10.1364/OE.27.032294.

Wan Mohd Ebtisyam Mustaqim Mohd Daniyal, Yap Wing Fen, Jaafar Abdullah, Amir Reza Sadrolhosseini, Silvan Saleviter, Nur Alia Sheh

Omar. Label-free optical spectroscopy for characterizing binding properties of highly sensitive nanocrystalline cellulose-graphene oxide based nanocomposite towards nickel ion. *Spectrochimica Acta Part A: Molecular and Biomolecular Spectroscopy*, 212, 25-31 (5/Apri/2019) IF 2.931, Q1. https://doi.org/10.1016/j.saa.2018.12.031.

2018:

Amir Reza Sadrolhosseini, Mohd Adzir Mahdi, Farideh Alizadeh, Suraya Abdul Rashid Laser Ablation Technique for Synthesis of Metal Nanoparticle. In: *Liquid, Laser Technology and its Applications*. Editor: Yufei Ma, IntechOpen, Chapter 4, 63-83 (20/Dec/2018). https://www.intechopen.com/books/laser-technology-and-its-applications/laser-ablation-technique-for-synthesis-of-metal-nanoparticle-in-liquid

Amir Reza Sadrolhosseini, Suraya Abdul Rashid, Suhaidi Shafie. Polypyrrole: Properties and Application. In: *Encyclopedia of Polymer Applications. Editor: Munmaya Mishra*, Vol III, CRC press Taylor & Francis, 2092-2103 (Jan/2019). https://www.routledgehandbooks.com/doi/10.4324/9781351019422-140000377.

Mahmoud Naseri, Amir Reza Sadrolhosseini, Yap Wing Fen, Mohd Adzir Mahdi. Measurement of Low Magnesium Concentration in Aqueous Solution. *Sensor and Material*, 30, 5, 1019-1026 (18/March/2018) IF:0.599 Q4. https://doi.org/10.18494/SAM.2018.1837.

Wan Mohd Ebtisyam Mustaqim Mohd Daniyal, Yap Wing Fen, Jaafar Abdullah, Amir Reza Sadrolhosseini, Silvan Saleviter, Nur Alia Sheh Omar. Exploration of Surface Plasmon Resonance for Sensing Copper Ion Based on Nanocrystalline Cellulose-Modified Thin Film. *Optics Express*, 26, 26, (24/Dec/2018), IF:3.669 Q1. https://doi.org/10.1364/OE.26.034880.

Soheila Akbari

Affiliation: Master Student in Photonics

Education: BSc. In Atomic and molecular Physics

Business Address: Magneto-plasmonic Lab, Laser and Plasma Research Institute, Shahid Beheshti University, Tehran, Iran.

Research and Professional Experience: Optical Sensors, Biophotonics, Plasmonic, Biosensors

Honors: Ranked within the top 10% of my graduating class

Publications from the Last 3 Years:

2021:

Akbari, S; Hamidi, SM; Eftekhari, H; Soheilian, A. Thermoplasmonic effect onto Toad physiology signals by plasmonic microchip structure. *Scientific Reports*, 11, 17287, 2021. IF:4.379 Q1. https://doi.org/10.1038/s41598-021-96640-w.

Hassan Tajarenejad, Mohammad Ali Ansari, Soheila Akbari, Hanieh Yazdanfar, Syedeh Mehri Hamidi. Optical neural stimulation using thermoplasmonic effect of gold nano-hexagon, *Biomedical Optics*, Accepted Impress.

S. Mehri Hamidi

Affiliation: Associate Professor

Education: PhD in Photonics

Business Address: Magneto-plasmonic Lab, Laser and Plasma Research Institute, Shahid Beheshti University, Daneshju Boulevard, 1983969411, Tehran, Iran

Research and Professional Experience: magneto-plasmonic; nano-photonics; neuroplasmonics; piezophotonics; Random lasers; atomic plasmonic.

Publications from the Last 3 Years:

Abdulfadhil, M; Gatea, H. Jawad, Hamidi, SM. "Detecting the thermoplasmonic effect using ellipsometry parameters for self-assembled gold nanoparticles within a polydimethylsiloxane matrix", *Applied Physics A: Materials science and processing*, 125, 103, (2019).

Aghamir, F; Hamidi, SM; Tehranchi, MM; Mirzaiee, R. "Combined Application of Imaging Methods for Estimating Soil Physicochemical Properties", *Eurasian Soil Science*, 52 (8), 926-934, 2019.

Asadishad, T; Sohrabi, F; Ghazimoradi, MH; Hamidi, SM; Anaghizi, SJ. "Detection of Nicotine Effect on Colon Cells in a Plasmonic Platform", *Journal of Lasers in Medical Sciences*, 11 (1), 8, 2020.

Cichelero, R; Oskuei, MA; Kataja, M; Hamidi, SM; Herranz, G. "Unexpected large transverse magneto-optic Kerr effect at quasi-normal incidence in magnetoplasmonic crystal"s", *Journal of Magnetism and Magnetic Materials*, (2019).

Ejbarah, RA; Jassim, JM; Haddawi, SF; Hamidi. SM. "Transition from incoherent to coherent random lasing by adjusting silver nanowires", *Applied Physics A*, 127 (6), 1-9, 2021.

Ejbarah, RA; Jassim, JM; Hamidi, SM. "Random laser action under picosecond laser pumping", *Optical and Quantum Electronics*, 52 (10), 1-8, 2020.

Ejbarah, RA; Jassim, JM; Hamidi, SM. "The effect of dye concentration and cell thickness on dye–polymer random laser action", *Optical and Quantum Electronics*, 53 (2), 1-9, 2021.

Fallah, H; Asadishad, T; Shafiei, M; Shokri, B; Javadianaghezi, S; Mohammed, WS; Hamidi, SM. "Utilizing ZnO Nanorods for CO gas detection by SPR technique", *Optics Communications*, 125490, 2020.

Ghasemi, M; Hamidi, SM; Dehzangi, A; Choudhury, PK. "Double-stadium Si-MZI racetrack microring resonator circuits: way to generate optical digital patterns", *JOSA B*, 37 (5), 1434-1446, 2020.

Ghasemi, M; Roostaei, N; Sohrabi, F; Hamidi, SM; Choudhury, PK. "Biosensing applications of all-dielectric SiO2-PDMS meta-stadium grating nanocombs", *Optical Materials Express*, 10 (4), 1018-1033, 2020.

Haddawi, SF; Hummud, H; Hamidi, SM. "Signature of plasmonic nanoparticles in multi-wavelength low power random lasing", *Optics and Laser technology*, 121, pp. 105770-105784, 2020.

Haddawi, SF; Humud, HR; Hamidi, SM. "Tunable low power piezo-plasmonic random laser under external voltage", *Optik*, 164482, 2020.

Haddawi, SF; Humud, HR; Monfared, SA; Hamidi, SM. "Two-dimensional plasmonic multilayer as an efficient tool for low power random lasing applications," *Waves in Random and Complex Media*, 1-10, 2021.

Haddawi, SF; Kodeary, AK; Shnan, NS; Humud, HR; Hamidi, SM. "Light emitting polymers in two dimensional plasmonic multi wavelength random laser", *Optik*, 231, 166437, 2021.

Haddawi, SF; Kodeary, AK; Shnan, NS; Roostaei, N; Hamidi, SM; Humud, HR. "Vertically emitted dual wavelength random laser based on light emitting polymers and silver nanowires double cavity structure", *Optik*, 219, 165256, 2020.

Haddawi, SF; Mirahmadi, M; Mbarak, H; Kodeary, AK; Ghasemi, M; Hamidi, SM. "Footprint of plexcitonic states in low-power green–blue plasmonic random laser", *Applied Physics A*, 125 (12), 843, 2019.

Hamidi, SM; Moradlou, R. "Tamm plasmon boosting Faraday rotation in a coupled resonator magneto-plasmonic structure", *Journal of Magnetism and Magnetic Materials*, 469, 364-372, (2019).

Hosseinpour, M; Zendehnam, A; Hamidi, SM; Marzdashti, HG. "Effects of different gas flow rates and non-perpendicular incidence angles of argon cold atmospheric-pressure plasma jet on silver thin film treatment", *Journal of Theoretical and Applied Physics*, 13 (4), 329-349, 2019.

Karaei Shiraz, A; Yazdanpanah Goharrizi, A; Hamidi, SM. "The electronic and optical properties of armchair germanene nanoribbons", *Physica E: Low-dimensional Systems and Nanostructures*, 107, 150-153, (2019).

Khajemiri, Z; Lee, D; Hamidi, SM; Kim, DS. "Rectangular plasmonic interferometer for high sensitive glycerol sensor", *Scientific Reports*, 9, 1378, (2019).

Kodeary, AK; Gatea, MA; Haddawi, SF; Hamidi, SM. "Tunable thermo-piezo-plasmonic effect on core/shell nanoparticles under laser irradiation and external electric field", *Optical and Quantum Electronics*, 52 (2), 112, 2020.

Kodeary, AK; Hamidi, SM. "Tunable Piezophotonic Effect on Core-Shell Nanoparticles Prepared by Laser Ablation in Liquids under External Voltage", *Journal of Nanotechnology*, 2019.

Kodeary, AK; Hamidi, SM; Moradlou, R. "Voltage controlled properties of piezo-magneto-plasmonic core/shell nanoparticles", *Nano-Structures & Nano-Objects*, 21, 100415, 2020.

Mahinroosta, T; Mehri Hamidi, S. "Strong Exciton–Plasmon Coupling in Waveguide-Based Plexcitonic Nanostructures", *physica status solidi* (b) 257 (12), 2000266, 2020.

Mbarak, H; Ghahrizjani, RT; Hamidi, SM; Mohajerani, E; Zaatar, Y. "Reversible and tunable photochemical switch based on plasmonic structure", *Scientific Reports*, 10 (1), 1-7, 2020.

Mbarak, H; Hamidi, SM; Belotelov, VI; Chernov, AI; Mohajerani, E; Zaatar, Y. "Surface lattice resonance-based magneto-plasmonic switch in NiFe patterned nano-structure", *Journal of Magnetism and Magnetic Materials*, 517, 167387, 2021.

Mbarak, H; Hamidi, SM; Belotelov, VI; Chernov, AI; Mohajerani, E; Zaatar, Y. "*Surface lattice resonance based magneto-plasmonic switch in NiFe patterned nano-structure*", arXiv preprint arXiv:2005.10913, (2020).

Mbarak, H; Hamidi, SM; Mohajerani, E; Zaatar, Y. "Electrically driven flexible 2D plasmonic structure based on a nematic liquid crystal", *Journal of physics D: Appl. Phys*, 52, pp. 415106-415114, 2019.

Mbarak, H; Kodeary, AK; Hamidi, SM; Mohajarani, E. "Control of nonlinear refractive index of AuNPs doped with nematic liquid crystal under external electric field", *Optik*, 163299, 2019.

Mirahmadi, M; Mahinroosta, T; Hamidi, SM. "Manipulating plasmon-exciton interactions in the plasmonic waveguide structure based on the dispersion relations concept", *Optical and Quantum Electronics*, 52 (6), 1-9, (2020).

Monfared, SA; Seifouri, M; Hamidi, SM; Mohseni, SM. "Two-dimensional graphene-plasmonic crystals for all-optical switch applications", *Optical and Quantum Electronics*, 52 (11), 1-16, 2020.

Mosleh, M; Ranjbaran, M; Hamidi, SM; Tehranchi, MM. "Ellipsometric spectroscopy of rubidium vapor cell at near-normal incidence", *Scientific Reports*, 10 (1), 1-9, 2020.

Nasiri, A; Hamidi, SM. "Bio compatible and highly sensitive two dimensional plasmonic sensor", *Optical and Quantum electronics*", 51, pp. 251-258, 2019.

Ranjbaran, M; Tehranchi, MM; Hamidi, SM; Khalkhali, SMH. "Relaxation time dependencies of optically detected magnetic resonance harmonics in highly sensitive Mx magnetometers", *Journal of Magnetism and Magnetic Materials*, 469, 522-530, (2019).

Roostaei, N; Mbarak, H; Monfared, SA; Hamidi, SM. "Plasmonic wideband and tunable absorber based on semi etalon nano structure in the visible region", *Physica Scripta*, 96 (3), 035805, 2021.

Roostaei, N; Shnan, NS; Hamidi, SM. "Red and Blue color production by flexible all-dielectric structure", *Optik*, 230, 166345, 2021.

Sadeghi, S; Hamidi, SM. "Bi:YIG@Au magneto-plasmonic core-shell nano-grating with robust, high magneto-optical Figure of merit", *Journal of magnetism and magnetic materials*, 493, pp. 165709-165715, 2020.

Sajede Saeidifard, Foozieh Sohrabi, Mohammad Hossein Ghazimoradi, Hamidi, SM; Shirin Farivar, Mohammad Ali Ansari. "Two-Dimensional Plasmonic Biosensing Platform: Cellular Activity Detection under Laser Stimulation", *Journal of applied physics*, 126 (10), 104701, 2019.

Sakineh Almasi Monfared, Mahmood Seifouri, Seyedeh Mehri Hamidi, Seyed Majid Mohseni. "Electro-optical switch based on one-dimensional graphene-plasmonic crystals", *optical Materials*, 115, 111051, 2021.

Shafei, M; Tehranchi, M; Hamidi, SM. "Switching Time Probing in Electric Field-Assisted Magnetization of PbZrTiO3/Cobalt Structure", *Journal of Superconductivity and Novel Magnetism*, Vol. 32, pp.1-6, 2019.

Shafei, M; Tehranchi, MM; Yazdi, HF; Hamidi, SM; Yusupov, R; Nikitin, S. "Experimental study and micro-magnetic modeling of magnetization dynamics in L10-FePt thin film", *Journal of Magnetism and Magnetic Materials*, 479, 12, (2019).

Sharifi, H; Hamidi, SM; Navi, K. "A design procedure for fan-out improvement in all-optical photonic crystal logic design", *Optical and Quantum Electronics*, 51 (10), 340, 2019.

Shilina, PV; Ignatyeva, DO; Kapralov, PO; Sekatskii, SK; Nur-E-Alam, M; Mikhail Vasiliev, Kamal Alameh, Venu Gopal Achanta, Yujun Song, Seyedeh Mehri Hamidi, Zvezdin, AK; Belotelov, VI. "Nanophotonic structures with optical surface modes for tunable spin current generation", *Nanoscale*, 13 (11), 5791-5799, 2021.

Shiraz, AK; Goharrizi, AY; Hamidi, SM. "Structural Stability and Electron Density Analysis of Doped Germanene: A First-Principles Study", *Materials Research Express*, 2019.

Shnan, NS; Roostaei, N; Hamidi, SM. "Tunable and reversible thermo-plasmonic hot spot imaging for temperature confinement", *Journal of Theoretical and Applied Physics*, 14 (4), 367-376, 2020.

Sohrabi, F; Asadishad, T; Ghazimoradi, MH; Mahinroosta, T; Saeidifard, S; Hamidi, SM. "Plasmophore Enhancement in Fibroblast Green Fluorescent Protein-Positive Cells Excited by Smoke, *ACS Omega*, 10.1021/acsomega.0c00496, 2020.

Sohrabi, F; Etezadi, D; Jahani, Y; Mohammadi, E; Ghadiani, B; Tamizifar, M; Hamidi, SM. "Blue-shift ultra sensitivity using rhombus-shaped plasmonic crystal on Si3N4 membrane", *Optical Materials Express*, 10 (7), 1649-1658, (2020).

Sohrabi, F; Etezadi, D; Perin, R; Jahani, Y; Mohammadi, E; Hamidi, SM. "Phase-sensitive optical neural recording of cerebellum tissue on a flexible interface", *Journal of Applied Physics*, 127 (11), 113101, 2020.

Sohrabi, F; Hamidi, SM**.** "Adjustable Plasmonic Bandgap in One-Dimensional Nanograting Based on Localized and Propagating Surface Plasmons", *International journal of optics and photonics*, 13 (2), 97-102, 2019.

Sohrabi, F; Hamidi, SM; Asgari, N; Ansari, M; Gachilou, T. "One dimensional photonic crystal as an efficient tool for in-vivo optical sensing of neural activity", *Optical materials*, 9, pp. 109275-109281, 2019.

Sohrabi, F; Hamidi, SM; Mohammadi, E. "Role of higher order plasmonic modes in one dimensional nanogratings", *Optical and quantum electronics*, 51, pp. 241-250, 2019.

Sohrabi, F; Jahani, Y; Sanchez-Mut, JV; Mohammadi, E; Barzegar, Z; Li, X; Hamidi, SM. "Membrane activity detection in cultured cells using phase-sensitive plasmonics", *Optics Express*, 28 (24), 36643-36655, 2020.

Sohrabi, F; Kiaei, M; Mahinroosta, T; Hamidi, SM. "Phase-Sensitive Pulse Sensor Using 2-D Active Plasmonics on Conformal Substrates", *IEEE Transactions on Electron Devices*, 67 (10), 4379-4384, 2020.

Sohrabi, F; Saeidifard, S; Ghasemi, M; Asadishad, T; Hamidi, SM; Hosseini, S. "Role of plasmonics in detection of deadliest viruses: a review," *The European Physical Journal Plus*, 136 (6), 1-71, 2021.

Tannaz Asadishad, Foozieh Sohrabi, Mohammad Hossein Ghazimoradi, Hamidi, SM; Saeed Javadi Anaghizi, Shirin Farivar,"Detection of Nicotine Effect on Colon Cells in a Plasmonic Platform", *Journal of Lasers in Medical Sciences*, 2019.

Chapter 2

Optical Constants of Noble Metal Nanostructures

Hira Joshi*, PhD
Department of Physics, Gargi College,
University of Delhi, Delhi, India

Abstract

This research is made to examine the optical constants of plasmonic nanostructures. The noble metals were thoroughly studied for their plasmonic properties in UV and visible range. It is found that optical response of nanostructures is mainly governed by dielectric permittivity of the metal and resonant wavelength depends on size, shape, composition and external matrix. Further, it is observed that the localised surface plasmon resonance (LSPR) dominates optical spectra of metal nanoparticles. During research the optical constants of noble metals (Au and Ag) were investigated theoretically, using Mie's formalism, in quasi static regime. Theoretically, it was evident that resonant wavelength changes with size of nanoparticle and external matrix. It is observed that LSPR is very sensitive tool for analyzing the environment or the matrix. In addition, the Optical absorption studies were performed on Ag and Au nanoparticles, of various sizes and keeping in different environments. It has been demonstrated that the optical absorption can be used to estimate dielectric constants of noble metal nanostructures.

Keywords: nanoparticles, LSPR, dielectric constant, absorption coefficient

* Corresponding Author's Email: hirajoshi123@gmail.com.

In: A Closer Look at Plasmonics
Editor: Christian Bosch
ISBN: 978-1-68507-709-9

Introduction

Surface plasmons are collective oscillations of conduction electrons, induced by electromagnetic fields [1-3]. When light is incident on metallic nanostructures, oscillations of conduction electrons are produced and at a particular value of wavelength of incident radiation, resonance occurs which is referred to as surface plasmon resonance (SPR) [4-10]. SPR, is an optical phenomenon in which collective oscillations of electrons in metals, result in enhancement of amplitude, energy, and electric field around the nanoparticle (np). SPR is also the reason for unique spectral response of Ag and Au. The SPR wavelength, extinction cross-section and local electric field enhancement are all highly dependent on the composition of the material, its size, shape and external environment [11-14]. Metal nanoparticles possess unique optical, electronic, chemical and magnetic properties. These properties can be tuned by varying the size, shape and composition of the nanoparticles. Noble metal nanoparticles exhibit a strong plasmon resonance band in the visible region. Noble metal nanostructures have various potential applications in surface enhanced Raman spectroscopy (SERS), chemical and biological sensors, imaging, photocatalysis, photovoltaics [15-20]. Noble metal nanoparticles such as Ag and Au, do exhibit sharp surface plasmon resonance generally referred to as Localized Surface Plasmon Resonance (LSPR) at their characteristic wavelengths (λ_{SPR}).

The tunability of λ_{SPR} depends on the size of nanoparticle and the surrounding matrix/medium [21]. Silver nanoparticles exhibit a high value efficiency of plasmon excitation when as compared to gold nanoparticles, as these interact more efficiently with visible light. In this chapter we have studied optical constants of Ag in detail, using dielectric constants from Baber [22].

Theoretical Explanations

When a small spherical metallic nanoparticle, is irradiated by electromagnetic light, the oscillating electric field causes the conduction electrons to oscillate coherently. The oscillation frequency (ω_p), is dependent on the density of electrons, the effective electron mass, also on the shape and size of the charge distribution. Optical response of materials is characterized by frequency dependent dielectric function, ε (ω). According to Drude model

$$\varepsilon(\omega) = 1 - \omega_p^2 / (\omega^2 + i\omega\Gamma) \quad (1)$$

$$\Gamma = \Gamma_o + A\, V_f / r \quad (2)$$

Here Γ and Γ_o are phenomenological and bulk damping constants respectively [23-24].

$$\varepsilon(\omega, r) = 1 - \omega_p^2 / [(\omega^2 + i\omega\,(\Gamma_o + A\, V_f / r)] \quad (3)$$

In 1908, Mie [25] proposed a solution of Maxwell's equations for spherical particles interacting with plane electromagnetic waves, which explains the origin of surface plasmons resonance (SPR) in the extinction spectra. Extinction crossection (which is sum of both absorption and scattering cross section), for metallic nanoparticles of sizes smaller than the wavelength of incident light, according to Mie's theory is reduced to:

$$C_{ext} = \frac{9\omega\epsilon_2 V(\epsilon_m)^{3/2}}{c[(\epsilon_1 + 2\epsilon_m)^2 + \epsilon_2{}^2]} \quad (4)$$

C_{ext} is extinction cross section for a metal sphere of volume (V) and radius (r) having dielectric function, of metal nanoparticle as $\epsilon(\omega) = (\epsilon_1 + i\epsilon_2)$, where ϵ_1 and ϵ_2 are real and imaginary parts of dielectric function of the metal under investigation, when placed in a surrounding medium with dielectric constant, ϵ_M, in quasi static limit. ω and c are angular frequency and velocity of light respectively.

Frohlich condition for maximum extinction cross section is achieved when

$$[\epsilon_1 + 2\,\epsilon_M)^2 + (\epsilon_2)^2] = \text{minimum}$$

and it is possible when $\epsilon_1 = -2\epsilon_M$, if ϵ_2 is small or if it is only weakly dependent on.

The extinction coefficient for Ag nanoparticles, using Mie's formalism, was determined for different sizes of nanoparticles by varying the size and the dielectric constant of the external matrix using equation (4). The dielectric constants for Ag used in the present calculations were taken from Baber [22].

Result and Discussion

Theoretical results are presented, for extinction spectra of spherical noble metal (Ag) nanoparticles, of various sizes and when embedded in different external matrices. Dielectric function, which plays a crucial role in optical response of the material, is frequency dependent only in Drude model. Extinction coefficient calculations, for Ag of different radii of spherical particles varying from 1nm –30nm, were performed and are shown in figure 1(a). It is observed that as the radius and hence, the volume of nanoparticles increases, intensity of extinction curve increases but resonant wavelength remains constant for a fixed surrounding matrix. This is, however, a very well-known fact for most of the metal nanoparticles [21]. Figure 1b shows variation of intensity of extinction with size indicating linear variation of intensity with size of nanoparticle.

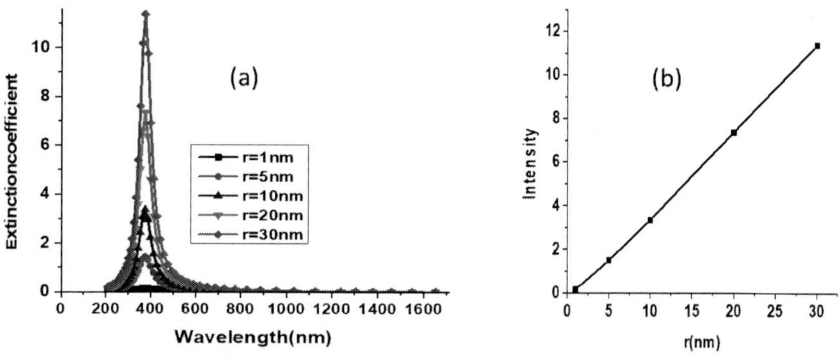

Figure 1. (a) Variation of extinction coefficient of Ag nanoparticle (b) Variation of Intensity of extinction coefficient with size of Ag nanoparticle.

Figure 2(a) shows variation of extinction coefficient with wavelength, for different surrounding media, for a fixed size of Ag nanoparticle. As the dielectric constant and hence, the refractive index of surrounding medium increases, resonance peak is red shifted. This redshift is used to find external medium in sensing devices [21]. Figure 2(b) shows variation of λ_{SPR} with external dielectric medium. The tuning of the resonance wavelength depends on the surrounding medium, hence, allowing the possibility of determining the dielectric constant of an unknown surrounding. In Mie's formalism, extinction spectra for spherical nanoparticles, is independent of size of nanoparticle as shown in figure 1.

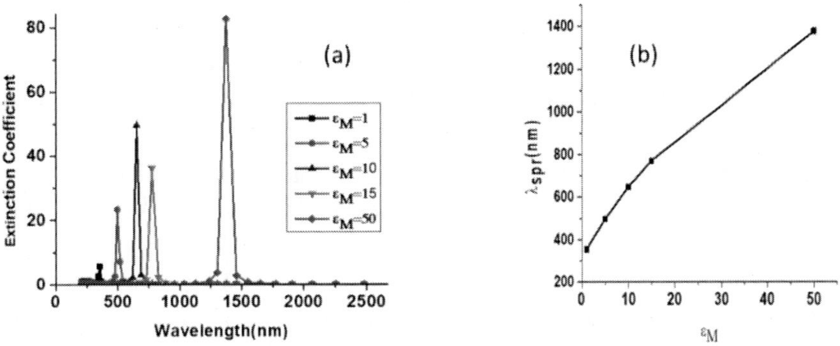

Figure 2. (a) Variation of extinction coefficient for 10 nm Ag nanoparticle of different dielectric matrix. (b) Variation of λ_{SPR} as a function of dielectric constant of external matrix.

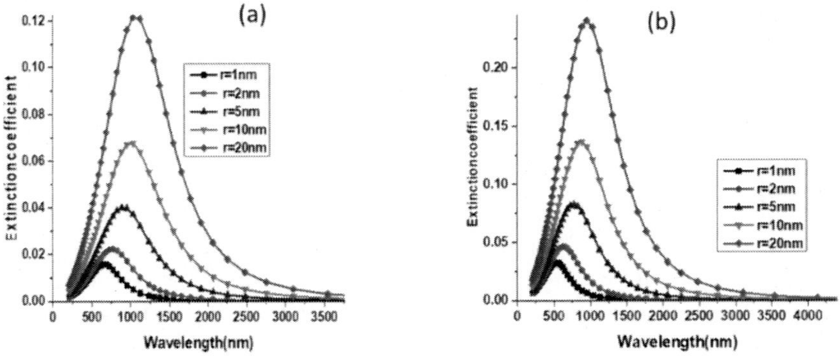

Figure 3. Extinction coefficient calculated using equation (4) for a gold sphere (a) and (b) for silver sphere in external matrix of refractive index as 1.33 when dielectric constants data [22].

In Modified Maxwell Garnet Mie (MMGM) theory, size of nanoparticle affects extinction spectra and it is shown in figure 3. Figure 3(a) shows variation of extinction spectra for Au nanoparticle of fixed external matrix refractive index as 1.33 and figure 3(b) shows for Ag nanoparticle. In this figure, size of nanoparticles in both Au & Ag are varying from 1nm to 20nm. It is shown that in Ag are sharper as compared to Au because Ag interacts more with visible light.

Conclusion

We have studied the extinction coefficient, which is the sum of scattering and absorption coefficient, for noble metal spherical nanoparticle, of size less than incident light in quasi-static limit, theoretically, using Mie and MMGM. It is shown that resonant wavelength depends on external medium in which spherical nanoparticle is placed and is also redshifted. This redshift in resonant wavelength can be used in sensors. It is also confrimed that resonant wavelength (λ_{SPR}), is independent of particle size but its intensity is directly dependent on the size of nanoparticle size in Mie's formalism but varies in MMGM. It is observed that LSPR is very sensitive tool for analyzing the environment or the matrix. With the help of this study, dielectric constants of noble metal nanostructures can be determined and can be used in sensors. Ag and Au nanoparticles embedded in various matrices can be used in the design of nonlinear optical devices.

References

[1] Maier, S. A. (2007). *Plasmonics: Fundamentals and Applications* (Springer Science Business Media LLC) (Maier 2007, 65 - 72).

[2] Jain, K., Eustis, S., El-Sayed, M. (2006). Optical properties of metal clusters (springer series in material science). *J. Phys. Chem. B.,* 110:18243.

[3] Sönnichen, C., Franzl, T., Wilk, T., von Plessen, G. and Fe, J. (2002). Plasmon resonances in large noble-metal clusters, *New J. Phys.*, 4: 93.1 - 93.8.

[4] Ghosh, S. K., Pal, T. (2007). Interparticle Coupling Effect on the Surface Plasmon Resonance of Gold Nanoparticles: From Theory to Applications. *Chem. Rev.*, 107: 4797 - 4862.

[5] Eustis, S., El-Sayed, M. A. (2006). Why gold nanoparticles are more precious than pretty gold: Noble metal surface plasmon resonance and its enhancement of the radiative and nonradiative properties of nanocrystals of different shapes. *Chem. Soc. Rev.,* 35:209 - 217.

[6] Link, Stephan and El-Sayed, Mostafa A. (1999). Size and Temperature Dependence of the Plasmon Absorption of Colloidal Gold Nanoparticles, *J. Phys. Chem. B.*, 103: 4212 - 4217.

[7] Klar, T., Perner, M., Grosse, S., von Plessen, G. W., Spirkl, W., Feldmann, J. (1998). "Surface-Plasmon Resonances in Single Metallic Nanoparticles", *Phys. Rev. Lett.*, 80:4249.

[8] Maiera, S. A. and Atwater, H. A. (2005). Plasmonics: Localization and guiding of electromagnetic energy in metal/dielectric structures. *J. of Appl. Phys.*, 98: 011101.

[9] Lakowicz, J. R. (2006). Plasmonics in Biology and Plasmon-Controlled Fluorescence. *Plasmonics,* 1: 5 - 33.

[10] Bohren, C. F. and Huffmann, D. R. (1983). *Absorption and Scattering of Light by Small Particles.* Wiley - Interscience, New York (Bohren 1983).

[11] Kolwas, K., Derkachova, A. and S. Demianiuk, S. (2006). The smallest free-electron sphere sustaining multipolar surface plasmon oscillation. *Comp. Mat. Sci.,* 34:337 - 341.

[12] Noguez, Cecilia. (2007). Surface Plasmons on Metal Nanoparticles: The Influence of Shape and Physical Environment. *J. Phys. Chem.* C., 111: 3806 - 3819.

[13] Mock, J. J., Barbic, M., Smith, D. R., D. A. Schultz, D. A. and Schultz, S. (2002). Shape effects in plasmon resonance of individual colloidal silver nanoparticles. *J. Chem. Phys.*, 116:6755 - 6759.

[14] Rupin, R. (1982). *Electromagnetic Surface Modes*, edited by A. D. Boardman, Wiley: Chichester (Rupin 1982).

[15] Laurent, G. N., Félidj, N., Aubard, J., Lévi, G., Krenn, J. R., Hohenau, A., Schider, G., Leitner, A., Aussenegg, F. R. (2005). Surface enhanced Raman scattering arising from multipolar plasmon excitation. *J. Chem. Phys.*, 122: 011102.

[16] Lee, P. C., Meisel, D. (1982). Adsorption and surface-enhanced Raman of dyes on silver and gold sols. *J. Phys. Chem.*, 86: 3391 - 3395.

[17] Yu, C. X., Irudayaraj, J. (2007). Multiplex Biosensor Using Gold Nanorods. *Anal. Chem.*, 79: 572 - 579.

[18] Wiederrecht, G. P., (2004). Near-field optical imaging of noble metal nanoparticles. *Eur. Phys. J. Appl. Phys.*, 28:3 - 18.

[19] Lindfors, K. T., Kalkbrenner, T. P., Stoller, P. and V. Sandoghdar, V. (2004). Detection and Spectroscopy of Gold Nanoparticles Using Supercontinuum White Light Confocal Microscopy. *Phys. Rev. Lett.,* 93: 037401.

[20] Arbouet, A., Christofilos, D., Fatti, N. D., Vallee, F., Huntzinger, J. R., Arnaud, L., Billaud, P., Broyer, M. (2004). Direct Measurement of the Single-Metal-Cluster Optical Absorption. *Phys. Rev. Lett.*, 93:127401.

[21] Joshi, H. and Annapoorni, S. (2020). Tuning Optical Properties in Nanocomposites. *Int. J. of Nanosc.,* 19(4): 1950026 - 32.

[22] Baber, S., Weaver, J. H., (2015). Optical constants of Cu, Ag, and Au revisited. *Applied Optics,* 54(3): 477 - 487.

[23] Kreibig, U., Vollmer, M. (1995). *Optical Properties of Metal Clusters,* Springer, Berlin Heidelberg New York (Kreibig 1995).

[24] Battie, Y., Resano, A- Garcia, Chaouli, N., Zhang, Y. and Nacini, A. (2014). Extended Maxwell-Garnet-Mie formulation applied to size dispersion of metallic nanoparticles embedded in host liquid matrix. *The Journal of chemical Physics,* 140:044705 - 9.

[25] Mie, G. (1908). *Annals of Physics*, 25(3):377 - 445.

Chapter 3

Effect of Magnetic Field on Plasmonics of Metal Nanostructures

Hira Joshi*, PhD

Department of Physics, Gargi College,
University of Delhi, Delhi, India

Abstract

Magnetoplasmonics of noble metal nanostructures using Mie's formalism have been investigated theoretically. Optical properties of noble metal structures such as resonant wavelength and extinction coefficients are investigated theoretically, in the presence of static external magnetic field. In this chapter, it is shown that the plasmonic nanostructures have increased magnetic response in visible range at low magnetic fields. Resonant wavelength for these nanostructures is demonstrated to be shifted towards lower wavelength (blue shift) in the presence of magnetic field(B). At low values of applied magnetic field, the spectral shift of resonant wavelength is towards the visible region. This critical value of magnetic field in case of Ag nanostructures is realized as 2 Tesla.

Keywords: LSPR, NPS, plasmonics, magnetic field, surface plasmons

* Corresponding Author's Email: hirajoshi123@gmail.com.

In: A Closer Look at Plasmonics
Editor: Christian Bosch
ISBN: 978-1-68507-709-9

Introduction

Magnetoplasmonic is result of excitations of plasmonics in the presence of magnetic field [1]. Surface plasmons (collective oscillations of conduction electrons) are very different in the presence of external magnetic field, named as surface magneto plasmons (SMPs). SMPs are well studied in different structures under various conditions [2-4]. Tuning of SMPs is possible by changing different parameters and external environment of metallic nanoparticles (nps). Many applications of magnetoplasmonics such as imaging [5], sensing [6], nanoantennas [7] and also in ultra fast switches [8] have already been reported. Applications of magnetic nanoparticles in cloaking [9], magnetic tweezers and separation [10] are also well demonstrated.

As is well known in magnetism, an external magnetic field applied to the metallic structures gives rise to magnetic dipoles. The magnetic moments so developed in nanoparticles result in the magnetic response of the system [11-14]. Shift in resonant wavelength can be seen with the variation in refractive index of the matrix [15]. The dielectric constant of matrix can be changed by choice of different matrices or by an external agents such as temperature, pressure or magnetic field. Here optical properties of noble metal structures such as resonant wavelength and extinction coefficient are investigated theoretically, in the presence of static external magnetic field. Using Mie's formalism for extinction coefficient,it is shown that the plasmonic nanostructures have increased magnetic response in visible range at low magnetic fields (~ 2T in Ag nanostructures which is not'' too large for realistic applications"[16]) and absorption spectrum is shifting towards lower wavelength (blue shift) as external magnetic field is enhanced.

Theory

According to Mie's theory (17), the polarisability of a spherical nanoparticle in the quasi static limit ($R<<\lambda$, R and λ are radius of np and incident light wavelength), is isotropic depending on scalar dielectric function of metal nps. Extinction coefficient in such a case is:

$$C_{ext} = \frac{9\omega\epsilon_2 V(\epsilon_m)^{3/2}}{c[(\epsilon_1+2\epsilon_m)^2+\epsilon_2{}^2]} \qquad (1)$$

V is volume, ω and c angular frequency and velocity of light. ϵ_m and ϵ are dielectric constants of external medium and metal np, ϵ being a complex function.

ϵ_1 and ϵ_2 are real and imaginary parts.

In presence of static magnetic field (B), dielectric function becomes a tensor and results in anisotropic polarizability [18, 19] given as:

$$\alpha(\omega) = 4\pi R3 \frac{[\tilde{\epsilon}(\omega) - \tilde{I}\epsilon_m]}{\tilde{\epsilon}(\omega) + 2\tilde{I}\epsilon_m} \tag{2}$$

On application of B in Z direction, diagonal terms of tensorial dielectric function [20] will be represented as:

$$\varepsilon_{xx} = \varepsilon_\infty - \varepsilon(\omega) \tag{3}$$

$$\varepsilon(\omega) = \frac{(\omega_p)^2}{(\omega)^2 + i\omega\Gamma} \text{ (Drude formula)} \tag{4}$$

ε_∞ is high frequency dielectric constant. Γ is damping constant.

Nondiagonal terms of dielectric function:

$$\varepsilon_{xy} = i\,\varepsilon(\omega) \frac{\omega_c}{\omega + i\Gamma} \tag{5}$$

$$\varepsilon(\omega) = \varepsilon_\infty \frac{(\omega_p)^2}{(\omega)^2 + i\omega\Gamma} \tag{6}$$

ω_c and ω_p cyclotron frequency and plasma frequency.

In noble metals cyclotron frequency is small as compared to plasma frequency [21]. Optical properties of Ag spherical nanoparticles are investigated when placed in static external magnetic field. The parameters necessary for calculation of extinction coefficient from equation 1 for Ag nanoparticles are:ω_p= 9.33 eV, A=1, Γ_o= 0.03eV, ε_∞=3.7 and ω_c = 94 μeV, B= 0.8 T, V_f =1.39 x10^6 (m/s) taken from Derkachova and Deepika [22, 23]. Results of these calculations for nanoparticles of size 10nm are shown in Figure 1 and 2.

Result and Discussion

Theoretical results are presented for extinction spectra of spherical noble metal nanoparticlesps when embedded in various external matrices. Extinction coefficient is calculated so as to show optical spectrum of noble metal(Ag) nanostructures in presence of B.

Without magnetic field study of extinction spectrum of noble metal nanostructures has been reported in [15]. When an external magnetic field is applied to Ag nanoparticle, optical extinction spectrum is shown in figure 1.

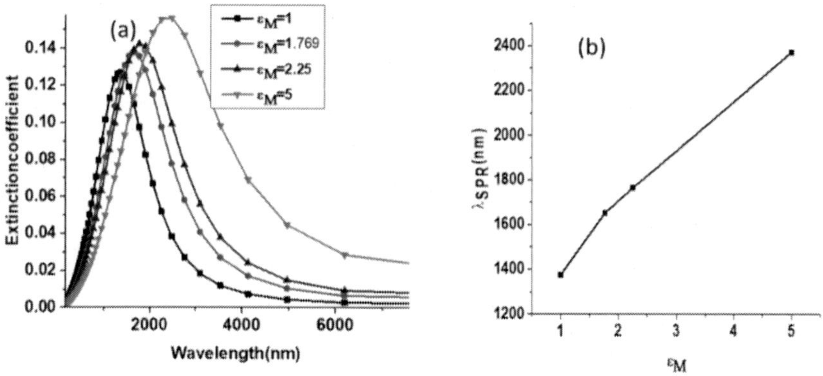

Figure 1. (a) Extinction spectra of Ag spherical np at B = 0.8 T and (**b**) Variation of peak wavelength for surface plasmons with ε_m, for B = 0.8T.

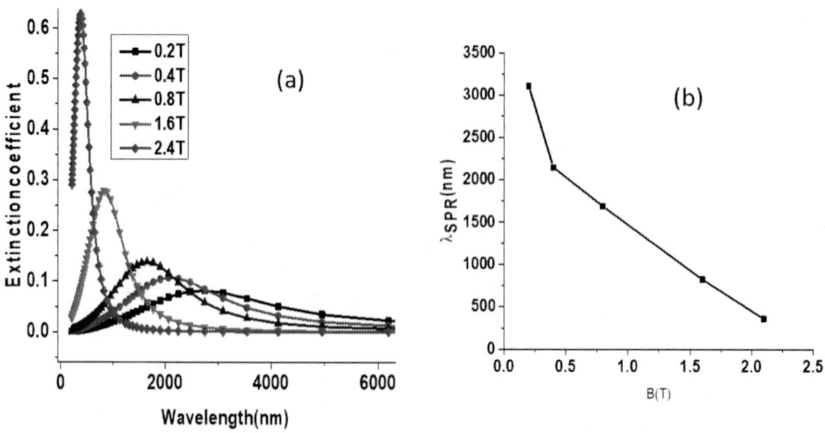

Figure 2. (a) Optical extinction spectra of Ag np at external matrix of refractive index 1.33 and (b) Variation of resonant wavelength with external magnetic field.

Extinction coefficient is calculated using equation 6 for dielectric function in Mie's formula for extinction at fixed B = 0.8 T. Effect of applied B is realized as resonant wavelength, broadness and peak intensity of absorption spectrum increases with ε_m. Figure 1(b) shows variation of resonant wavelength with ε_m thereby confirming, red shift and utility of Ag nanostructures in sensing.

Figure 2(a) shows extinction spectra of Ag spherical nanoparticle for different strengths of applied magnetic field. As strength of B is increased, resonant wavelength is showing a decreasing trend i.e., it is blue shifted. At a particular value of external magnetic field = 2T, resonant wavelength for this nanostructure coincides with the resonant value for Ag nanoparticle of same size. Also the sharpness and intensity of absorption spectrum increases with increase in value of B.

In figure 2(b), variation of resonant wavelength with external magnetic field is shown for a fixed size and refractive index of matrix. It shows decrease in resonant value of absorption spectrum with increase in B and hence confirming blue shift in the absorption spectrum of Ag nanostructure.

Optical absorption studies performed on Ag metal nanostructures exhibit sharp plasmonic resonance using the Mie scattering calculations based on change in the dielectric constant of the surrounding medium. Resonant wavelength is found to be red shifted as the external medium refractive index is increased both with B and without [15]. With increase in strength of external magnetic field, the resonant wavelength is found to be shifted towards lower wavelength (blue shift). At a particular value of B ~ 2T (which is not so large as quoted in several papers), the resonant wavelength shifts towards the visible region so as to realize applications.

Conclusion

In this chapter magnetic plasmonic response of noble metal nanoparticles was realized and optical absorption spectrum for Ag nanoparticles in external magnetic field was investigated theoretically. It is shown that absorption spectrum of noble metal nanostructures strongly depends on the static external magnetic field .Spectral shift of absorbance is shown towards blue end (blue shift). Also it is demonstrated that resonant wavelength, for Ag nanoparticles at small value of external magnetic field ~ 2T, approaches towards its value lying in visible spectral range. This magnetic response of noble metal nanostructures will lead to the origin of many unique properties of noble metals (e.g., Ag) nanostructures in the visible range at low magnetic fields. It is concluded that such investigated magneto plasmonic nanostructures can also be utilized in the construction of nanoactive devices like NSOM and in biomedical applications such as MRI, hyperthermia and biosensors.

References

[1] Makyskov, L. S. (2015). Magneto-Plasmonics and Resonant Interaction of Light with Dynamic Magnetisation in Metallic and All –Magneto-Dielectric Nanostructures. *Nanomaterials,* 5(2):577 - 613.

[2] Maccaferri, N. et al. (2013). Tuning the Magneto-Optical response of Nanosize Ferromagnetic Ni Disks Using the Phase of Localised Plasmons. *Phys. Rev. Lett.*, 111:167401 - 5.

[3] Cabllero, B., Garcia-Martin, A., Cuevas, J. C. (2016). Hybrid Magnetoplasmonics Crystals Boost the Performance of Nanohole Arrays as Plasmonic Sensor. *ACS Photonics,* 3:203 - 208.

[4] Armelles, G. et al. (2013). Plasmonic Enhancement of the third order nonlinear optical nanoresonators. *Opt. Express.* 21(22): 27356 - 27370.

[5] Liu, N., Tang, M. L., Hentschel, M., H. Giessen, H., Alivisatos, A. P. (2011). Nanoantenna-enhanced gasensing in a single tailored nanofocus. *Nat. Mater.*, 10: 631 - 636.

[6] Pendry, J. B. (2000). Negative refraction makes a perfect lens, *Phys. Rev. Lett.*, 85: 3966 - 3969.

[7] Bardhan, R., Lal, S., Joshi, A., Halas, N. J. (2011). Theranostic Nanoshells: From Probe Design to Imaging and Treatment of Cancer. *Acc. Chem. Res.,* 44: 936 - 946.

[8] Chai, Z., Hu, X., Wang, F., Niu, X., Xie, J., Gong, Q. (2017). Ultrafast All-Optical Switching, *Adv. Opt. Mat.*, 5(7): 1600665.

[9] Xie, J., Liu, G., Edens, H. S., Chen, H., Ai., X. (2011) Surface-Engineered Magnetic Nanoparticle Platforms for Cancer Imaging and Therapy, *Chem. Res.*, 44: 883 - 892.

[10] Sarkar, R. and Rybenkov, V. V. (2016). Guide to Magnetic Tweezers and Their Applications, *Frontiers in physics*, 4(28): 1.

[11] Lodewijks, K. et al. (2014). Magnetoplasmonic design rules for active magneto-optics. *Nano Lett.,* 14: 7207 - 7214.

[12] Sepúlveda, B., Calle, A., Lechuga, L. M., Armelles, G. (2006). Highly sensitive detection of biomolecules with the magneto-optic surfaceplasmon-resonance sensor, *Opt. Lett.*, 31: 1085 - 1087.

[13] Maccaferri, N. et al. (2013). Tuning the Magneto-Optical Response of Nanosize Ferromagnetic Ni Disks Using the Phase of Localized Plasmons. *Phys. Rev. Lett.*, 111 (16): 167401 - 05.

[14] Andrey, K., Sarychev, Gennady, S., Shalev Vladimir, M. (2006). Magnetic plasmon resonance, *Physical Review E,* 73: 036609 - 19.

[15] Joshi, H. and Annapoorni, S. (2020). Tuning Optical Properties in Nanocomposites. *Int. J. of Nanosc.,* 19(4): 1950026 - 32.

[16] Temnov, V. V., Armelles, G., Woggon, U., Guzatov, D., Cebollada, A., Garcia-Martin, A., Garcia-Martin, J. M., Thomay, T., Leitenstorfer, A., Bratschitsch, R. Active magneto-plasmonics in hybrid metal–ferromagnet structure, *Nat. Photonics,* 4 (2010) 107 - 111.

[17] Mie, M. (1908). Beitrzur Optik trMedien, speziell kolloidaler Metall osungen, *Ann. Phys.*, 330: 377 - 445.

[18] Sepúlveda, B., González-Díaz, J. B., García-Martín, A., Lechuga, L. M., Armelles, G. (2010). Plasmon-induced magneto-optical activity in nanosized gold disks. *Phys. Rev. Lett.,* 104: 147401 - 4.

[19] Smith, D. A. and Stokes, K. L. (2006). Discrete dipole approximation for magneto-optical scattering calculations, *Opt. Express,* 14(12): 5746 - 5764.

[20] Armelles, G., Cebollada, A., García-Martín, González, M. U. (2013). Magneto-plasmonics: Combining Magnetic and Plasmonic Functionalities, *Adv. Opt. Mater.*, 1(1): 1.

[21] Duchs, G., Rikken, G., Grenet, T., Wyder, P. (2001). Magnetotransverse Scattering of Surface Plasmon Polaritons, *Phys. Rev. Lett.*, 87: 127402.

[22] Kolwas, K. and Derkachova, A. (2010). Plasmonic abilities of gold and silver spherical nanoantennas in terms of size dependent multipolar resonance frequencies and Plasmon damping rates. *Opto-Electronics Review,* 18(4): 421 - 437.

[23] Deepika, G., Chauhan, Prashant., Varshney, Anshu., Singh, D. B. and Sajal, Vivek. (2015). Enhanced absorption of surface plasma wave by metal nano-particles in the presence of external magnetic field. *J. Phys. D. APPl. Phys.,* 48: 345103 - 7.

Chapter 4

Notes on Surface Plasmonics in Electromagnetics and Quantum Optics

Jian Qi Shen*
Centre for Optical and Electromagnetic Research,
College of Optical Science and Engineering,
State Key Laboratory of Modern Optical Instrumentations,
Zijingang Campus, Zhejiang University, Hangzhou, China

Abstract

In metal optics, including metallic or plasmonic micro/nano structure electromagnetics, electromagnetic or optical surface plasmonic waves (e.g., planar-interface supported surface plasmon polaritons and nanaosphere-sustained localized surface plasmons) have captured a lot of interest and attention of a number of researchers in various research fields, including both fundamental physical principles and application-oriented areas (e.g., materials science and biotechnology). In this chapter, we will study some aspects of classical and quantum optical effects in surface plasmonics. For the first, we will consider the free-electron Drude model for describing metal optical properties and point out that there are two mysteries in the Drude model (i.e., though the local field correction and plasma oscillation effect have not been taken into account in the Drude model, yet the dielectric constants for metals obtained in such a model still agree well with the experimental observations). We resolve this mystery that seems to be puzzling, i.e., it can be found that the local field correction and the plasma oscillation effect cancel each other out exactly, no matter whether the free electron gas is in one-, two- or three-dimensional space. Since the negative permittivity is required in surface plasmonics, we suggest some alternative mechanisms for achieving tunable negative permittivity in three- and four-level quantum-coherent atomic vapor (e.g., alkali

* Corresponding Author's Email: jqshen@zju.edu.cn.

In: A Closer Look at Plasmonics
Editor: Christian Bosch
ISBN: 978-1-68507-709-9

metal atomic vapor). Then some scenarios of tunable prism coupler (based on such mechanisms relevant to quantum coherence) for exciting surface plasmon polaritons will be reviewed. The field quantization of surface plasmon modes is also considered and some new quantum field theoretical effects (including quantum-vacuum effects) will be suggested. These quantum effects include atom trapping and nanosphere quantum levitation. For example, because of the interaction between nanaosphere-sustained localized surface plasmons and quantized microcavity modes, the metallic nanosphere trapping (or quantum levitation) caused by the nanosphere-sustained localized surface plasmonic modes in a quantum microcavity will be suggested. The concept of metal-surface plasmonic wave modes can be extended to optical parity-time symmetric interface, where the electric permittivities of two electromagnetic media are complex conjugate. Based on this, two new parity-time symmetries (i.e., a generalized optical parity-time symmetry and a supersymmetric parity-time invariance) will also be suggested. We expect that some of the considerations in this chapter (such as tunable quantum-coherent atomic vapor, quantum surface plasmonic waves, and generalized parity-time symmetries) would open up readers' train of thought and allow them to propose more surface plasmonic effects, providing more theoretical principles and new working mechanisms for future surface plasmonics experiments as well as electromagnetic or photonic plasmonic device design.

Keywords: Drude-Lorentz model, quantum-coherent plasmonics, atomic phase coherence, quantum interference, electric and magnetic surface plasmon waves, surface plasmon field quantization, parity-time symmetric optics

PACS: 73.20.Mf, 12.38.Mh, 41.20.Jb, 52.25.-b, 52.25.Dg, 42.50.Gy, 42.50.Nn, 32.10.Dk, 33.15.Kr, 78.20.Ci

1. Introduction

Metal optics, surface plasmonics or micronano metal photonics is an essentially important research field in applied electromagnetism and modern optics ranging from fundamental principles to application-oriented technologies [1, 2, 3] (and references therein). As we know, the Drude-Lorentz model is one of the theoretical foundations of optical materials involved in the above studies (i.e., metallic materials and plasmas in surface plasmonics and micronano metal photonics). It should be pointed out that the Drude model for the electric permittivity of metals and plasmas is in good agreement with optical observations. But it should also be pointed out that such an electric permittivity of metals and plasmas does not take account of the well-known *local field correction* and *plasma oscillation effect.* Why does the Drude model, in which the effects of local field and plasma oscillation have been ignored, can still be consistent well with the experimental results? In this chapter, we find that these two effects (local field and plasma oscillation) can just cancel each other out, and therefore, the two effects cannot be exhibited in the Drude type electric permittivity for metals and plasmas.

In this chapter, some relevant aspects such as quantum coherent effects [4, 5, 6, 7, 8, 9] and quantum field effects (including quantum-vacuum effects) [10, 11, 12] in surface plasmonics will be reviewed or studied in details. In the literature, quantum vacuum fields and modes in cavity and electromagnetic media can exhibit various interesting behaviors [10, 11, 12, 13, 14]. For a tunable optical medium consisting of three-level systems, we consider a pumped quantum-dot (or atomic-vapor) electric-dipole allowed transition aiming at achieving an incoherent-gain negative permittivity and based on this, we suggest a scenario of incoherent-gain prism coupler system, which can give rise to effects of *surface-plasmon-like wave amplification* and *enhanced total reflection* [15]. Because of the quantum coherence or quantum-dot or atomic level phase coherence [7], the *surface-plasmon-like electromagnetic modes* are sensitive to the frequency detuning, e.g., the dispersion would be much more stronger than that in conventional metal-dielectric interface by about seven orders of magnitude [15]. Since the novel effect of *dispersion-sensitive enhancement in tunable reflection* can be exhibited, we expect that such an optical gain-assisted prism coupler would find potential applications for new quantum optical and photonic device design, such as photonic transistors and optical switching that is of dispersion sensitivity.

We also consider a system of four-level quantum-coherence assisted tunable prism coupler for generating surface plasmon resonance effect [16]. As is known, the optical response or characteristics of a four-level atomic system can be quantum-coherently controlled via tunable quantum (constructive or destructive) interference [17, 18, 19] between atomic transition pathways [6]. This quantum interference between atomic transition pathways can lead to electromagnetically induced transparency (EIT) and various interesting interference-switchable (classical and quantum) optical behavior in atomic media [20]. We will study a system of EIT-prism coupling (based on the knowledge of conventional surface-plasmon-polariton prism coupler) will be suggested. This scenario can be used for realizing tunable reflection spectrum because of quantum destructive and constructive interference occurring in the atomic transition pathways [21]. It should be noted that the four-level EIT atomic vapor in the present EIT-prism coupler can have an electric permittivity with a relatively large imaginary part when the quantum destructive interference (between the atomic transitions) takes place and then the effect of surface-plasmon-like resonance, in which the incident light can be coupled into surface electromagnetic modes (evanescent waves) on the interface between the prism dielectric and the atomic vapor cell, can be exhibited. Such an effect of controllable optical characteristics (based on the atomic level-transition quantum interference in the EIT-prism system) may find some possible applications in design of new electromagnetic, photonic or quantum optical devices. These might include the photonic logic gates, optical switching and photonic transistors [16] and some devices that control light with light on chips [22].

As an illustrative example of artificial design of electromagnetic or acoustic media, parity-time symmetry [23, 24, 25, 26, 27, 28] in quantum mechanics can be generalized to optical parity-time symmetry (acoustic parity-time symmetry can be found in some previous references [29, 30]), and hence the parity-time symmetric electromagnetics (optics) or parity-time symmetric acoustics can be established. The parity-time symmetry, of which the Hamiltonian is non-Hermitian but has real eigenvalues [23, 24], is difficult to be realized in quantum mechanics, but it is relatively easy to be achieved in optics and acoustics. In this chapter, we study the parity-time symmetric electromagnetic surface wave and its generalization, including the supersymmetric parity-time invariance.

We are concerned in the present chapter primarily on six topics in surface plasmonics. These classical and quantum topics include: i) The Drude-Lorentz model and its mysteries in local field effect and plasma oscillation effect (in Section 2); ii) The tunable negative permittivity assisted by three- and four-level quantum-coherent atomic vapor or semiconductor quantum-dot materials and its application in surface plasmon resonance (in

Sections 3 and 4); iii) Field quantization for surface plasmonic modes (in Section 5); iv) Single-interface surface plasmonic wave and its electromagnetic spin current density distribution (in Sections 6 and 7); v) Ordinary and generalized electromagnetic parity-time symmetric surface wave modes (in Sections 8 and 9); vi) Comparison of the equations between the Maxwell electromagnetic field and the elastic field (in Section 10). It has been my research work during the last seven years. These topics would be of interest to some readers in electromagnetics and quantum optics, particularly in metal optics and surface plasmonics.

Personally, I think the most important issues in this chapter include: i) The exact elimination in local field and plasma oscillation effects in the Drude model for metals and plasmas (in Section 2); ii) Leptons and quarks are the elementary particles without internal structures seen from the Landé factors of magnetic dipole moments (in Section 5); iii) The comparison in the dynamical equations and Lagrangians between the elastic wave and the electromagnetic wave (in Section 10).

2. The Drude-Lorentz Model and Its Two Mysteries

2.1. The Drude-Lorentz Model and the Relevant Issues

The Drude model for free electron gas is essentially important in metal optics and plasma physics (relevant to optical and electromagnetic properties). In addition to the Drude (metal-optics) model, there is the Lorentz (polarized-dielectric) model, which describes the electric dipole oscillators in dielectrics such as ionic crystals, atomic and molecular crystals [31]. Since the former (Drude model) is a special case of the latter (Lorentz model), we refer to these two models together as the "Drude-Lorentz model".

The Drude-Lorentz model is a dielectric polarization model, where the electromagnetic materials are assumed to be composed of electric dipole oscillators, which are driven by an incident propagating electromagnetic wave [31]. In a metal, the movement of heavy positive ions can be negligible, whereas the free electrons move out of equilibrium, and such ion-electron systems can be regarded as electric dipoles. In ionic crystals, seen from the aspect of optical or electromagnetic characteristics, the periodic arrangement of positive and negative ions can be reduced to electric dipoles. In atomic and molecular crystals, the outermost valence electrons can also deviate from their equilibrium positions under the action of an external electric field, so that each atom or molecule in an external electric field behaves as an electric dipole (seen from the aspect of optical characteristics).

2.1.1. Radiation Damping for a Charged Particle and Dirac Equation of Charged Particle in Acceleration

In the Drude-Lorentz model, the equation of the charge in an electric dipole oscillator is given by $m\ddot{x} + m\gamma\dot{x} + m\omega_0^2 x = qE$ with the driving electric field $E = E_0 \mathrm{e}^{-\mathrm{i}\omega t}$. Here, $-m\omega_0^2 x$ is an elastic restoring force and $-m\gamma\dot{x}$ is a damping force (proportional to the velocity $\dot{x}$, where dot denotes the time derivative). Since the damping force contains only the first derivative with respect to time, it breaks the invariance under the inversion of time, and so the effect caused by the damping force is irreversible. The damping force is proportional to the first power of the velocity of the charged particle. In general, such a regular rule can be interpreted in two ways: i) The Stokes viscosity formula $f_{\mathrm{Stokes}} = 6\pi\eta a v$ with η the viscous coefficient, a the particle radius and v the particle velocity; ii) Radiative damping force of charged particles $f = \frac{m}{\tau}v$.

The theoretical proof of the Stokes viscosity formula can be found in some textbooks of theoretical mechanics [32]. Here, we shall address the radiative damping force of charged particles: According to the classical electrodynamics, the Larmor radiation power of a charged particle in acceleration is given by $P = \tau_{\mathrm{c}} m\dot{v}^2$, where the radiation-damping characteristic time scale $\tau_{\mathrm{c}} = \frac{e^2}{6\pi\epsilon_0 c^3 m}$ and $\dot{v}$ is the particle acceleration [31]. This radiation power can be derived by using $P = \int \mathbf{S}\cdot\mathbf{n}R^2 \mathrm{d}\Omega$ with $\mathrm{d}\Omega$ a solid angle element and $\mathbf{S}$ a Poynting vector ($\mathbf{S} = \frac{1}{\mu_0}\mathbf{E}\times\mathbf{B}$). Here, $\mathbf{E}$ and $\mathbf{B}$ are the radiative electric and magnetic induction fields of an electrically charged particle in accelerated motion. We shall consider two typical cases, i.e., periodic oscillation in velocity and exponential decay in velocity: i) Since the Larmor radiation power of a charged particle in acceleration is $P = \tau_{\mathrm{c}} m\dot{v}^2$ with the radiation-damping characteristic time scale $\tau_{\mathrm{c}} = \frac{e^2}{6\pi\epsilon_0 c^3 m}$ [31], in a periodic (oscillatory) motion, we need to rewrite it as $P = \tau_{\mathrm{c}} m\dot{v}^*\dot{v}$, where $v = v_0 \mathrm{e}^{-\mathrm{i}\omega t}$ (in the complex form). Then we have $\dot{v}^*\dot{v} = \omega^2 v^* v$ and the Larmor radiation power is $P = \omega^2 \tau_{\mathrm{c}} m v^* v$. By defining the radiation-damping force f through $P = fv^*$, we have $f = \frac{P}{v^*} = \omega^2\tau_{\mathrm{c}}mv$. Thus, in the equation $m\ddot{x} + m\gamma\dot{x} + m\omega_0^2 x = qE$ can be changed to be $m\ddot{x} + m(\omega^2\tau_{\mathrm{c}})\dot{x} + m\omega_0^2 x = qE$. Thus, $\gamma = \omega^2\tau_{\mathrm{c}} = \omega^2\frac{e^2}{6\pi\epsilon_0 c^3 m}$. This result of γ is also given in some previous references (e.g., Ref. [33]); ii) According to the mechanical equations $m\dot{v} = f_{\mathrm{damp}}$ and $-f_{\mathrm{damp}}v = P$, we have $-m\dot{v}v = \tau_{\mathrm{c}}m\dot{v}^2$, i.e., $\dot{v}(\dot{v} + \frac{1}{\tau_{\mathrm{c}}}v) = 0$. There are two solutions: $\dot{v} = 0$ (uniform motion in a straight line) and $\dot{v} = -\frac{1}{\tau_{\mathrm{c}}}v$, i.e., $v = v_0\mathrm{e}^{-\frac{t}{\tau_{\mathrm{c}}}}$. Now by substituting $\dot{v} = -\frac{1}{\tau_{\mathrm{c}}}v$ and $P = \tau_{\mathrm{c}}m\dot{v}^2$ into $-f_{\mathrm{damp}}v = P$, we can obtain the radiative damping force $f_{\mathrm{damp}} = -\frac{m}{\tau_{\mathrm{c}}}v$ for a charged particle. However, it should be pointed out that the derivation for the above radiative damping force formula is not rigorous, because there are some insurmountable inherent contradictions in the radiation damping effect in classical electrodynamics, so such a method is presented only in an instructive sense and it cannot be taken seriously.

Since the Larmor radiation power of an accelerated charged particle is given by $P = \tau_{\mathrm{c}}m\dot{v}^2$, Dirac constructed a phenomenological relativistic equation of the charged particle:

$$\frac{\mathrm{d}p^\nu}{\mathrm{d}\tau} = \tau_{\mathrm{c}}\left(\frac{\mathrm{d}^2p^\nu}{\mathrm{d}\tau^2} + \frac{p^\nu}{(m_0c)^2}\frac{\mathrm{d}p_\mu}{\mathrm{d}\tau}\frac{\mathrm{d}p^\mu}{\mathrm{d}\tau}\right). \tag{2.1}$$

Since $p_\nu p^\nu = (m_0c)^2$, by multiplying p_ν on two sides of Eq. (2.1), the term on the left-handed side is $\mathrm{L} = p_\nu\frac{\mathrm{d}p^\nu}{\mathrm{d}\tau}$ and the term on the right-handed side is $\mathrm{R} = \tau_{\mathrm{c}}p_\nu(\frac{\mathrm{d}^2p^\nu}{\mathrm{d}\tau^2} + \frac{\mathrm{d}p_\mu}{\mathrm{d}\tau}\frac{\mathrm{d}p^\mu}{\mathrm{d}\tau}) = \tau_{\mathrm{c}}\frac{\mathrm{d}}{\mathrm{d}\tau}(p_\nu\frac{\mathrm{d}p^\nu}{\mathrm{d}\tau})$. We have $\frac{\mathrm{d}}{\mathrm{d}\tau}(p_\nu\frac{\mathrm{d}p^\nu}{\mathrm{d}\tau}) = \frac{1}{\tau_{\mathrm{c}}}(p_\nu\frac{\mathrm{d}p^\nu}{\mathrm{d}\tau})$. From this equation, we can obtain the solution: $p_\nu\frac{\mathrm{d}p^\nu}{\mathrm{d}\tau} = 0$ or $p_\nu\frac{\mathrm{d}p^\nu}{\mathrm{d}\tau} \propto \mathrm{e}^{\frac{t}{\tau_{\mathrm{c}}}}$. But the latter form contradicts the relation $p_\nu p^\nu = (m_0c)^2$.

It should be noted that there is no external force in Eq. (2.1). By adding the external force f^ν that causes the acceleration of the electric charge, the Dirac equation for the accelerated charge is

$$\frac{\mathrm{d}p^\nu}{\mathrm{d}\tau} = \tau_{\mathrm{c}}\left(\frac{\mathrm{d}^2p^\nu}{\mathrm{d}\tau^2} + \frac{p^\nu}{(m_0c)^2}\frac{\mathrm{d}p_\mu}{\mathrm{d}\tau}\frac{\mathrm{d}p^\mu}{\mathrm{d}\tau}\right) + f^\nu, \tag{2.2}$$

where the 4-D electromagnetic force is $f^\nu = qF^{\nu\lambda}U_\lambda = \frac{q}{m_0}F^{\nu\lambda}p_\lambda$. Here, $F^{\nu\lambda} = \partial^\nu A^\lambda - \partial^\lambda A^\nu$ and $p_\nu f^\nu = 0$ (orthogonality). Once there is an external force f^ν, the acceleration term $\frac{\mathrm{d}p^\nu}{\mathrm{d}\tau} \neq 0$, and $\tau_{\mathrm{c}}\frac{\mathrm{d}^2p^\nu}{\mathrm{d}\tau^2}$ and

$\tau_{\rm c}\frac{p^{\nu}}{(m_{0}c)^{2}}\frac{{\rm d}p_{\mu}}{{\rm d}\tau}\frac{{\rm d}p^{\mu}}{{\rm d}\tau}$ are unavoidably nonzero, i.e., the electric charge in acceleration will emit the electromagnetic radiation.

Now we shall consider another problem, though it is not related intimately to the present problem of Drude-Lorentz model, yet it is still an essential issue for damping of motion of charged particles: Does the electric charge in acceleration emit the electromagnetic radiation when it is in a gravitational field? For example, when a free charged particle is in a freely-falling elevator (in a gravitational field), the free charge is at rest seen from the observer in the elevator (local inertial frame), so it does not radiate the electromagnetic radiation. But to an observer in a ground frame of reference, this charged particle accelerates in the gravitational field and would emit the electromagnetic radiation. This constitutes a contradiction. Let us solve this problem: In the gravitational field, the equation (2.2) should be replaced by

$$\frac{{\rm D}p^{\nu}}{{\rm d}\tau}=\tau_{\rm c}\left(\frac{{\rm D}^{2}p^{\nu}}{{\rm d}\tau^{2}}+\frac{p^{\nu}}{(m_{0}c)^{2}}\frac{{\rm D}p_{\mu}}{{\rm d}\tau}\frac{{\rm D}p^{\mu}}{{\rm d}\tau}\right), \tag{2.3}$$

where the covariant derivative is given by $\frac{{\rm D}p^{\nu}}{{\rm d}\tau}=\frac{{\rm d}p^{\nu}}{{\rm d}\tau}+\Gamma^{\nu}{}_{\alpha\beta}U^{\beta}p^{\alpha}$. Here, $\Gamma^{\nu}{}_{\alpha\beta}$ is the Levi-Civita affine connection. It can be found that there is a solution $\frac{{\rm D}p^{\nu}}{{\rm d}\tau}\equiv 0$ (and hence $\frac{{\rm D}^{2}p^{\nu}}{{\rm d}\tau^{2}}=0$), i.e., it seems that an accelerated charge in the gravitational field does not emit electromagnetic radiation. We can further interpret the solution $\frac{{\rm D}p^{\nu}}{{\rm d}\tau}\equiv 0$ as follows: In the equation (2.2), $\frac{{\rm d}p^{\nu}}{{\rm d}\tau}\neq 0$ because of the presence of the external force f^{ν}. But in the equation (2.3), there is no such an external force term and we should have $\frac{{\rm D}p^{\nu}}{{\rm d}\tau}\equiv 0$. (This was the question I pondered and the conclusion I reached in 1997–1999 when I was a physics-department undergraduate student.) Of course, we should also point out that this may be only a result of a first-order approximation, i.e., there would also be a possibility of $\frac{{\rm D}^{2}p^{\nu}}{{\rm d}\tau^{2}}\neq 0$ (For example, when $\frac{{\rm D}^{2}p^{\nu}}{{\rm d}\tau^{2}}\neq 0$, i.e., the charged particle moves with varying acceleration, which can possibly be caused by a tidal force, i.e., $\frac{{\rm D}^{2}p^{\nu}}{{\rm d}\tau^{2}}\sim f^{\nu}$, where the tidal force f^{ν} is determined by the Riemannian curvature tensor. If the tidal force can be ignored, the solution must be the form $\frac{{\rm D}p^{\nu}}{{\rm d}\tau}\equiv 0$). In the literature, within the framework of the theory of general relativity, by calculating the energy flow density (Poynting vector) $\mathbf{E}\times\mathbf{H}$ of a charged particle in a gravitational field, Wang has shown that such a particle really does not emit electromagnetic radiation [34].

In the theory of general relativity, the accelerated mass can emit the gravitational radiation. This means that there is also a relativistic equation of motion like the equation (2.2), where the characteristic time scale $\tau_{\rm c}$ is determined by the theory of gravity. In the theories of gravity-gauge unification (e.g., Kaluza-Klein 5-D or non-Abelian Kaluza-Klein theory), the Dirac damping equation (2.2) emerges from this unified theory.

2.1.2. The Drude-Lorentz Model

Now we return to the equation $m\ddot{x} + m\gamma\dot{x} + m\omega_0^2 x = qE$. The displacement x must be $x = x_0 {\rm e}^{-{\rm i}\omega t}$ and one can have $(-\omega^2 + \omega_0^2 - {\rm i}\gamma\omega)x = \frac{q}{m}E$. The solution is $x = \frac{q}{m} \cdot \frac{1}{-\omega^2+\omega_0^2-{\rm i}\gamma\omega}E$ [31]. In general, an electric dipole consists of a heavy positive ion and an electron (in, e.g., a metal). Since the displacement of heavy positive ions is small, we will ignore their motion and we consider the electron displacement x only. Then the electric polarization is given by $P = n_{\rm d}qx = \frac{n_{\rm d}q^2}{m} \cdot \frac{1}{-\omega^2+\omega_0^2-{\rm i}\gamma\omega}E$, where m and q are the mass and the electric charge of an electron; $n_{\rm d}$ denotes the total electric dipole number per unit volume in the medium; In a metal, $n_{\rm d}$ can equal the total free electron number per unit volume. The medium polarization vector $\mathbf{P}$ is

$$\mathbf{P} = \frac{n_{\rm d}q^2}{\epsilon_0 m} \cdot \frac{1}{-\omega^2 + \omega_0^2 - {\rm i}\gamma\omega}\epsilon_0\mathbf{E} = n_{\rm d}\beta_{\rm pol}\epsilon_0\mathbf{E}. \tag{2.4}$$

In the electromagnetics, as is well known, the electric polarization vector $\mathbf{P} = \chi\epsilon_0\mathbf{E}$, where χ is the electric susceptibility of the medium. The relative permittivity is $\epsilon_{\rm r} = 1 + \chi$, where the explicit expression for the susceptibility is given by

$$\chi = n_{\rm d}\beta_{\rm pol} = \frac{n_{\rm d}q^2}{\epsilon_0 m} \cdot \frac{1}{-\omega^2 + \omega_0^2 - {\rm i}\gamma\omega} = \frac{\omega_{\rm p}^2}{\omega_0^2 - \omega^2 - {\rm i}\gamma\omega}. \tag{2.5}$$

Here, the plasma frequency square is $\omega_{\rm p}^2 = \frac{n_{\rm d}q^2}{\epsilon_0 m} = \frac{n_{\rm d}e^2}{\epsilon_0 m}$. Now the relative electric permittivity of the medium in the Drude-Lorentz model is of the form [31]

$$\epsilon_{\rm r} = 1 + n_{\rm d}\beta_{\rm pol} = 1 + \frac{\omega_{\rm p}^2}{\omega_0^2 - \omega^2 - {\rm i}\gamma\omega}. \tag{2.6}$$

Now we shall consider the plasma permittivity. The plasma is an ionized gas, where the electric-dipole (intrinsic) oscillation frequency square $\omega_0^2 = 0$ (in the Drude model). The relative electric permittivity of the plasma is given by

$$\epsilon_{\rm r} = 1 - \frac{\omega_{\rm p}^2}{\omega(\omega + {\rm i}\gamma)}, \qquad \epsilon_{\rm r} = 1 - \frac{\omega_{\rm p}^2}{\omega^2} \quad (\gamma \simeq 0), \tag{2.7}$$

where in the second relation the damping rate γ is negligibly small (in a dilute plasma or in a metal when the driving electromagnetic wave frequency ω is large, e.g., $\omega \simeq 10^{15}$ rad/s. In general, the damping rate γ has the order of magnitude of $10^{13} \sim 10^{14}$ rad/s in a metal). This electric permittivity (2.7) of the plasma has been successfully used in ionized gases [35, 36].

If there is a background permittivity $\epsilon_{\rm b}$, i.e., the electric displacement vector $\mathbf{D} = \epsilon_0\mathbf{E} + \mathbf{P}_{\rm b} + \mathbf{P} = \epsilon_{\rm b}\epsilon_0\mathbf{E} + \mathbf{P} = \epsilon_{\rm b}\epsilon_0\mathbf{E} + n_{\rm d}\beta_{\rm pol}\epsilon_0\mathbf{E} = (\epsilon_{\rm b} + n_{\rm d}\beta_{\rm pol})\epsilon_0\mathbf{E} = \epsilon_{\rm r}\epsilon_0\mathbf{E}$, we have

$$\epsilon_{\rm r}(\omega) = \epsilon_{\rm b} + n_{\rm d}\beta_{\rm pol} = \epsilon_{\rm b}\left(1 + \frac{n_{\rm d}\beta_{\rm pol}}{\epsilon_{\rm b}}\right) = \epsilon_{\rm b} - \frac{\omega_{\rm p}^2}{\omega(\omega + {\rm i}\gamma)}. \tag{2.8}$$

The background permittivity $\epsilon_{\rm b}$ is ϵ_∞ or $\epsilon_{\rm r}(\infty)$, which is caused by residual polarization relevant to the positively charged background of ionic nuclei [37].

We shall give an illustrative example of application of the electric permittivity (2.8) in a metal:

In the Drude model for a free electron gas, the electric permittivity is given by $\epsilon_{\rm r} = \epsilon_\infty - \frac{\omega_{\rm p}^2}{\omega(\omega+{\rm i}\gamma)}$, where γ is the damping rate. On the other hand, according to the Ampère circuital law in the Maxwell equations, we have $\epsilon_{\rm r} = \epsilon_\infty + \frac{{\rm i}\sigma(\omega,\gamma)}{\epsilon_0\omega}$. As is well known in the textbooks of electromagnetics, this effective permittivity can be obtained as follows: The curl equation $\nabla \times \mathbf{H} = \frac{\partial \mathbf{D}}{\partial t} + \mathbf{J}_{\rm f}$ in the metal can be rewritten as

$$\begin{aligned}\frac{\partial \mathbf{D}}{\partial t} + \mathbf{J}_{\rm f} &= \epsilon_\infty\epsilon_0\frac{\partial \mathbf{E}}{\partial t} + \sigma\mathbf{E} = \epsilon_\infty\epsilon_0\frac{\partial \mathbf{E}}{\partial t} + \sigma\frac{\rm i}{\omega}\frac{\partial \mathbf{E}}{\partial t} \\ &= \epsilon_\infty\epsilon_0\left(1 + \frac{\sigma}{\epsilon_\infty\epsilon_0}\frac{\rm i}{\omega}\right)\frac{\partial \mathbf{E}}{\partial t} = \epsilon_{\rm eff}\epsilon_0\frac{\partial \mathbf{E}}{\partial t},\end{aligned} \tag{2.9}$$

where the effective dielectric "constant" is given by

$$\epsilon_{\text{eff}} = \epsilon_{\text{r}} = \epsilon_{\infty} + \frac{\text{i}}{\omega}\frac{\sigma}{\epsilon_0}. \tag{2.10}$$

Then the dynamical electric conductivity in a metal is given by $\sigma(\omega,\gamma) = \text{i}\frac{\omega_{\text{p}}^2\epsilon_0}{\omega+\text{i}\gamma} = \text{i}\frac{n_{\text{e}}e^2}{m_{\text{e}}(\omega+\text{i}\gamma)}$. This formula can be used to classify the perfect conductor (pc) and the superconductor (sc). Our result can be listed as follows:

$$\begin{aligned}
\sigma_{\text{pc}} &= \lim_{\gamma\to 0}\lim_{\omega\to 0}\sigma(\omega,\gamma) = \lim_{\gamma\to 0}\lim_{\omega\to 0}\left[\text{i}\frac{n_{\text{e}}e^2}{m_{\text{e}}(\omega+\text{i}\gamma)}\right] \\
&= \lim_{\gamma\to 0}\frac{n_{\text{e}}e^2}{m_{\text{e}}\gamma} \to \text{real infinite number}, \\
\sigma_{\text{sc}} &= \lim_{\omega\to 0}\lim_{\gamma\to 0}\sigma(\omega,\gamma) = \lim_{\omega\to 0}\lim_{\gamma\to 0}\left[\text{i}\frac{n_{\text{e}}e^2}{m_{\text{e}}(\omega+\text{i}\gamma)}\right] \\
&= \lim_{\omega\to 0}\left(\text{i}\frac{n_{\text{e}}e^2}{m_{\text{e}}\omega}\right) \to \text{imaginary infinite number}.
\end{aligned} \tag{2.11}$$

From the above result, it can be seen that at static frequency, the electric conductivity of a perfect conductor is a real infinite number, whereas the electric conductivity of a superconductor is an imaginary infinite number.

It should be stressed that the perfect conductor is an ideal concept, which may not exist in nature. But some materials at low frequencies can be identified with an analog of perfect conductors. For example, in a metal, the typical damping rate is $\gamma \simeq 10^{14}$ rad/s, and at terahertz and microwave frequencies, we should choose the first case in Eq. (2.11) because $\omega \ll \gamma$. Now we obtain $\sigma_{\text{pc}} \to \frac{n_{\text{e}}e^2}{m_{\text{e}}\gamma} = \frac{n_{\text{e}}e^2}{m_{\text{e}}\epsilon_0}\frac{1}{\gamma\omega}\epsilon_0\omega$. Here the dimensionless factor $\frac{n_{\text{e}}e^2}{m_{\text{e}}\epsilon_0}\frac{1}{\gamma\omega}$ is very large (larger than 10^6), where $\frac{n_{\text{e}}e^2}{m_{\text{e}}\epsilon_0} \simeq 10^{32}\ \text{rad}^2/\text{s}^2$. Thus, we do not need to choose $\gamma \to 0$ and the present metal at terahertz and microwave frequencies can still be considered to be a "perfect conductor".

As far as the superconductor is concerned, the damping rate is $\gamma = 0$ because of electron-phonon coupling and Cooper pair formation. Thus, we should choose the second case in Eq. (2.11) because $\omega \gg \gamma = 0$. In such a superconductor, the electric permittivity is $\epsilon_{\text{r}} = \epsilon_{\infty} - \frac{n_{\text{e}}e^2}{\epsilon_0 m_{\text{e}}}\frac{1}{\omega^2}$ and then we have

$$k^2 = \epsilon_r \frac{\omega^2}{c^2} = \left(\epsilon_\infty - \frac{n_e e^2}{\epsilon_0 m_e} \frac{1}{\omega^2}\right) \frac{\omega^2}{c^2} = \epsilon_\infty \frac{\omega^2}{c^2} - \frac{n_e e^2}{\epsilon_0 m_e c^2}$$

$$\Rightarrow \quad (\hbar k)^2 = \frac{(\hbar\omega)^2}{\frac{c^2}{\epsilon_\infty}} - \frac{n_e e^2 \epsilon_\infty \hbar^2}{\epsilon_0 m_e c^4} \frac{c^2}{\epsilon_\infty}. \tag{2.12}$$

This relation can be rewritten as $p^2 = \frac{E^2}{c'^2} - m_{\text{eff}}^2 c'^2$, where

$$c'^2 = \frac{c^2}{\epsilon_\infty}, \qquad m_{\text{eff}}^2 = \frac{n_e e^2 \epsilon_\infty \hbar^2}{\epsilon_0 m_e c^4}. \tag{2.13}$$

It seems that the photon in a superconductor carries an effective rest mass, which is also related to the symmetry breaking of the Higgs mechanism in elementary particle physics and quantum field theory.

2.1.3. Discussion for the Drude-Lorentz Model

This electric permittivity (2.8) in metals is in good agreement with experimental observations [37, 38]. In metal silver, the plasma frequency $\omega_p = 1.40 \times 10^{16}$ rad/s and the damping rate $\gamma = 3.23 \times 10^{13}$ rad/s; In metal gold, the plasma frequency $\omega_p = 1.38 \times 10^{16}$ rad/s and the damping rate $\gamma = 1.075 \times 10^{14}$ rad/s [38]. When $\hbar\omega$ is in the range $[1.0, 4.0]$ eV (i.e., ω is in the range $[1.5, 6.0] \times 10^{15}$ rad/s) for Ag and when $\hbar\omega$ is in the range $[1.0, 2.0]$ eV (i.e., ω is in the range $[1.5, 3.0] \times 10^{15}$ rad/s) for Au, Johnson and Christy demonstrated experimentally that the metal permittivity of Drude model agrees very well with such noble metals [38]. But it should be pointed out that the *internal band transition* in the metal limits the applicability of the Drude model in the bands of higher frequencies (i.e., much larger than the visible frequencies) [37]. In any case, we can conclude that the Drude model for metal optics agrees well with the experimental results, except for some additional complications (e.g., internal band transition [38]).

However, the story does not end here, since the above Drude-Lorentz model is not complete. We need to consider two important physical factors of *collective motion* or *collective behavior* of charged particles that are always present: "local field correction" and "plasma oscillation" (in metals and plasmas). Since 2003, I have been wondering why researchers in metal optics and surface plasmonics did not take into account the *local field correction effect of adjacent electric dipoles* and the *restoring force effect due to plasma oscillation of free electron gas* when studying the Drude-model dielectric coefficients of plasmas and metals.

As is well known, in metals and plasmas, there is free electron gas. In general, due to thermal fluctuation, the local electric neutrality imbalance in free electrons and heavy positive ions is always present, so that a local electric field is established, and the magnitude of such an established electric field strength is proportional to the distance from the equilibrium position, so that the electrons in this local space are subject to an electric field force similar to an elastic restoring force, and the behavior of electron moving in this local space is similar to that of a harmonic oscillator, of which the oscillation frequency is the plasma frequency ω_{p}. This is the plasma oscillation effect that was first considered by Langmuir in 1929 [39].

Let us evaluate the thermal fluctuation contribution to driving the plasma oscillation: In an ordinary plasma such as arc discharge and ionosphere (in the upper atmosphere of the Earth [36]), the plasma frequency ω_{p} is 6×10^{11} and 6×10^{7} rad/s, respectively, and ω_{p} is 2×10^{5} rad/s in the Solar wind ($r = 1$ AU from the Sun) and ω_{p} is 6×10^{11} rad/s in Tokamak (a circular vessel in magnetic-confinement controlled nuclear fusion) [40]. The thermally-excited plasmon energy $\varepsilon_{\mathrm{plasmon}} = \hbar\omega_{\mathrm{p}}$ in these plasmas is less than $\sim 10^{-23}$ J. According to statistical physics, the order of magnitude of average thermal energy for each particle is $\varepsilon_{\mathrm{thermal}} \simeq \frac{3}{2}k_{\mathrm{B}}T$. Clearly, the average thermal energy $\varepsilon_{\mathrm{thermal}}$ has many orders of magnitude larger than the plasmon energy $\varepsilon_{\mathrm{plasmon}}$, i.e., the plasma oscillation modes of free electrons can be easily thermally excited. In a metal, the plasma frequency ω_{p} has the order of magnitude of 10^{16} rad/s; in inertial confinement fusion (ICF), the plasma frequency ω_{p} is 6×10^{15} rad/s [40]. The plasmon energy in the metal is $\varepsilon_{\mathrm{plasmon}} = \hbar\omega_{\mathrm{p}} \simeq (0.6 \sim 1) \times 10^{-18}$ J. In a metal, the thermal energy $\varepsilon_{\mathrm{thermal}} \simeq \frac{3}{2}k_{\mathrm{B}}T$ is much less than the plasmon energy, and it seems that the quantized plasmons (including the plasmon at quantum vacuum level) cannot be excited thermally or established. But when we study the optical or electromagnetic properties of a metal, there must be a weak probe light, which can still drive the plasma oscillation. The displacement (the distance from equilibrium) of an ion driven by the light has the order of magnitude $s \sim \frac{qE}{m}(\frac{1}{f})^{2}$ with f the wave frequency. If the incident electromagnetic wave electric field is $E \simeq 10^2$ $\mathrm{V/m}$, the displacement s is $10^{-16} \sim 10^{-15}$ m for an free electron in the metal. The displacement of the heavy positive ion is negligibly small. The displacement ($10^{-16} \sim 10^{-15}$ m) of electrons estimated by this method is very small, but as a slight perturbation, it must exist as a correction to the wavefunction of free electrons in the metal. Therefore, plasma oscillation, both classical and quantal, must be present in any metals and plasmas. In short, as long as there is an external electric field driving the electron gas layer in the plasma, there will be a new electric field (in the plasma) in the opposite direction to the external electric field. The strength of the electric field is proportional to the distance from the equilibrium position of the electron layer, and the electric field force provided by it acts as a restoring force, pulling the electrons back to the equilibrium position. So the plasma oscillation effect is inevitable in free electron gas.

Now we return to the local field correction in the electric permittivity of metals and plasmas [41]. The local field effect is caused by the dipole-dipole interaction (mutual polarization between the neighboring electric dipoles in the medium) [41]. The theoretical formalism describing such a dipole-dipole interaction is the Clausius-Mossotti relation (also called the Lorenz-Lorentz relation, or the Clausius-Mossotti-Lorenz-Lorentz relation). If the local field effect is not taken into account, the relative permittivity of a medium is the well known formula $\epsilon_{\mathrm{r}} = 1 + n_{\mathrm{d}}\beta_{\mathrm{pol}}$, where n_{d} and β_{pol} denote the electric dipole number density and the microscopic electric polarizability of each dipole, respectively. If, however, we consider the local field correction, the relative permittivity of the medium is $\epsilon_{\mathrm{r}} = \epsilon_{\mathrm{r}}(n_{\mathrm{d}}\beta_{\mathrm{pol}}) = 1 + n_{\mathrm{d}}\beta_{\mathrm{pol}} + ...$, which is a nonlinear function of $n_{\mathrm{d}}\beta_{\mathrm{pol}}$. To be precise, from the Clausius-Mossotti relation, in a plasma, the relative permittivity is given by $\epsilon_{\mathrm{r}} = 1 + \frac{n_{\mathrm{d}}\beta_{\mathrm{pol}}}{1-\frac{n_{\mathrm{d}}\beta_{\mathrm{pol}}}{3}}$ [41]. Such a nonlinear function of n_{d} represents the mutual interaction between neighboring dipoles in the medium.

In the Drude model, the effects of both *local field* and *plasma oscillation* will unavoidably modify the electric permittivity of plasmas and metals. To our surprise, in previous references and monographs, the modification caused by local field and plasma oscillation in the permittivity of plasmas and metals was not taken into account. Why is the Drude model still consistent with the experimental data even when these two factors were not taken into account?

2.2. A Mystery of Local Field Correction in the Drude-Lorentz Model

Now we shall analyze the concept of "local field effect" caused by mutual polarization of neighboring atoms, molecules or electric dipoles in an electric medium [41, 42, 43, 44].

In the above, when the local field effect was not taken into account, we have used the two relations

$$\mathbf{P} = \chi\epsilon_0\mathbf{E}, \qquad \mathbf{P} = n_{\mathrm{d}}\beta_{\mathrm{pol}}\epsilon_0\mathbf{E}, \tag{2.14}$$

where $\mathbf{E}$ is in general an applied external electric field. We did not make a distinction between the local fields $\mathbf{E}_{\mathrm{local}}$ (acting upon the electric dipoles in the electric medium) and the external applied fields $\mathbf{E}_{\mathrm{external}}$ [41]. But as a matter of fact, the correct relations that need to replace Eq. (2.14) should be listed as follows

$$\mathbf{P} = \chi\epsilon_0\mathbf{E}_{\mathrm{external}}, \qquad \mathbf{P} = n_{\mathrm{d}}\beta_{\mathrm{pol}}\epsilon_0\mathbf{E}_{\mathrm{local}}, \tag{2.15}$$

because in a dielectric, in addition to the external *average macroscopic field* $\mathbf{E}_{\mathrm{external}}$, there is a polarized electric field $\mathbf{E}_{\mathrm{p}}$ (caused by the dipole-dipole interaction) for any given electric dipoles [41, 44]. Such a dipole-dipole interaction originating from the polarization of neighboring electric dipoles occurs in the medium with closely packed atoms, molecules and electric dipoles [41], and therefore a distinction between the local electric fields (at the electric dipoles in the medium) and the external applied fields must be made [41, 44]. The internal local electric field $\mathbf{E}_{\mathrm{local}}$ acting on the electric dipoles can be expressed as the difference between two terms, namely, $\mathbf{E}_{\mathrm{local}} = \mathbf{E}_{\mathrm{external}} - \mathbf{E}_{\mathrm{p}}$, where $\mathbf{E}_{\mathrm{p}}$ is the contribution made by the electric dipoles (this can be addressed in the "average continuum approximation" characterized by the electric polarization $\mathbf{P}$, where $\mathbf{E}_{\mathrm{p}}$ in this medium is given by $\mathbf{E}_{\mathrm{p}} = -\frac{\mathbf{P}}{3\epsilon_0}$ [41]). Thus, the local internal field is $\mathbf{E}_{\mathrm{local}} = \mathbf{E}_{\mathrm{external}} + \frac{\mathbf{P}}{3\epsilon_0}$ [41, 42, 43, 44]. Let the microscopic electric polarizability of an electric dipole be β_{pol}. Then the polarization vector is $\mathbf{P} = n_{\mathrm{d}}\beta_{\mathrm{pol}}\epsilon_0\left(\mathbf{E}_{\mathrm{external}} + \frac{\mathbf{P}}{3\epsilon_0}\right)$, where $\mathbf{E}_{\mathrm{external}}$ is the external average macroscopic field strength and n_{d} is the electric dipole number density (the total number of dipoles per unit volume). With the help of $\mathbf{P} = \epsilon_0 \chi \mathbf{E}_{\mathrm{external}}$, one can obtain the susceptibility $\chi = \frac{n_{\mathrm{d}}\beta_{\mathrm{pol}}}{1-\frac{n_{\mathrm{d}}\beta_{\mathrm{pol}}}{3}}$ [41, 42, 43, 44], which manifests the relation between the electric susceptibility χ (the macroscopic parameter) and the polarizability β_{pol} (the microscopic parameter). The above analysis can lead to the Clausius-Mossotti relation: $n_{\mathrm{d}}\beta_{\mathrm{pol}} = \frac{3(\epsilon_{\mathrm{r}}-1)}{\epsilon_{\mathrm{r}}+2}$ [42, 43], where ϵ_{r} denotes the relative permittivity of the medium. This relation holds in many dilute gases [41]. This suggests that local field effect does exist in substances made up of atoms and molecules. Of course, in these substances, there is no motion of plasma oscillation, partly because free electrons are absent in them. But after careful consideration, we think this interpretation may not be valid because, after all, positive and negative charges are still separated from each other, so there is still an induced electric field between them, in the direction opposite to the external field. So we need to look for other explanations. According to quantum theory, when an external electric field drives the atomic energy level transition from a lower energy state to an upper state, the electric dipole moment of the atom originates from the electric-dipole allowed transition dipole matrix element, which is different from the separation phenomenon of positive and negative charges (i.e., the classical ordinary cases). In such a situation, the effect related to plasma oscillation does not appear in atomic and molecular gases (and substances).

Here is our summary relevant to the local field correction, which can lead to the Clausius-Mossotti relation [45]:

$$\mathbf{E}_{\mathrm{local}} = \mathbf{E}_{\mathrm{external}} + (-\mathbf{E}_{\mathrm{neighboring}}),$$
$$\mathbf{E}_{\mathrm{neighboring}} = -\frac{\mathbf{P}}{3\epsilon_0}, \qquad \mathbf{P} = n_{\mathrm{d}}\beta_{\mathrm{pol}}\epsilon_0 \mathbf{E}_{\mathrm{local}}, \tag{2.16}$$

where the dimensionless quantity $n_{\mathrm{d}}\beta_{\mathrm{pol}}$ based on the Lorentz model is given by

$$n_{\mathrm{d}}\beta_{\mathrm{pol}} = \frac{n_{\mathrm{d}}e^2}{\epsilon_0 m} \cdot \frac{1}{\omega_0^2 - \omega^2 - \mathrm{i}\gamma\omega} = \frac{\omega_{\mathrm{p}}^2}{\omega_0^2 - \omega^2 - \mathrm{i}\gamma\omega}. \tag{2.17}$$

The electric permittivity is $\epsilon_{\mathrm{r}} = 1 + \chi$, where χ is defined through $\mathbf{P} = \chi\epsilon_0 \mathbf{E}_{\mathrm{external}}$. In Eq. (2.17), the plasma frequency square is $\omega_{\mathrm{p}}^2 = \frac{n_{\mathrm{d}}e^2}{\epsilon_0 m}$.

It follows from the relations given in Eq. (2.16) that the electric polarization $\mathbf{P}$ under the condition of local field effect is given by

$$\begin{aligned}\mathbf{P} &= \chi\epsilon_0\mathbf{E}_{\text{external}} = \chi\epsilon_0\left(\mathbf{E}_{\text{local}} + \mathbf{E}_{\text{neighboring}}\right) \\ &= \chi\epsilon_0\left(\mathbf{E}_{\text{local}} - \frac{\mathbf{P}}{3\epsilon_0}\right) = \chi\epsilon_0\left(\frac{\mathbf{P}}{n_{\text{d}}\beta_{\text{pol}}\epsilon_0} - \frac{\mathbf{P}}{3\epsilon_0}\right). \qquad (2.18)\end{aligned}$$

Then the electric susceptibility χ is of the form

$$\chi = \frac{1}{\frac{1}{n_{\text{d}}\beta_{\text{pol}}} - \frac{1}{3}} = \frac{n_{\text{d}}\beta_{\text{pol}}}{1 - \frac{n_{\text{d}}\beta_{\text{pol}}}{3}}. \qquad (2.19)$$

From this equation, one can have $\chi\left(1 - \frac{n_{\text{d}}\beta_{\text{pol}}}{3}\right) = n_{\text{d}}\beta_{\text{pol}}$ and $\chi = n_{\text{d}}\beta_{\text{pol}}(1 + \frac{\chi}{3})$, i.e., $n_{\text{d}}\beta_{\text{pol}} = \frac{\chi}{1+\frac{\chi}{3}}$. Then it can yield

$$\frac{\epsilon_{\text{r}} - 1}{\epsilon_{\text{r}} + 2} = \frac{n_{\text{d}}\beta_{\text{pol}}}{3}. \qquad (2.20)$$

This is called the Clausius-Mossotti relation. In the history, Mossotti in 1850 and Clausius in 1879 suggested that the ratio $\frac{\epsilon_{\text{r}}-1}{\epsilon_{\text{r}}+2}$ is proportional to the number density, n_{d}, of the matters (substances) [41]. It can also be identified as the Lorenz-Lorentz relation when $\epsilon_{\text{r}} = n_{\text{ref}}^2$ is substituted (1880), where n_{ref} denotes the refractive index of the medium [41].

Substituting the result (2.17) into Eq. (2.19), the electric susceptibility is

$$\chi = \frac{\frac{\omega_{\text{p}}^2}{\omega_0^2-\omega^2-\text{i}\gamma\omega}}{1 - \frac{\frac{\omega_{\text{p}}^2}{\omega_0^2-\omega^2-\text{i}\gamma\omega}}{3}} = \frac{\omega_{\text{p}}^2}{\left(\omega_0^2 - \frac{\omega_{\text{p}}^2}{3}\right) - \omega^2 - \text{i}\gamma\omega}. \qquad (2.21)$$

Comparing it with Eq. (2.5), it seems that the proper oscillation frequency square, ω_0^2, of the electric-dipole oscillator is shifted by the local field effect, i.e., $\omega_0^2 \rightarrow \omega_0^2 - \frac{\omega_{\text{p}}^2}{3}$.

If there is a background permittivity ϵ_{b}, then all the vacuum permittivity ϵ_0 should be replaced by $\epsilon_{\text{b}}\epsilon_0$ and the plasma frequency square ω_{p}^2 is replaced with $\frac{\omega_{\text{p}}^2}{\epsilon_{\text{b}}}$. The absolute permittivity ϵ and the relative permittivity

ϵ_{r} is given by

$$\epsilon = \epsilon_{\mathrm{b}}\epsilon_0\left(1 + \frac{\frac{\omega_{\mathrm{p}}^2}{\epsilon_{\mathrm{b}}}}{\left(\omega_0^2 - \frac{\omega_{\mathrm{p}}^2}{3\epsilon_{\mathrm{b}}}\right) - \omega^2 - \mathrm{i}\gamma\omega}\right)$$

$$\Rightarrow \quad \epsilon_{\mathrm{r}} = \epsilon_{\mathrm{b}} + \frac{\omega_{\mathrm{p}}^2}{\left(\omega_0^2 - \frac{\omega_{\mathrm{p}}^2}{3\epsilon_{\mathrm{b}}}\right) - \omega^2 - \mathrm{i}\gamma\omega}. \tag{2.22}$$

This is the result in the Lorentz model, where we have taken account of both local field effect and background polarization.

As far as the plasmas and metals are concerned, we choose $\omega_0^2 = 0$ for the free electron gases. The relative permittivity is given by

$$\epsilon_{\mathrm{r}} = \epsilon_{\mathrm{b}} - \frac{\omega_{\mathrm{p}}^2}{\frac{\omega_{\mathrm{p}}^2}{3\epsilon_{\mathrm{b}}} + \omega^2 + \mathrm{i}\gamma\omega}. \tag{2.23}$$

This is the result in the Drude model, where we have taken account of both local field effect and background polarization.

It should be noted that the permittivity $\epsilon_{\mathrm{r}} = \epsilon_{\mathrm{b}} - \frac{\omega_{\mathrm{p}}^2}{\omega^2+\mathrm{i}\gamma\omega}$ without the local field effect is consistent with the experimental data of Ag and Au [38] in the frequency ω range $[1.5, 6.0] \times 10^{15}$ rad/s (for Ag) and $[1.5, 3.0] \times 10^{15}$ rad/s (for Au). Since the plasma frequency $\omega_{\mathrm{p}} = 1.40 \times 10^{16}$ rad/s (for Ag) and $\omega_{\mathrm{p}} = 1.38 \times 10^{16}$ rad/s (for Au), the term $\frac{\omega_{\mathrm{p}}^2}{3\epsilon_{\mathrm{b}}}$ caused by the local field correction has the same order of magnitude as ω^2 (or even much larger than ω^2), and this local field correction will then drastically modify the permittivity. Then the permittivity (2.23) will violate the experimental measurements of Johnson and Christy [38]. This leads to a paradox, i.e., the optical parameter measurement experiments have shown that the local field effect was not manifested in the dielectric coefficients $\epsilon_{\mathrm{r}}(\omega)$ of metals. But as a matter of fact, the local field effect has a solid theoretical basis and has been tested in some substances, e.g., the permittivity formula with local field correction (or the Clausius-Mossotti relation) works well in rarefied gases and is valid approximately in liquids and solids [41]. However, why does the local field effect disappear in metals? This became a mystery in the metal optics and surface plasmonics.

Now we shall consider the plasma permittivity, where the local field effect is involved. The result is given by

$$n_{\mathrm{d}}\beta_{\mathrm{pol}} = -\frac{\omega_{\mathrm{p}}^2}{\omega^2 + \mathrm{i}\gamma\omega}, \qquad \chi = \frac{\omega_{\mathrm{p}}^2}{-\frac{\omega_{\mathrm{p}}^2}{3} - \omega^2 - \mathrm{i}\gamma\omega} \tag{2.24}$$

with $\omega_{\mathrm{p}}^2 = \frac{n_{\mathrm{d}}e^2}{\epsilon_0 m_{\mathrm{e}}}$ (m_{e} is the electron mass). When the damping rate γ is negligibly small, the plasma permittivity is given by

$$\epsilon_{\mathrm{r}} = 1 - \frac{\omega_{\mathrm{p}}^2}{\frac{\omega_{\mathrm{p}}^2}{3} + \omega^2}. \tag{2.25}$$

This permittivity is different from $\epsilon_{\mathrm{r}} = 1 - \frac{\omega_{\mathrm{p}}^2}{\omega^2}$ given in Eq. (2.7). In Eq. (2.25), when $\omega^2 < \frac{2}{3}\omega_{\mathrm{p}}^2$ the relative permittivity $\epsilon_{\mathrm{r}} < 0$, and when $\omega^2 > \frac{2}{3}\omega_{\mathrm{p}}^2$ the relative permittivity $\epsilon_{\mathrm{r}} > 0$. In other words, in theory, local field effects severely modify the dielectric coefficients. However, in practice, such corrections are only shown in atomic and molecular gases and liquids, but not in metals and plasmas. This is our content of the mystery of local field correction in the Drude-Lorentz model.

2.3. A Mystery of Plasma Oscillation in the Drude-Lorentz Model

We shall address the concept of *plasma oscillation* in plasma [44, 46, 47]. As is well known, plasma is of (globally) electric neutrality. However, because of the fluctuation in its density, the electrons and heavy ions would deviate from each other. This, therefore, leads to an electric field between the layers of moving electrons and stationary heavy ions. The magnitude of this electric field (caused by the collective oscillation of electron layers in plasma [47]) is proportional to the (geometric) displacement of the electrons from their equilibrium positions. This established electric field gives rise to a force (acting on the free electrons) like an elastic restoring force.

The *plasma oscillation* in the space with various dimensions ($d = 1, 2, 3$) are different. We shall for the first consider the *plasma oscillation* in the 1-D space [39]. Since there is always thermal motion fluctuation and other mechanical reasons (driven by external fields), the free electrons in plasmas and metals will unavoidably be out of equilibrium, the electric neutrality at any local positions cannot be preserved [39, 44]. Let us consider a simple model: specifically, we assume the very heavy positive-charge ions cannot move (i.e., the motional amplitudes are negligibly small), but nearly all the free electrons, which were in the spatial region $[0, a]$ of x coordinate, are forced to move to one side [39, 40], say, the right side at the position $x = a$.

We shall now calculate the electric field strength $x = a - \varepsilon$, where ε is an infinitesimal positive parameter. This electric field is produced by the heavy positive charge ions on the left and the free electron charges on the right [39, 40], which are located at $x = a$. Then the electric field is given by $E = \frac{\sigma}{2\epsilon_0}$ (pointed to the positive x direction) produced by either positive charges (which is on the left) or negative charges (which is on the right), where σ is the surface charge density of the charged layers [44]. Then at the position $x = a - \varepsilon$, the total electric field strength created by both positive charges (which is on the left) and negative charges (which is on the right) is $E = \frac{\sigma}{\epsilon_0}$. As a result, the produced electric field strength (because the positive and negative charge positions do not coincide) is $E_x = \frac{\sigma_{\rm s}}{\epsilon_0}$, where the electric charge surface density $\sigma_{\rm s} = n_{\rm e} q \delta x$. Here, $q > 0$ and δx is the difference between positions of positive and negative charges. Therefore, the local electric field because of the collective motion of *free electron gas layer* in the plasma is given by

$$E_x = \frac{n_{\rm e} q}{\epsilon_0} \delta x. \tag{2.26}$$

Note that the electron charge is $-q$ (< 0). The electric field exerted on an electron is $F_x = -qE_x = -\frac{n_{\rm e} q^2}{\epsilon_0} \delta x$. Then the 1-D plasma oscillation of free electrons *driven by the layer of charge out of equilibrium* is governed by

$$\frac{{\rm d}^2 \delta x}{{\rm d}t^2} = -\frac{n_{\rm e} q^2}{\epsilon_0 m_{\rm e}} \delta x, \qquad \text{or} \qquad \frac{{\rm d}^2 \delta \mathbf{r}}{{\rm d}t^2} = -\frac{n_{\rm e} q^2}{\epsilon_0 m_{\rm e}} \delta \mathbf{r}. \tag{2.27}$$

Now we shall consider the 3-D *plasma oscillation*. The produced electric field is $E_i = \frac{n_{\rm e} q}{3\epsilon_0} \delta r_i$ with $i = x, y, z$ [i.e., $\delta r_i = (\delta x, \delta y, \delta z)$]. In this case, the closed surface integral of the electric field (in the Cartesian coordinate system) is of the form

$$\oint_{\mathrm{S}} \mathbf{E} \cdot \mathrm{d}\mathbf{S} = E_x(\delta y \delta z) + E_y(\delta z \delta x) + E_z(\delta x \delta y)$$
$$= \frac{n_{\mathrm{e}} q}{3\epsilon_0}(3\delta x \delta y \delta z) = \frac{n_{\mathrm{e}} q}{\epsilon_0}\delta V = \frac{\delta Q}{\epsilon_0}. \tag{2.28}$$

It should be pointed out that, in the 3-D spherical polar coordinate system, the above mechanism can alternatively be treated. As we know, in the 3-D space, the Coulomb electric field is $\mathbf{E} = \frac{\delta Q}{4\pi\epsilon_0(\delta r)^2}\mathbf{e}_r$, where the electric charge is $\delta Q = n_{\mathrm{e}} q \delta V = n_{\mathrm{e}} q \frac{4\pi}{3}(\delta r)^3$ and δr denotes the distance from the electrons to their equilibrium positions, namely, now the electric field is $\mathbf{E} = \frac{n_{\mathrm{e}} q}{3\epsilon_0}\delta r \mathbf{e}_r$. The equation of motion $\frac{\mathrm{d}^2 \delta \mathbf{r}}{\mathrm{d}t^2} = \frac{-q}{m_{\mathrm{e}}}\mathbf{E}$, where $-q$ is the electron electric charge, can be rewritten as

$$\frac{\mathrm{d}^2 \delta \mathbf{r}}{\mathrm{d}t^2} = -\frac{n_{\mathrm{e}} q^2}{3\epsilon_0 m_{\mathrm{e}}}\delta \mathbf{r}. \tag{2.29}$$

Here, $\delta \mathbf{r}$ represents the displacement vector (in the radial $\mathbf{r}$ direction). It follows from the equation (2.29) that $\frac{n_{\mathrm{e}} q^2}{3\epsilon_0 m_{\mathrm{e}}}$ can be identified as the frequency square of the 3-D plasma oscillation. To the best of our knowledge, in general, only the 1-D plasma oscillation was discussed in the literature, i.e., the frequency square of 1-D plasma oscillation is given by $\frac{n_{\mathrm{e}} q^2}{\epsilon_0 m_{\mathrm{e}}}$ [44, 46, 47]. It should be stressed that the 3-D plasma oscillation frequency in the 3-D space is smaller than the oscillation frequency of the plasma oscillation in the 1-D space.

Then each charged particle is governed by the dynamical equation $\frac{\mathrm{d}^2 \delta \mathbf{r}}{\mathrm{d}t^2} = -\frac{n_{\mathrm{e}} q^2}{3\epsilon_0 m_{\mathrm{e}}}\delta \mathbf{r}$, where $\delta \mathbf{r}$ denotes the displacement vector (in the radial direction). This equation can be written as $\ddot{\mathbf{x}} + \omega_{_{3\mathrm{D}}}^2 \mathbf{x} = 0$ (here, $\mathbf{x} \equiv \delta \mathbf{r}$), where the frequency square $\omega_{_{3\mathrm{D}}}^2$ of the 3-D *plasma oscillation* is given by

$$\omega_{_{3\mathrm{D}}}^2 = \frac{n_{\mathrm{e}} e^2}{3\epsilon_0 m_{\mathrm{e}}} = \frac{\omega_{\mathrm{p}}^2}{3}. \tag{2.30}$$

When the damping term and the driving term are taken into account, $\ddot{\mathbf{x}} + \gamma \dot{\mathbf{x}} + \omega_{_{3\mathrm{D}}}^2 \mathbf{x} = \frac{q\mathbf{E}}{m_{\mathrm{e}}}$ with the external applied electromagnetic wave electric field $\mathbf{E} = \mathbf{E}_0 \mathrm{e}^{-\mathrm{i}\omega t}$. The displacement vector of an electron must be $\mathbf{x} = \mathbf{x}_0 \mathrm{e}^{-\mathrm{i}\omega t}$ and we have $(-\omega^2 + \omega_{_{3\mathrm{D}}}^2 - \mathrm{i}\gamma\omega)\mathbf{x} = \frac{q}{m_{\mathrm{e}}}\mathbf{E}$. By using the same formalism as that given in Eqs. (2.4) and (2.5), we have

$$n_{\mathrm{e}}\beta_{\mathrm{pol}}=\frac{\omega_{\mathrm{p}}^{2}}{\omega_{\mathrm{3D}}^{2}-\omega^{2}-\mathrm{i}\gamma\omega},\qquad \epsilon_{\mathrm{r}}=1+n_{\mathrm{e}}\beta_{\mathrm{pol}}=1+\frac{\omega_{\mathrm{p}}^{2}}{\frac{\omega_{\mathrm{p}}^{2}}{3}-\omega^{2}-\mathrm{i}\gamma\omega}. \tag{2.31}$$

When the damping rate γ is ignored (in a dilute plasma), the electric permittivity ϵ_{r} is

$$\epsilon_{\mathrm{r}}=1+n_{\mathrm{e}}\beta_{\mathrm{pol}}=1+\frac{\omega_{\mathrm{p}}^{2}}{\frac{\omega_{\mathrm{p}}^{2}}{3}-\omega^{2}}. \tag{2.32}$$

This is different from $\epsilon_{\mathrm{r}}=1-\frac{\omega_{\mathrm{p}}^{2}}{\omega^{2}}$ (in the dilute plasma) given in Eq. (2.7). The permittivity (2.32) can be rewritten as

$$\epsilon_{\mathrm{r}}=\frac{\omega^{2}-\frac{4}{3}\omega_{\mathrm{p}}^{2}}{\omega^{2}-\frac{1}{3}\omega_{\mathrm{p}}^{2}}. \tag{2.33}$$

When $\omega^{2}<\frac{1}{3}\omega_{\mathrm{p}}^{2}$ and $\omega^{2}>\frac{4}{3}\omega_{\mathrm{p}}^{2}$, the permittivity $\epsilon_{\mathrm{r}}>0$, and when $\frac{4}{3}\omega_{\mathrm{p}}^{2}>\omega^{2}>\frac{1}{3}\omega_{\mathrm{p}}^{2}$, the permittivity $\epsilon_{\mathrm{r}}<0$. This is completely different from the case, in which the plasma oscillation is not taken into account.

In the above, we have shown the contribution of the plasma oscillation in the electrically neutral metals and plasmas (ionic gases): The displacement of electron layer leads to a restoring force exerted on the freely moving electrons and this gives rise to the electron collective oscillation effect, i.e., the electrons can oscillate backwards and forwards at a certain frequency around the heavy positive charge ions [44]. We obtain this oscillation frequency as $\frac{\omega_{\mathrm{p}}}{\sqrt{d}}$, where d denotes the dimension number of spaces (this is our result obtained here). Therefore, the equation of motion of a free electron $\ddot{\mathbf{x}}+\gamma\dot{\mathbf{x}}=\frac{q\mathbf{E}}{m_{\mathrm{e}}}$ in the Drude model (for a metal or plasma) is no longer valid. Instead, the correct equation of motion of a free electron should be of the form $\ddot{\mathbf{x}}+\gamma\dot{\mathbf{x}}+\omega_{\mathrm{3D}}^{2}\mathbf{x}=\frac{q\mathbf{E}}{m_{\mathrm{e}}}$, which has the same form as the equation of motion of charged particles in the Lorentz model.

2.4. The Solution to the Two Mysteries in the Drude-Lorentz Model

Let me briefly describe my research experience on the puzzle of effects related to local field and plasma oscillation in the Drude model: I have focused on the contribution of *local field effect* and *plasma oscillation effect* to the Drude model of plasmas and metals for many years. But I staggered in time when I considered them, and so I have not treated these two effects *together* for these years (2003–2020): specifically, since 2004 to 2012, when I studied electromagnetically induced transparency (EIT) and the negative refractive indices of atomic gases and quantum-dot solids (where the local field correction to the dielectric coefficients was required) [48, 49], I was always wondering why the authors of monographs and literature of electromagnetics and optics did not take the *local field correction* into account in the Drude model for plasmas and metals. In the first half of 2016, I studied a highly ionized gas effect in the problem of dark matter in galaxies, where the negative dielectric coefficient might be involved: If the local field effect is taken into account, the permittivity of galactic plasma (highly ionized gas) is given by $\epsilon_{\mathrm{r}} = 1 - \frac{\omega_{\mathrm{p}}^2}{\frac{\omega_{\mathrm{p}}^2}{3} + \omega^2}$. When $\omega^2 \ll \omega_{\mathrm{p}}^2$, the quasi-static electric permittivity $\epsilon_{\mathrm{r}} \rightarrow -2$ for the highly ionized gas. But such a negative permittivity was surprising to me and then I wondered whether I need again to take the local field effect into consideration in the plasma dielectric coefficient. It made me scratch my head. I also consulted some colleagues who taught the courses of electromagnetism and electromagnetic wave theory, but I did not get an exact conclusion about such a local field effect. As far as the effect of plasma oscillation is concerned, although I was familiar with this mechanism before 2003 and I knew that it must exist in metals and plasmas and that it gives rise to elastic restorative forces on free electrons, I had never taken them seriously (I was not sure whether they actually affect the dielectric coefficients of plasmas and metals). I suspected that this elastic restorative force of the electron does not actually cause a correction in the dielectric coefficient of ionic gases, in part because the displacement of the electrons driven by the plasma oscillation is so small. Then I presumed that within the framework of the Drude model, the issue of plasma oscillation contribution to the permittivity may not matter. I therefore took no further consideration of the plasma oscillation effect in the Drude model since then. In January 2021, I studied the equivalent dark matter effect [50] within the framework of gravitational gauge theory, where my theoretical mechanism requires the mass term of an elastic displacement (shear wave) of spacetime crystal in order to truncate the (effective) density of equivalent dark matter [50] at large (galactic) scales and to avoid the total mass of dark matter becoming infinity. Since

I studied a possibility of introducing a mass term of the elastic vibration of crystals for many years after 2013 (based on Huang's lattice dynamics for studying the optical properties of ionic crystals [51]), so as to generalize Huang's theory, I transplanted this idea into my theoretical mechanism of so-called equivalent dark matter. However, the introduction of a mass term for elastic vibrations in solids requires a specific mechanism. At first (in January 2021), I thought there would be an elastic restorative force between the large-displacement (relatively free) charges in a metal or an ionic crystal versus the small-displacement charges (background), but I was not sure whether it was correct or not. In February 2021, it occurred to me that the plasma oscillation in plasma physics has just such an *elastic restorative force effect.* As a result, I had to face the plasma oscillation head on and was convinced again that the plasma oscillation effect must be taken into account in the dielectric coefficients of the Drude model for plasmas and metals. Therefore, I am even more confused about why the authors of textbooks and literature did not consider the correction to the dielectric coefficient originating from the plasma oscillation in the Drude model of plasmas and metals. In May 2021, I suddenly realized that these two mysteries (the contribution of *local field effect* and *plasma oscillation effect* in the Drude-model permittivity) might be the answer to each other, that is, *the influences of local field effect and plasma oscillation effect on the Drude model are just mutually eliminated.* Then immediately I did a careful calculation for these two effects and found that they were truly cancelled out by each other!

Although this is simply a problem in electromagnetism, metal optics and plasma physics, what prompted me to consider such mysteries was that I was working on the concept of equivalent dark matter [50]. Within the framework of gravitational gauge theory [52, 53], the spin current density tensor of vector fields such as electromagnetic field and spacetime-crystal elastic field may serve as a gravitating source of the gravitational gauge field. For the first, in 2015–2016, I suggested that the electromagnetic spin current density tensor $S^{\lambda\mu\nu} = -\left(F^{\lambda\mu}A^{\nu} - F^{\lambda\nu}A^{\mu}\right)$ (which agrees with the relation $\partial_{\lambda}S^{\lambda\mu\nu} = -\left(T^{\mu\nu} - T^{\nu\mu}\right)$) might lead to an emergent equivalent dark matter in galaxies. Since electrons have far less mass than protons and nuclei, they are more likely to be ejected or evaporated from stars and galaxies at high temperatures, making the stars and galaxies positively charged, and the resulting spin current of the electromagnetic field around the stars and galaxies may be a source of gravity in the gravitational gauge theory. The above $S^{\lambda\mu\nu}$ is written in the natural units. Actually, the electric permittivity should be introduced into the spin current $S^{\lambda\mu\nu}$ in practice. Then in this mechanism, the local field effect in the permittivity of the galactic plasma should be considered. For example, the permittivity under the condition of local field effect is given by $\epsilon_{\mathrm{r}} = 1 - \frac{\omega_{\mathrm{p}}^2}{\frac{\omega_{\mathrm{p}}^2}{3} + \omega^2}$. For the highly ionized gas, $\omega^2 \ll \omega_{\mathrm{p}}^2$, the quasi-static electric permittivity will be given by $\epsilon_{\mathrm{r}} \to -2$. But as a matter of fact, the electromagnetic spin current density tensor $S^{\lambda\mu\nu}$ is not of gauge symmetry, and cannot be an independent source of gravity. So I abandoned this mechanism and expected to find another vector field. Since the spacetime crystal elastic wave equation is similar to the electromagnetic field equation in mathematical structure, in 2018–2020, I used the so-called spatiotemporal elastic wave driven by *spacetime lattice defect entropy* to substitute for the electromagnetic field in the aforementioned mechanism, i.e., the spin current of spatiotemporal elastic wave in this new mechanism can give rise to a gravitational source in the gravitational gauge theory [52, 53]. In this case, a rest mass term of the spatiotemporal elastic wave should also be introduced. Fortunately, the effect similar to plasma oscillation or Langmuir oscillation can lead to such a mass term in the present elastic wave equation. Thus, I thought again about whether the plasma oscillation needs to be included in the Drude model in practical problems of applied physics. This is the background of my research on the mystery, i.e., why the traditional Drude model for plasma ionic gas and metal electron gas does not take into account the local field and plasma oscillation effects but can still agree very well with experimental observations.

It should be pointed out that the dimensionless quantity $n_{\mathrm{e}}\beta_{\mathrm{pol}}$ obtained in Eq. (2.31) has involved the plasma oscillation effect, which is exhibited by ω_{3D}^2, but it does not take the local field effect into account. If the local field effect is also involved, the complete expression for the susceptibility χ of this plasma is given by

$$\chi = \frac{n_{\mathrm{e}}\beta_{\mathrm{pol}}}{1 - \frac{n_{\mathrm{e}}\beta_{\mathrm{pol}}}{3}} = \frac{\frac{\omega_{\mathrm{p}}^2}{\omega_{\mathrm{3D}}^2 - \omega^2 - \mathrm{i}\gamma\omega}}{1 - \frac{\frac{\omega_{\mathrm{p}}^2}{\omega_{\mathrm{3D}}^2 - \omega^2 - \mathrm{i}\gamma\omega}}{3}} = \frac{\omega_{\mathrm{p}}^2}{\left(\omega_{\mathrm{3D}}^2 - \frac{\omega_{\mathrm{p}}^2}{3}\right) - \omega^2 - \mathrm{i}\gamma\omega}. \quad (2.34)$$

Here, the terms $-\frac{\omega_{\rm p}^2}{3}$ and $\omega_{\rm 3D}^2$ are the effects of local field correction and the plasma-oscillation "elastic" restoring force, respectively. Since $\omega_{\rm 3D}^2 - \frac{\omega_{\rm p}^2}{3} = 0$ [i.e., the result (2.30)], we have the electric susceptibility and the permittivity of the present plasma:

$$\chi = -\frac{\omega_{\rm p}^2}{\omega^2 + {\rm i}\gamma\omega}, \qquad \epsilon_{\rm r} = 1 - \frac{\omega_{\rm p}^2}{\omega^2 + {\rm i}\gamma\omega}. \tag{2.35}$$

The effects of the local field correction and the plasma oscillation in the permittivity of plasma are eliminated by each other and the final result is the same as the ordinary form without involving the contributions of both local field correction and plasma oscillation. So we can explain why the dielectric coefficients of metals and plasmas agree very well with the experimental results, even when the two effects (i.e., both local field correction and plasma oscillation) were not taken into account.

It should also be pointed out that the permittivity formula with local field correction (or the Clausius-Mossotti relation) works well in rarefied gases and is valid approximately in liquids and solids [41]. This means that the contribution of the local field effect is not cancelled out by that of plasma oscillation in such dilute gases, liquids and solids, because in these substances, the plasma oscillation may be absent. The premise of the occurrence of plasma oscillation is that the electron gas as a whole deviates from its equilibrium position, creating an electric field that drives the electron's periodic recovery motion (under the action of restoring force). But in an atomic or molecular medium, the electric field resulting from such an overall deviation is not as ideal as in a plasma. Even if, at the very least, the so-called plasma oscillation exists in atomic and molecular media, the local field effect still cannot be eliminated by the plasma oscillation. This can be interpreted as follows:

In atomic and molecular gases (such as two-level and three-level alkali metallic atomic gases), the microscopic electrical polarizability and dielectric coefficients originate from energy level transitions. In order to calculate the dielectric coefficients of the atomic and molecular gases, it is necessary to solve the density matrix equation, which is a Schrödinger equation. This equation is associated with the first derivative with respect to time, so in the denominator of such dielectric coefficients, the frequency appears ω in a linear form. For example, the microscopic electric polarizability of a two-level atomic system $\{|a\rangle, |b\rangle\}$ is $\beta_{\rm e} = -\frac{|\wp_{ab}|^2}{\epsilon_0\hbar}\frac{1}{\omega - \omega_{ab} + {\rm i}\gamma_{ab}}$.

By taking into account the local field effect, the dielectric coefficient of this atomic vapor is $\varepsilon_{\mathrm{r}} = 1 + \frac{N\beta_{\mathrm{e}}}{1-\frac{N\beta_{\mathrm{e}}}{3}}$, where N denotes the number density of atoms. From this formula, it follows that this would not enable us to make the effects of local field correction and plasma oscillation cancel each other out in atomic and molecular media.

2.5. The Local Field Effect and the Plasma Oscillation in One and Two Dimensions

In the above, we have considered the local field effect and plasma oscillation in the 3-D space. Now we shall also address the local field effect and the plasma oscillation in one and two dimensions.

From the equations (2.16) and (2.18), we have

$$\begin{aligned}
\mathbf{E}_{\mathrm{local}} &= \mathbf{E}_{\mathrm{external}} + (-\mathbf{E}_{\mathrm{neighboring}}), \\
\mathbf{E}_{\mathrm{neighboring}} &= -\frac{\mathbf{P}}{d\epsilon_0}, \qquad \mathbf{P} = n_{\mathrm{d}}\beta_{\mathrm{pol}}\epsilon_0\mathbf{E}_{\mathrm{local}}, \\
\mathbf{P} &= \chi\epsilon_0\mathbf{E}_{\mathrm{external}} = \chi\epsilon_0\left(\mathbf{E}_{\mathrm{local}} + \mathbf{E}_{\mathrm{neighboring}}\right) \\
&= \chi\epsilon_0\left(\mathbf{E}_{\mathrm{local}} - \frac{\mathbf{P}}{d\epsilon_0}\right) = \chi\epsilon_0\left(\frac{\mathbf{P}}{n_{\mathrm{d}}\beta_{\mathrm{pol}}\epsilon_0} - \frac{\mathbf{P}}{d\epsilon_0}\right)
\end{aligned} \tag{2.36}$$

Then the electric susceptibility χ in the d-dimensional space is of the form

$$\chi = \frac{1}{\frac{1}{n_{\mathrm{d}}\beta_{\mathrm{pol}}} - \frac{1}{d}} = \frac{n_{\mathrm{d}}\beta_{\mathrm{pol}}}{1 - \frac{n_{\mathrm{d}}\beta_{\mathrm{pol}}}{d}}. \tag{2.37}$$

Substituting the result (2.17) into Eq. (2.37), the electric susceptibility is

$$\chi = \frac{\frac{\omega_{\mathrm{p}}^2}{\omega_0^2-\omega^2-\mathrm{i}\gamma\omega}}{1 - \frac{\frac{\omega_{\mathrm{p}}^2}{\omega_0^2-\omega^2-\mathrm{i}\gamma\omega}}{d}} = \frac{\omega_{\mathrm{p}}^2}{\left(\omega_0^2 - \frac{\omega_{\mathrm{p}}^2}{d}\right) - \omega^2 - \mathrm{i}\gamma\omega}. \tag{2.38}$$

We have shown that the 3-D plasma oscillation frequency square is $\omega_{_{3\mathrm{D}}}^2 = \frac{n_{\mathrm{e}}e^2}{3\epsilon_0 m_{\mathrm{e}}} = \frac{\omega_{\mathrm{p}}^2}{3}$ given in Eq. (2.30). In the preceding subsection, we have shown that the 1-D and 2-D plasma oscillation frequency squares are $\omega_{_{1\mathrm{D}}}^2 = \omega_{\mathrm{p}}^2$ and $\omega_{_{2\mathrm{D}}}^2 = \frac{\omega_{\mathrm{p}}^2}{2}$, respectively. In a word, the d-D plasma oscillation frequency square should be of the form

$$\omega_{_{d\mathrm{D}}}^2 = \frac{\omega_{\mathrm{p}}^2}{d}. \tag{2.39}$$

Then the (intrinsic or proper) oscillation frequency ω_0 of the electric dipole oscillator in Eq. (2.38) should be chosen as $\omega_0 = \omega_{_{\rm dD}} = \frac{\omega_{\rm p}}{\sqrt{d}}$. Thus, the term $\omega_0^2 - \frac{\omega_{\rm p}^2}{d} = 0$ in Eq. (2.38), and the susceptibility is still of the form $\chi = \frac{\omega_{\rm p}^2}{-\omega^2 - {\rm i}\gamma\omega}$ (ordinary form). In the ordinary (conventional) form for the susceptibility, neither the local field nor the plasma oscillation has been taken into account. If the local field effect in Eq. (2.36) and the plasma oscillation term ω_0^2 given in Eq. (2.38) are taken into account together, their contributions cancel out precisely, and the result (of the electric permittivity) still reduces to the ordinary form.

From the equation (2.37), one can have $\chi\left(1 - \frac{n_{_{\rm d}}\beta_{\rm pol}}{d}\right) = n_{_{\rm d}}\beta_{\rm pol}$ and $\chi = n_{_{\rm d}}\beta_{\rm pol}(1 + \frac{\chi}{d})$, i.e., $n_{_{\rm d}}\beta_{\rm pol} = \frac{\chi}{1+\frac{\chi}{d}}$. Then it can yield

$$\frac{\epsilon_{\rm r} - 1}{\epsilon_{\rm r} + d - 1} = \frac{n_{_{\rm d}}\beta_{\rm pol}}{d}. \tag{2.40}$$

This relation suggested by us can be referred to as the Clausius-Mossotti relation in the d-dimensional space. When $d = 1$ and $d = 2$, we suggest that the Clausius-Mossotti relation is given by

$$\frac{\epsilon_{\rm r} - 1}{\epsilon_{\rm r}} = n_{_{\rm d}}\beta_{\rm pol} \quad (d = 1); \qquad \frac{\epsilon_{\rm r} - 1}{\epsilon_{\rm r} + 1} = \frac{n_{_{\rm d}}\beta_{\rm pol}}{2} \quad (d = 2). \tag{2.41}$$

It should be pointed out that the electric field induced by the neighboring dipoles $\mathbf{E}_{\rm neighboring} = -\frac{\mathbf{P}}{d\epsilon_0}$ given in Eq. (2.36) and the Clausius-Mossotti relation (2.41) have so far never been rigorously proved. In the next subsection, we shall show that these relations hold true in one- and two-dimension spaces.

2.6. A Proof of Induced Electric Fields and the Clausius-Mossotti Relations in One- and Two-Dimensional Spaces

We will give a proof of induced electric fields and the Clausius-Mossotti relations in one- and two-dimensional spaces. But for the first, we shall list the conventional proof for the case of 3-D space [54, 55]. This helps us to develop the proofs in two- and one-dimensional spaces.

i) The electric field induced by neighboring dipoles and the Clausius-Mossotti relation in 3-D space. For the first, we shall consider a dielectric in a 3-D space, e.g., a nanometer sphere whose length scale is much less than the incident electromagnetic wavelength. Then the Laplace equation $\nabla^2\Phi = 0$ for an electric scalar potential Φ is applicable to this issue. By using the solution with rotational symmetry $\Phi(r,\theta) = \sum_{l=0}^{\infty}[A_l r^l + B_l r^{-(l+1)}]P_l(\cos\theta)$, the electric scalar potential Φ outside and inside the nanometer sphere is given by [54, 55]

$$\begin{aligned}\Phi_{\text{out}}(r,\theta) &= -E_0 r\cos\theta + \frac{\epsilon_{\text{in}} - \epsilon_{\text{out}}}{\epsilon_{\text{in}} + 2\epsilon_{\text{out}}} E_0 \frac{a^3}{r^2}\cos\theta \quad (r > a); \\ \Phi_{\text{in}}(r,\theta) &= -\frac{3\epsilon_{\text{out}}}{\epsilon_{\text{in}} + 2\epsilon_{\text{out}}} E_0 r\cos\theta \quad (r \le a). \quad (2.42)\end{aligned}$$

Here, E_0 is the applied external electric field, which is parallel to the z axis. The parameters ϵ_{in} and ϵ_{out} are the electric permittivities of nanosphere and of the medium outside the nanosphere. The nanosphere radius is a.

It should be stressed that such a nanosphere, which is illuminated by an incident electromagnetic wave, can be considered to be an *equivalent electric dipole* [55]. Now we shall consider such a dipole model. Let an electric dipole lie on the z axis with its center at the origin $r = 0$, and the positions of the positive and negative point charges (with the same magnitude of charge q) are $(0, 0, +\frac{d}{2})$ and $(0, 0, -\frac{d}{2})$, respectively. Here, d is the distance between the positive and negative point charges. The electric potentials of the positive and negative point charges are given by $\Phi_+ = +\frac{q}{4\pi\epsilon_0}\frac{1}{r_+}$ and $\Phi_- = -\frac{q}{4\pi\epsilon_0}\frac{1}{r_-}$, respectively, where the distance between the observation point $\mathbf{r}$ and the positive charge is $r_+ = \sqrt{r^2 + \frac{d^2}{4} - rd\cos\theta} \simeq r(1 - \frac{d}{2r}\cos\theta)$ and the distance between the observation point $\mathbf{r}$ and the negative charge is $r_- = \sqrt{r^2 + \frac{d^2}{4} + rd\cos\theta} \simeq r(1 + \frac{d}{2r}\cos\theta)$. Then $r_- - r_+ = d\cos\theta$ (up to the first order of d). Now the electric scalar potential Φ of this equivalent electric dipole, which is the embodiment of the electric polarization distribution on the nanosphere, is given by

$$\begin{aligned}\Phi &= \Phi_+ + \Phi_- = \frac{q}{4\pi\epsilon_0}\left(\frac{1}{r_+} - \frac{1}{r_-}\right) \\ &\simeq \frac{q}{4\pi\epsilon_0}\frac{r_- - r_+}{r^2} = \frac{q}{4\pi\epsilon_0}\frac{d\cos\theta}{r^2} = \frac{1}{4\pi\epsilon_0}\frac{\mathbf{p}\cdot\mathbf{r}}{r^3}, \quad (2.43)\end{aligned}$$

where $\mathbf{p}\cdot\mathbf{r} = qdr\cos\theta$ and the (induced) electric dipole moment vector $\mathbf{p} = qd\mathbf{e}_z$ (on the nanosphere).

The outer solution in Eq. (2.42) can be written as an electric potential of a dipole [55], i.e., the outer solution is

$$\begin{aligned}\Phi_{\text{out}}(r,\theta) &= -E_0 r\cos\theta + \frac{\epsilon_{\text{in}}-\epsilon_{\text{out}}}{\epsilon_{\text{in}}+2\epsilon_{\text{out}}}E_0 a^3\frac{\cos\theta}{r^2} \\ &= -E_0 r\cos\theta + \frac{1}{4\pi\epsilon_{\text{out}}}\left(\frac{\epsilon_{\text{in}}-\epsilon_{\text{out}}}{\epsilon_{\text{in}}+2\epsilon_{\text{out}}}4\pi\epsilon_{\text{out}}E_0 a^3\right)\frac{\cos\theta}{r^2}. \quad (2.44)\end{aligned}$$

By comparing the second term in the result of Eq. (2.44) with $\Phi = \frac{qd}{4\pi\epsilon_{\text{out}}}\frac{\cos\theta}{r^2} = \frac{p}{4\pi\epsilon_{\text{out}}}\frac{\cos\theta}{r^2}$ obtained in Eq. (2.43), one can easily obtain the magnitude p of the *equivalent electric dipole moment* lying on the z axis [55]:

$$p = \frac{\epsilon_{\text{in}}-\epsilon_{\text{out}}}{\epsilon_{\text{in}}+2\epsilon_{\text{out}}}4\pi\epsilon_{\text{out}}E_0 a^3. \quad (2.45)$$

We assume the outside medium permittivity $\epsilon_{\text{out}} = \epsilon_0$. Note that E_0 is the local field E_{local} whose vector is $\mathbf{E}_{\text{local}}$ (parallel to the z axis). Then from the equation (2.45), the *equivalent electric dipole moment vector* $\mathbf{p}$ is

$$\mathbf{p} = \frac{\epsilon_{\text{in}}-\epsilon_0}{\epsilon_{\text{in}}+2\epsilon_0}4\pi\epsilon_0\mathbf{E}_{\text{local}}a^3 = \frac{\epsilon_{\text{in}}-\epsilon_0}{\epsilon_{\text{in}}+2\epsilon_0}3\epsilon_0\mathbf{E}_{\text{local}}V_{\text{sphere}} \quad (2.46)$$

with the nanosphere volume $V_{\text{sphere}} = \frac{4\pi}{3}a^3$. Since there is an equivalent dipole moment $\mathbf{p}$ in the nanosphere volume V_{sphere}, the polarization $\mathbf{P}$ is given by

$$\mathbf{P} = \frac{\mathbf{p}}{V_{\text{sphere}}} = \frac{\epsilon_{\text{in}}-\epsilon_0}{\epsilon_{\text{in}}+2\epsilon_0}3\epsilon_0\mathbf{E}_{\text{local}}. \quad (2.47)$$

On the other hand, the microscopic electric polarizability β_{pol} is defined through $\mathbf{P} = n_{\text{d}}\beta_{\text{pol}}\epsilon_0\mathbf{E}_{\text{local}}$. Then a relation yields

$$\frac{n_{\text{d}}\beta_{\text{pol}}}{3} = \frac{\epsilon_{\text{in}}-\epsilon_0}{\epsilon_{\text{in}}+2\epsilon_0}. \quad (2.48)$$

This is the relation (2.20), which is referred to as the Clausius-Mossotti relation or the Lorenz-Lorentz relation.

Since now $\frac{\epsilon_{\text{in}}-\epsilon_0}{\epsilon_0} \neq n_{\text{d}}\beta_{\text{pol}}$, this means $\mathbf{E}_{\text{local}}$ is not equal to the external field $\mathbf{E}_{\text{external}}$. Define $\mathbf{E}_{\text{neighboring}} = \mathbf{E}_{\text{external}} - \mathbf{E}_{\text{local}}$, where $\mathbf{P} = (\epsilon_{\text{in}}-\epsilon_0)\mathbf{E}_{\text{external}}$ and $\mathbf{P} = \frac{\epsilon_{\text{in}}-\epsilon_0}{\epsilon_{\text{in}}+2\epsilon_0}3\epsilon_0\mathbf{E}_{\text{local}}$. From these two relations, we have

$$\begin{aligned}&(\epsilon_{\text{in}}-\epsilon_0)\mathbf{E}_{\text{external}} = \frac{\epsilon_{\text{in}}-\epsilon_0}{\epsilon_{\text{in}}+2\epsilon_0}3\epsilon_0\mathbf{E}_{\text{local}} \Rightarrow \mathbf{E}_{\text{external}} = \frac{3\epsilon_0}{\epsilon_{\text{in}}+2\epsilon_0}\mathbf{E}_{\text{local}} \\ \Rightarrow\quad &\mathbf{E}_{\text{neighboring}} = \mathbf{E}_{\text{external}} - \mathbf{E}_{\text{local}} = -\frac{\epsilon_{\text{in}}-\epsilon_0}{\epsilon_{\text{in}}+2\epsilon_0}\mathbf{E}_{\text{local}} = -\frac{\mathbf{P}}{3\epsilon_0}. \quad (2.49)\end{aligned}$$

The relation $\mathbf{E}_{\text{neighboring}} = -\frac{\mathbf{P}}{3\epsilon_0}$ is a well-known result in the dielectric sphere polarization [54], i.e., a case of our obtained result $\mathbf{E}_{\text{neighboring}} = -\frac{\mathbf{P}}{d\epsilon_0}$ with $d = 3$ (dimension number of space).

Inspired by the above topic of dielectric sphere polarization in the three-dimensional space [54, 55], we will study the polarization of one-dimensional dielectric and two-dimensional circular cross section in a nanorod.

ii) The electric field induced by neighboring dipoles and the Clausius-Mossotti relation in 1-D space. We shall consider the problem of one-dimensional dielectric polarization. The present one-dimensional dielectric is illuminated by the electromagnetic wave, of which the wavelength is much longer than the length of the one-dimensional dielectric. The one-dimensional Laplace equation is $\frac{\partial^2 \Phi}{\partial x^2} = 0$, of which the solution is given by

$$\begin{aligned} \Phi_{\text{out}}^{(\text{L})} &= -E_0 x + C^{(\text{L})} \qquad (x < -a), \\ \Phi_{\text{out}}^{(\text{R})} &= -E_0 x + C^{(\text{R})} \qquad (x > +a); \\ \Phi_{\text{in}} &= -Ax + C^{(\text{in})} \qquad (-a \le x \le +a), \end{aligned} \tag{2.50}$$

where E_0 is the electric field of the incident wave. Here we assume the one-dimensional dielectric lies on the x axis with its center at the origin ($x = 0$). The length of the present one-dimensional dielectric is $2a$ and its permittivity is ϵ_{in}; The electric permittivity when $x > a$ and $x < -a$ is ϵ_{out}. By using the boundary condition (the electric displacement vector $\mathbf{D}$ in the x axis is continuous at $x = \pm a$), we have $-\epsilon_{\text{out}} \frac{\partial \Phi_{\text{out}}^{(\text{L})}}{\partial x}\Big|_{x=-a} = -\epsilon_{\text{in}} \frac{\partial \Phi_{\text{in}}}{\partial x}\Big|_{x=-a}$ and $-\epsilon_{\text{out}} \frac{\partial \Phi_{\text{out}}^{(\text{R})}}{\partial x}\Big|_{x=+a} = -\epsilon_{\text{in}} \frac{\partial \Phi_{\text{in}}}{\partial x}\Big|_{x=+a}$, i.e., we have $\epsilon_{\text{out}} E_0 = \epsilon_{\text{in}} A$, namely, $A = \frac{\epsilon_{\text{out}}}{\epsilon_{\text{in}}} E_0$. It should be noted that the electric scalar potential should also be continuous at $x = \pm a$, i.e.,

$$\begin{aligned} E_0 a + C^{(\text{L})} &= \frac{\epsilon_{\text{out}}}{\epsilon_{\text{in}}} E_0 a + C^{(\text{in})}, \\ -E_0 a + C^{(\text{R})} &= -\frac{\epsilon_{\text{out}}}{\epsilon_{\text{in}}} E_0 a + C^{(\text{in})}. \end{aligned} \tag{2.51}$$

From these relations, we can obtain

$$C^{(\text{in})} - C^{(\text{L})} = \left(1 - \frac{\epsilon_{\text{out}}}{\epsilon_{\text{in}}}\right) E_0 a,$$
$$C^{(\text{in})} - C^{(\text{R})} = \left(\frac{\epsilon_{\text{out}}}{\epsilon_{\text{in}}} - 1\right) E_0 a. \tag{2.52}$$

For convenience, we can assume $C^{(\text{in})} = 0$. Thus, we have

$$C^{(\text{L})} = -\left(1 - \frac{\epsilon_{\text{out}}}{\epsilon_{\text{in}}}\right) E_0 a, \qquad C^{(\text{R})} = -\left(\frac{\epsilon_{\text{out}}}{\epsilon_{\text{in}}} - 1\right) E_0 a. \tag{2.53}$$

It should be pointed out that the present one-dimensional dielectric illuminated by the electromagnetic wave can be identified as an equivalent electric dipole. We shall now consider this problem. We assume the dipole lies on the x axis. At the endpoints, the electric charge surface density is $+\sigma_{\text{s}}$ at $x = +\frac{d}{2}$ and $-\sigma_{\text{s}}$ at $x = -\frac{d}{2}$. In the direction of positive x axis, the electric potential produced by the positive charge at $x = +\frac{d}{2}$ is $\Phi_+(x) = -\frac{\sigma_{\text{s}}}{2\epsilon_{\text{out}}}(x - \frac{d}{2})$ and the potential produced by the negative charge at $x = -\frac{d}{2}$ is $\Phi_-(x) = +\frac{\sigma_{\text{s}}}{2\epsilon_{\text{out}}}(x + \frac{d}{2})$, where $x > \frac{d}{2}$ (i.e., here we consider the electric scalar potential of right-handed side of the one-dimensional dielectric). The total electric potential (produced by the present dipole) is $\Phi = \Phi_+ + \Phi_- = \frac{\sigma_{\text{s}} d}{2\epsilon_{\text{out}}}$. If the one-dimensional dielectric illuminated by the electromagnetic wave can really be viewed as an electric dipole, the constant $C^{(\text{R})}$ in the solution $\Phi^{(\text{R})}_{\text{out}} = -E_0 x + C^{(\text{R})}$ on the right-handed side given in Eq. (2.50) should be equal to the potential $\frac{\sigma_{\text{s}} d}{2\epsilon_{\text{out}}}$ produced by the equivalent electric dipole. Thus, we have

$$-\left(\frac{\epsilon_{\text{out}}}{\epsilon_{\text{in}}} - 1\right) E_0 a = \frac{\sigma_{\text{s}} d}{2\epsilon_{\text{out}}}$$
$$\Rightarrow \quad -\left(\frac{\epsilon_{\text{out}}}{\epsilon_{\text{in}}} - 1\right) E_0 \frac{2a\delta S}{2} = \frac{(\sigma_{\text{s}}\delta S) d}{2\epsilon_{\text{out}}}, \tag{2.54}$$

where δS is an infinitesimal area (cross section parallel to the y-z plane) of the two endpoints at $x = \pm a$ (for simplicity, $\frac{d}{2} = a$). Clearly, $\mathcal{V} = 2a\delta S$ is the volume of this one-dimensional dielectric with length $2a$ and $\sigma_{\text{s}}\delta S$ is the electric charge (at $x = +\frac{d}{2}$) of the present equivalent electric dipole. Thus, $p = (\sigma_{\text{s}}\delta S) d$ is the electric dipole moment lying on the x axis. Then the result in Eq. (2.54) can be rewritten as

$$-\left(\frac{\epsilon_{\rm out}}{\epsilon_{\rm in}}-1\right)E_0\mathcal{V}=\frac{p}{\epsilon_{\rm out}}\Rightarrow P=\frac{p}{\mathcal{V}}=-\left(\frac{\epsilon_{\rm out}}{\epsilon_{\rm in}}-1\right)\epsilon_{\rm out}E_0,\tag{2.55}$$

where P denotes the polarization in the present one-dimensional dielectric. We assume $\epsilon_{\rm out}=\epsilon_0$. Note that here E_0 is the local field $E_{\rm local}$. Now the electric polarization is

$$\mathbf{P}=\frac{\epsilon_{\rm in}-\epsilon_0}{\epsilon_{\rm in}}\epsilon_0\mathbf{E}_{\rm local}.\tag{2.56}$$

On the other hand, $\mathbf{P}=n_{\rm d}\beta_{\rm pol}\epsilon_0\mathbf{E}_{\rm local}$. Thus, we have

$$n_{\rm d}\beta_{\rm pol}=\frac{\epsilon_{\rm in}-\epsilon_0}{\epsilon_{\rm in}}.\tag{2.57}$$

This is the Clausius-Mossotti relation in one-dimensional space. The above formalism is a proof for the relation (2.57) when the space dimension number $d=1$. This result has been given in Eq. (2.41).

By using the polarization $\mathbf{P}=(\epsilon_{\rm in}-\epsilon_0)\mathbf{E}_{\rm external}$ and $\mathbf{P}=\frac{\epsilon_{\rm in}-\epsilon_0}{\epsilon_{\rm in}}\epsilon_0\mathbf{E}_{\rm local}$, we can obtain

$$(\epsilon_{\rm in}-\epsilon_0)\mathbf{E}_{\rm external}=\frac{\epsilon_{\rm in}-\epsilon_0}{\epsilon_{\rm in}}\epsilon_0\mathbf{E}_{\rm local}$$

$$\Rightarrow\quad \mathbf{E}_{\rm neighboring}=\mathbf{E}_{\rm external}-\mathbf{E}_{\rm local}=-\frac{\epsilon_{\rm in}-\epsilon_0}{\epsilon_{\rm in}}\mathbf{E}_{\rm local}=-\frac{\mathbf{P}}{\epsilon_0}.\tag{2.58}$$

The result $\mathbf{E}_{\rm neighboring}=-\frac{\mathbf{P}}{\epsilon_0}$ is applicable in the one-dimensional dielectric. This is a case of our obtained result $\mathbf{E}_{\rm neighboring}=-\frac{\mathbf{P}}{d\epsilon_0}$ with $d=1$ (dimension number of space).

iii) The electric field induced by neighboring dipoles and the Clausius-Mossotti relation in 2-D space.

We consider a cylindrical dielectric or a nanorod whose length scale is much less than the incident electromagnetic wavelength. The Laplace equation of the electric scalar potential in the circular cross section is given by $\frac{\partial^2\Phi}{\partial r^2}+\frac{1}{r}\frac{\partial\Phi}{\partial r}+\frac{1}{r^2}\frac{\partial^2\Phi}{\partial\phi^2}=0$ (in the cylindrical coordinate system, where the spatial variables are (r,ϕ,z)). It should be noticed that r is the 2-D radial distance and that the nanorod lies on the z axis. The solution can be written as $\Phi={\rm e}^{\pm im\phi}R_m(r)$. Then the Laplace equation for the radial component is

$$\frac{\partial^2R_m(r)}{\partial r^2}+\frac{1}{r}\frac{\partial R_m(r)}{\partial r}-\frac{m^2}{r^2}R_m(r)=0.\tag{2.59}$$

Let the radial function $R_m(r) = C_m r^n$. By substituting this radial function into the Laplace equation (2.59), we can obtain $n(n-1)r^{n-2} + nr^{n-2} - m^2 r^{n-2} = 0$, and we have $n = \pm m$. Then the solution of the radial function is $R_m = C_{(+m)} r^{+m} + C_{(-m)} r^{-m}$.

There is another solution that may be easily missing. When $m = 0$, let $\mathcal{Y} = \frac{\partial R_m(r)}{\partial r}$. Thus, since $\frac{\partial \mathcal{Y}}{\partial r} + \frac{1}{r}\mathcal{Y} = 0$, $\frac{\mathrm{d}\mathcal{Y}}{\mathcal{Y}} = -\frac{\mathrm{d}r}{r}$ and $\ln \mathcal{Y} = -\ln(C^{-1} r)$, we have $\mathcal{Y} = \frac{C}{r}$ and hence $R(r) = C \ln r + C'$. This function is not analytic at $r = 0$ (it is the electric potential of an electrically charged thin wire).

If there is an applied electric field pointing to the direction parallel to the x axis (the 2-D cross section plane is the x-y plane), the electric scalar potential at infinity is $-E_0 r \cos\phi$. It can be found that only the solution of the case of $m = 1$ can be continuous (at $r \to +\infty$) with the applied electric scalar potential $-E_0 r \cos\phi$. Thus, we consider the case of $m = 1$ only. The electric scalar potential outside and inside the cylindrical dielectric is

$$\begin{aligned} \Phi_{\text{out}}(r,\phi) &= \left(-E_0 r + \frac{C_{-1}^{(\text{out})}}{r}\right)\cos\phi \qquad (r > a); \\ \Phi_{\text{in}}(r,\phi) &= C_{+1}^{(\text{in})} r \cos\phi \qquad (r \le a). \end{aligned} \tag{2.60}$$

The electric scalar potential should be continuous across the cylinder wall at $r = a$ and thus we have $-E_0 a + \frac{C_{-1}^{(\text{out})}}{a} = C_{+1}^{(\text{in})} a$. Note that at the interface (cylinder wall), the normal component of the electric displacement vector must be continuous, i.e., the normal direction of the electric displacement vector $\mathbf{D}$ should also be equal across the cylinder wall at $r = a$ and then we have $\epsilon_{\text{out}}\big(-E_0 - \frac{C_{-1}^{(\text{out})}}{a^2}\big) = \epsilon_{\text{in}} C_{+1}^{(\text{in})}$, where ϵ_{out} is the permittivity of the medium outside the cylinder and ϵ_{in} is the permittivity of the cylindrical dielectric. We list these two boundary condition equations:

$$\begin{aligned} -E_0 a + \frac{C_{-1}^{(\text{out})}}{a} &= C_{+1}^{(\text{in})} a, \\ \epsilon_{\text{out}}\Big(-E_0 - \frac{C_{-1}^{(\text{out})}}{a^2}\Big) &= \epsilon_{\text{in}} C_{+1}^{(\text{in})}. \end{aligned} \tag{2.61}$$

There are two coefficients $C_{+1}^{(\text{in})}$ and $C_{-1}^{(\text{out})}$ to be determined. The result is given by

$$C_{+1}^{(\mathrm{in})} = -2\frac{\epsilon_{\mathrm{out}}}{\epsilon_{\mathrm{out}} + \epsilon_{\mathrm{in}}}E_0,$$
$$C_{-1}^{(\mathrm{out})} = -\frac{(\epsilon_{\mathrm{out}} - \epsilon_{\mathrm{in}})a^2}{\epsilon_{\mathrm{out}} + \epsilon_{\mathrm{in}}}E_0. \quad (2.62)$$

Now from the equation (2.60), we have the electric scalar potential solution:

$$\Phi_{\mathrm{out}} = E_0\left(-r - \frac{(\epsilon_{\mathrm{out}} - \epsilon_{\mathrm{in}})}{\epsilon_{\mathrm{out}} + \epsilon_{\mathrm{in}}}\frac{a^2}{r}\right)\cos\phi \qquad (r > a);$$
$$\Phi_{\mathrm{in}} = -2E_0\frac{\epsilon_{\mathrm{out}}}{\epsilon_{\mathrm{out}} + \epsilon_{\mathrm{in}}}r\cos\phi \qquad (r \le a). \quad (2.63)$$

In what follows, we shall show that the above electric potential Φ_{out} can be rearranged as the potential of an *equivalent electric dipole* that lies on the x axis, namely, the 2-D dielectric in the circular cross section can be considered to be an effective dipole when the 2-D dielectric is illuminated by the applied light (with the electric field E_0). We need the potential $\Phi(r) = C\ln r$ for a line charge. Let the equivalent dipole lie at the x axis with its center at $x = 0$. In the x-y plane, the position coordinates of $+q$ and $-q$ of the present equivalent dipole are $(+\frac{d}{2}, 0)$ and $(-\frac{d}{2}, 0)$, respectively. Then their potentials are given by $\Phi_+ = C_+\ln r_+$ and $\Phi_- = C_-\ln r_-$, respectively, where $r_+ = \sqrt{r^2 + \frac{d^2}{4} - rd\cos\phi} \simeq r(1 - \frac{d}{2r}\cos\phi)$ and $r_- = \sqrt{r^2 + \frac{d^2}{4} + rd\cos\phi} \simeq r(1 + \frac{d}{2r}\cos\phi)$. If $C_- = -C_+$, we have

$$\begin{aligned}\Phi(r) &= \Phi_+ + \Phi_- = C_+(\ln r_+ - \ln r_-) = C_+\ln\frac{r_+}{r_-}\\ &= C_+\ln\frac{r(1 - \frac{d}{2r}\cos\phi)}{r(1 + \frac{d}{2r}\cos\phi)}\\ &\simeq C_+\ln\left(1 - \frac{d}{r}\cos\phi\right) \simeq -C_+\frac{d}{r}\cos\phi. \end{aligned} \quad (2.64)$$

Let us consider a charged cylinder with a line charge density λ_{e}. According to the Gauss law, $\Phi = -\frac{\lambda_{\mathrm{e}}}{2\pi\epsilon_0}\ln r$, i.e., $C_+ = -\frac{\lambda}{2\pi\epsilon_0}$ and $C_- = +\frac{\lambda}{2\pi\epsilon_0}$. Thus, in Eq. (2.64), the electric scalar potential is $\Phi(r) = \Phi_+ + \Phi_- = \frac{\lambda_{\mathrm{e}}d}{2\pi\epsilon_0}\frac{\cos\phi}{r}$.

Since we have assumed the nanometer-sized circular cross section in the cylinder can be identified as an equivalent dipole lying on the x axis, the second term $-E_0a^2\frac{(\epsilon_{\rm out}-\epsilon_{\rm in})}{\epsilon_{\rm out}+\epsilon_{\rm in}}\frac{\cos\phi}{r}$ in the solution $\Phi_{\rm out} = E_0\left(-r-\frac{(\epsilon_{\rm out}-\epsilon_{\rm in})}{\epsilon_{\rm out}+\epsilon_{\rm in}}\frac{a^2}{r}\right)\cos\phi \quad (r>a)$ of the cylinder given in Eq. (2.63) should equal the potential, $\frac{\lambda_{\rm e}d}{2\pi\epsilon_0}\frac{\cos\phi}{r}$, of the present equivalent dipole. Here, let $\epsilon_{\rm out}=\epsilon_0$. Then we have the following result:

$$\begin{aligned} -E_0a^2\frac{(\epsilon_0-\epsilon_{\rm in})}{\epsilon_0+\epsilon_{\rm in}}\frac{\cos\phi}{r} &= \frac{\lambda_{\rm e}d}{2\pi\epsilon_0}\frac{\cos\phi}{r} \\ \Rightarrow \quad -E_0\frac{\pi a^2h}{\pi}\frac{(\epsilon_0-\epsilon_{\rm in})}{\epsilon_0+\epsilon_{\rm in}} &= \frac{\lambda_{\rm e}hd}{2\pi\epsilon_0}, \end{aligned} \tag{2.65}$$

where h denotes the nanometer circular cross section thickness in the direction of the height of the cylinder and πa^2h is the volume $\mathcal{V}$ of this nanometer circular cross section. Since $\lambda_{\rm e}h$ is the positive charge at $x=+\frac{d}{2}$ in the equivalent electric dipole lying on the x axis with its center at the coordinate origin, $\lambda_{\rm e}hd$ is the magnitude (p) of this dipole moment. Now the equation (2.65) can be rewritten as

$$-E_0\mathcal{V}\frac{\epsilon_0-\epsilon_{\rm in}}{\epsilon_0+\epsilon_{\rm in}} = \frac{p}{2\epsilon_0} \quad \Rightarrow \quad P=\frac{p}{\mathcal{V}} = -2\epsilon_0E_0\frac{\epsilon_0-\epsilon_{\rm in}}{\epsilon_0+\epsilon_{\rm in}}. \tag{2.66}$$

Here, as a matter of fact, E_0 is the local field, of which the direction is parallel to the x axis. Thus, the induced (equivalent) electric dipole moment vector is $\mathbf{P}=-2\epsilon_0\frac{\epsilon_0-\epsilon_{\rm in}}{\epsilon_0+\epsilon_{\rm in}}\mathbf{E}_{\rm local}$. On the other hand, the microscopic electric polarizability $\beta_{\rm pol}$ is defined through $\mathbf{P}=n_{\rm d}\beta_{\rm pol}\epsilon_0\mathbf{E}_{\rm local}$. Then a relation yields

$$\frac{n_{\rm d}\beta_{\rm pol}}{2} = \frac{\epsilon_{\rm in}-\epsilon_0}{\epsilon_{\rm in}+\epsilon_0}. \tag{2.67}$$

This is the Clausius-Mossotti relation in two-dimensional space. The above formalism is a proof for the relation (2.67) when the space dimension number $d=2$. This result has already been given in Eq. (2.41).

With the help of the definition of the polarization vector $\mathbf{P}=(\epsilon_{\rm in}-\epsilon_0)\mathbf{E}_{\rm external}$ and $\mathbf{P}=2\frac{\epsilon_{\rm in}-\epsilon_0}{\epsilon_{\rm in}+\epsilon_0}\epsilon_0\mathbf{E}_{\rm local}$ given in Eq. (2.66), we can obtain

$$\begin{aligned} &(\epsilon_{\rm in}-\epsilon_0)\mathbf{E}_{\rm external} = 2\frac{\epsilon_{\rm in}-\epsilon_0}{\epsilon_{\rm in}+\epsilon_0}\epsilon_0\mathbf{E}_{\rm local} \Rightarrow \mathbf{E}_{\rm external} = \frac{2\epsilon_0}{\epsilon_{\rm in}+\epsilon_0}\mathbf{E}_{\rm local} \\ &\Rightarrow \quad \mathbf{E}_{\rm neighboring} = \mathbf{E}_{\rm external}-\mathbf{E}_{\rm local} = \frac{\epsilon_0-\epsilon_{\rm in}}{\epsilon_{\rm in}+\epsilon_0}\mathbf{E}_{\rm local} = -\frac{\mathbf{P}}{2\epsilon_0}, \end{aligned} \tag{2.68}$$

where in the last step, the relation $\mathbf{P} = -2\epsilon_0 \frac{\epsilon_0 - \epsilon_{\rm in}}{\epsilon_0 + \epsilon_{\rm in}} \mathbf{E}_{\rm local}$ (in the two-dimensional space) has been substituted. The result $\mathbf{E}_{\rm neighboring} = -\frac{\mathbf{P}}{2\epsilon_0}$ is applicable in the two-dimensional dielectric (to be more precise, it is the dielectric in the two-dimensional space).

Now from the results (2.49), (2.58) and (2.68), the induced neighboring electric field is given by

$$\mathbf{E}_{\rm neighboring} = -\frac{\mathbf{P}}{d\epsilon_0}, \tag{2.69}$$

where d denotes the space dimension number ($d = 1, 2, 3$). Other important relations include the electric susceptibility and the Clausius-Mossotti relation. The electric susceptibility χ in the d-dimensional space is given in Eq. (2.37) and the Clausius-Mossotti relation in the d-dimensional space is given in Eq. (2.40).

2.7. Two-Dimensional Surface Plasmon Resonance in a Metallic Nanorod

We have obtained the electric scalar potential distribution (2.63) in a nanorod (in the cylindrical coordinate system, where the spatial variables are (r, ϕ, z)):

$$\begin{aligned} \Phi_{\rm out} &= -E_0 r\cos\phi - E_0 \frac{\epsilon_{\rm out} - \epsilon_{\rm in}}{\epsilon_{\rm out} + \epsilon_{\rm in}} \frac{a^2}{r} \cos\phi \quad (r > a); \\ \Phi_{\rm in} &= -2E_0 \frac{\epsilon_{\rm out}}{\epsilon_{\rm out} + \epsilon_{\rm in}} r\cos\phi \qquad (r \le a). \end{aligned} \tag{2.70}$$

Here, E_0 is the driving (stimulating) field and $-E_0 r\cos\phi$ is the external potential. The term $-E_0 \frac{\epsilon_{\rm out} - \epsilon_{\rm in}}{\epsilon_{\rm out} + \epsilon_{\rm in}} \frac{a^2}{r} \cos\phi$ in the electric scalar potential outside the cylinder and the electric potential $-2E_0 \frac{\epsilon_{\rm out}}{\epsilon_{\rm out} + \epsilon_{\rm in}} r\cos\phi$ inside the cylinder are the excited electric scalar potentials on this nanorod system. If the dielectric cylinder is an insulator, the magnitude of the excited electric scalar potential is small. If, however, the dielectric cylinder is made of a metal, such excited electric scalar potentials would be very strong, because the metal permittivity $\epsilon_{\rm in}$ can be negative, and the denominator $\epsilon_{\rm out} + \epsilon_{\rm in}$ in the above excited electric scalar potential can be zero at a certain frequency. Then the electric scalar potentials $\Phi_{\rm out}$ and $\Phi_{\rm in}$ given in Eq. (2.70) will be extremely large. This can be referred to as an effect of "surface plasmon resonance".

In the Drude model, the relative permittivity of a metallic nanorod is $\frac{\epsilon_{\rm in}}{\epsilon_0} = \epsilon^{(\rm r)}_{\rm metal} = \epsilon_{\rm b} - \frac{\omega_{\rm p}^2}{\omega^2}$, which can be negative. The relative permittivity of the medium outside the metallic nanorod is $\epsilon^{(\rm r)}_{\rm out} = \frac{\epsilon_{\rm out}}{\epsilon_0} > 0$, which is not sensitive to the frequency and can be considered to be a constant. The condition of surface plasmon resonance is given by

$$\epsilon_{\rm b} - \frac{\omega_{\rm p}^2}{\omega^2} + \epsilon^{(\rm r)}_{\rm out} = 0 \qquad \Rightarrow \qquad \omega = \frac{\omega_{\rm p}}{\sqrt{\epsilon_{\rm b} + \epsilon^{(\rm r)}_{\rm out}}}. \tag{2.71}$$

This is the electromagnetic wave frequency chosen when the surface plasmon resonance occurs in a metallic nanorod.

2.8. Magnetic Clausius-Mossotti Relation

In the derivation of electric Clausius-Mossotti relation, we used the quasi-static electric field equations: $\nabla \cdot \mathbf{E} = 0$, $\nabla \times \mathbf{E} = 0$ and $\mathbf{E} = -\nabla \Phi$ with Φ the electric scalar potential, because the size of the dielectric structure with which we are concerned is much smaller than the incident electromagnetic wavelength. Now we consider the magnetic Clausius-Mossotti relation when the condition of quasi-static magnetic field is also fulfilled. In this case, the curl equation $\nabla \times \mathbf{H} = \frac{\partial \mathbf{D}}{\partial t}$, where $\frac{\partial \mathbf{D}}{\partial t}$ can be ignored, can reduce to $\nabla \times \mathbf{H} = 0$. Then a magnetic scalar potential $\Phi_{\rm m}$ can be defined, i.e., $\mathbf{H} = -\nabla \Phi_{\rm m}$. By using $\nabla \cdot \mathbf{B} = 0$ with $\mathbf{B} = \mu_0 \mu_{\rm r} \mathbf{H}$, in the homogeneous magnetic medium, we have $\nabla \cdot \mathbf{H} = 0$. Thus, *the quasi-static magnetic field equations (i.e.,* $\nabla \cdot \mathbf{H} = 0$, $\nabla \times \mathbf{H} = 0$ *and* $\mathbf{H} = -\nabla \Phi_{\rm m}$*) hold true.* In the fashion similar to the case of electrostatic field, the following relations can also be obtained:

$$\begin{aligned}
\mathbf{H}_{\rm local} &= \mathbf{H}_{\rm external} + (-\mathbf{H}_{\rm neighboring}), \\
\mathbf{H}_{\rm neighboring} &= -\frac{\mathbf{M}}{d}, \qquad \mathbf{M} = n_{\rm d} \beta_{\rm pol} \mathbf{H}_{\rm local}, \\
\mathbf{M} &= \chi_{\rm m} \mathbf{H}_{\rm external} = \chi_{\rm m} \left(\mathbf{H}_{\rm local} + \mathbf{H}_{\rm neighboring} \right) \\
&= \chi_{\rm m} \left(\mathbf{H}_{\rm local} - \frac{\mathbf{M}}{d} \right) = \chi_{\rm m} \left(\frac{\mathbf{M}}{n_{\rm d} \beta_{\rm pol}} - \frac{\mathbf{M}}{d} \right),
\end{aligned} \tag{2.72}$$

where $n_{\rm d}$ is the number density of the magnetic dipoles and $\beta_{\rm pol}$ is the microscopic magnetic polarizability; $\mathbf{M}$ is the magnetic polarization (magnetization) and $\chi_{\rm m}$ is the magnetic susceptibility. Then the magnetic susceptibility $\chi_{\rm m}$ in the d-dimensional space is given by

$$\chi_{\rm m} = \frac{1}{\frac{1}{n_{\rm d}\beta_{\rm pol}} - \frac{1}{d}} = \frac{n_{\rm d}\beta_{\rm pol}}{1 - \frac{n_{\rm d}\beta_{\rm pol}}{d}}. \tag{2.73}$$

From the equation (2.73), we can have $\chi_{\rm m}\left(1 - \frac{n_{\rm d}\beta_{\rm pol}}{d}\right) = n_{\rm d}\beta_{\rm pol}$ and $\chi_{\rm m} = n_{\rm d}\beta_{\rm pol}(1 + \frac{\chi_{\rm m}}{d})$, i.e., $n_{\rm d}\beta_{\rm pol} = \frac{\chi_{\rm m}}{1+\frac{\chi_{\rm m}}{d}}$. Then it can yield

$$\frac{\mu_{\rm r} - 1}{\mu_{\rm r} + d - 1} = \frac{n_{\rm d}\beta_{\rm pol}}{d}, \tag{2.74}$$

where $\chi_{\rm m} = \mu_{\rm r} - 1$ has been substituted. *This is the magnetic Clausius-Mossotti relation in the d-dimensional space.* When $d = 1$ and $d = 2$, the magnetic Clausius-Mossotti relation is given by

$$\frac{\mu_{\rm r} - 1}{\mu_{\rm r}} = n_{\rm d}\beta_{\rm pol} \quad (d = 1); \qquad \frac{\mu_{\rm r} - 1}{\mu_{\rm r} + 1} = \frac{n_{\rm d}\beta_{\rm pol}}{2} \quad (d = 2). \tag{2.75}$$

Now we shall return to the magnetic polarizability relation $\mathbf{M} = n_{\rm d}\beta_{\rm pol}\mathbf{H}_{\rm local}$. As is known, the electric and magnetic polarization energy densities are given by $\mathcal{H}_{\rm e} = -\mathbf{P}\cdot\mathbf{E}$ and $\mathcal{H}_{\rm m} = -\mathbf{M}\cdot\mathbf{B}$, respectively. Thus, it seems that the magnetic polarizability relation that corresponds to the well known electric polarizability relation $\mathbf{P} = n_{\rm d}\beta_{\rm pol}\epsilon_0\mathbf{E}_{\rm local}$ must be $\mathbf{M} = n_{\rm d}\beta_{\rm pol}\frac{1}{\mu_0}\mathbf{B}_{\rm local}$. But it should be pointed out that the local magnetic field is determined by the local magnetic induction field ($\mathbf{H}_{\rm local} = \frac{\mathbf{B}_{\rm local}}{\mu_0}$). Therefore, $\mathbf{M} = n_{\rm d}\beta_{\rm pol}\frac{1}{\mu_0}\mathbf{B}_{\rm local}$ *reduces to the magnetic polarizability relation* $\mathbf{M} = n_{\rm d}\beta_{\rm pol}\mathbf{H}_{\rm local}$.

In the issue of surface plasmons on the metallic nanospheres, researchers were concerned only with the electric field effect, i.e., the magnetic field effect is in general not taken into account. Why did they ignore the magnetic field effect? This can be interpreted as follows: The electric scalar potential Φ outside and inside the nanometer-sized sphere is given by Eq. (2.42): $\Phi_{\rm out}(r,\theta) = -E_0 r\cos\theta + \frac{\epsilon_{\rm in}-\epsilon_{\rm out}}{\epsilon_{\rm in}+2\epsilon_{\rm out}}E_0\frac{a^3}{r^2}\cos\theta$ with $r > a$ and $\Phi_{\rm in}(r,\theta) = -\frac{3\epsilon_{\rm out}}{\epsilon_{\rm in}+2\epsilon_{\rm out}}E_0 r\cos\theta$ with $r \leq a$, where a is the radius of the nanometer-sized sphere; E_0 is the applied external electric field, which is parallel to the z axis. Here, $-E_0 r\cos\theta$ in $\Phi_{\rm out}(r,\theta)$ is the electric scalar potential of the incident electromagnetic wave; $-\frac{\epsilon_{\rm out}}{\epsilon_{\rm in}+2\epsilon_{\rm out}}E_0\frac{a^3}{r^2}\cos\theta$ in $\Phi_{\rm out}(r,\theta)$ and $-\frac{3\epsilon_{\rm out}}{\epsilon_{\rm in}+2\epsilon_{\rm out}}E_0 r\cos\theta$

in $\Phi_{\text{in}}(r,\theta)$ [54, 55] are the induced electric field effect by the illuminating light on the nanometer-sized sphere. Clearly, in the magnetic scalar potential Φ_{m} on the nanometer-sized sphere, there are also such magnetic scalar potentials caused by the incident electromagnetic wave and the magnetically induced effect. *Since the surface plasmon resonance (drastic excitation) occurs when $\epsilon_{\text{in}} + 2\epsilon_{\text{out}} = 0$ (surface plasmon resonance condition), the induced surface-plasmon electric scalar potentials $-\frac{\epsilon_{\text{out}}}{\epsilon_{\text{in}}+2\epsilon_{\text{out}}}E_0\frac{a^3}{r^2}\cos\theta$ in $\Phi_{\text{out}}(r,\theta)$ and $-\frac{3\epsilon_{\text{out}}}{\epsilon_{\text{in}}+2\epsilon_{\text{out}}}E_0 r\cos\theta$ in $\Phi_{\text{in}}(r,\theta)$ caused by the illuminating light are very strong* (because the denominator vanishes and the factor $\frac{1}{\epsilon_{\text{in}}+2\epsilon_{\text{out}}}$ becomes divergent theoretically). When $\epsilon_{\text{in}}+2\epsilon_{\text{out}} = \epsilon_{\text{b}} - \frac{\omega_{\text{p}}^2}{\omega^2} + 2\frac{\epsilon_{\text{out}}}{\epsilon_0} = 0$, the surface plasmon resonance frequency is $\omega = \frac{\omega_{\text{p}}}{\sqrt{\epsilon_{\text{b}}+2\frac{\epsilon_{\text{out}}}{\epsilon_0}}}$. Since, however, at such frequencies, $\mu_{\text{in}} + 2\mu_{\text{out}} \neq 0$, the magnetic energy density $\frac{1}{2}\mu_{\text{r}}\mu_0\mathbf{H}^2$ is negligibly small compared with the surface-plasmon energy density $\frac{1}{2}\epsilon_{\text{r}}\epsilon_0\mathbf{E}^2$. Thus, in the issue of surface plasmons on the metallic nanospheres, the magnetic field effect can in general be ignored.

2.9. The Relationship and Difference between Plasma Oscillation and Langmuir Oscillation

In the above, we have addressed the plasma oscillation. It should be pointed out that we need to distinguish between the *plasma oscillation* and the *Langmuir oscillation* in a plasma. Though these two kinds of oscillations were studied by Langmuir, and they can be derived in a same set of equations of non-magnetized plasma, yet they are different in physical essence. We shall briefly discuss this problem (i.e., the connection and difference between the *plasma oscillation* and the *Langmuir oscillation*):

As is well known, the set of equations of non-magnetized plasma includes three equations: $\frac{\partial n_{\text{e}}}{\partial t} + \nabla\cdot(n_{\text{e}}\mathbf{u}_{\text{e}}) = 0$ (electric charge conservation), $m_{\text{e}}n_{\text{e}}\frac{\mathrm{d}\mathbf{u}_{\text{e}}}{\mathrm{d}t} + \nabla p_{\text{e}} = -n_{\text{e}}e\mathbf{E}$ (Newton-Euler fluid equation of motion) and $\nabla\cdot\mathbf{E} = \frac{(n_{\text{p}}-n_{\text{e}})e}{\epsilon_0}$ (electrostatic Gauss law) [39], where n_{e} is the electron number density, $-e$ is the electron charge ($-e < 0$), $\mathbf{u}_{\text{e}}$ is the electron velocity, p_{e} is the thermal pressure of the electron gas, and n_{p} is the positive ion number density. We ignore the motion of the positive ions, because they are heavy in mass. The plasma densities can be written as $n_{\text{p}} = n_{\text{p}}^{(0)} + n_{\text{p}}^{(1)}$

and $n_{\rm e} = n_{\rm e}^{(0)} + n_{\rm e}^{(1)}$, where $n_{\rm p}^{(0)}$ and $n_{\rm e}^{(0)}$ denote the unperturbed number densities of positive ions and electrons, respectively, and $n_{\rm p}^{(1)}$ and $n_{\rm e}^{(1)}$ denote the small perturbed (fluctuated) number densities of positive ions and electrons, respectively. Then the electrostatic Gauss law can be rewritten as

$$\nabla \cdot \mathbf{E} = \frac{(n_{\rm p}^{(0)} - n_{\rm e}^{(0)})e}{\epsilon_0} + \frac{(n_{\rm p}^{(1)} - n_{\rm e}^{(1)})e}{\epsilon_0}. \tag{2.76}$$

Here, there are two source terms in the electrostatic Gauss law. The first source term $\frac{(n_{\rm p}^{(0)} - n_{\rm e}^{(0)})e}{\epsilon_0}$ and the second source term $\frac{(n_{\rm p}^{(1)} - n_{\rm e}^{(1)})e}{\epsilon_0}$ in Eq. (2.76) lead to the plasma oscillation and the Langmuir oscillation, respectively. As an illustrative example, we shall address the source term of the plasma oscillation. Since the electrons are much lighter than the positive ions (i.e., the motion of positive ions can be ignored and the electrons can move freely at relatively high acceleration and high velocity), for the first, we assume the electron number density $n_{\rm e}^{(0)}$ can be driven overall to the side as a whole (for example, they are driven by thermal factors or by an applied external electric field), and there is then an excess of positive charges with the number density $n_{\rm p}^{(0)}$ in space, which creates a new electric field in the direction opposite to the applied external field, pulling the electrons back to their original equilibrium positions. This leads to the plasma oscillation, of which the exact source term is $\frac{n_{\rm p}^{(0)}}{\epsilon_0}$. The difference between the *plasma oscillation* and the *Langmuir oscillation* lies in that: i) The plasma oscillation is a collective behavior, but the Langmuir oscillation is not; ii) The plasma oscillation frequency in d-dimensional space is $\omega_d^{(\rm plasma)} = \frac{n_{\rm e} e^2}{d\epsilon_0 m_{\rm e}}$, whereas the Langmuir oscillation frequency is always the same frequency $\omega^{(\rm Langmuir)} = \frac{n_{\rm e} e^2}{\epsilon_0 m_{\rm e}}$, no matter how many space dimensions it is in. Only in one dimension can they have the same frequency, $\frac{n_{\rm e} e^2}{\epsilon_0 m_{\rm e}}$, of oscillation.

To close this section, let us give a brief overview of the important result:

The two mysteries in local field correction and plasma oscillation in the Drude-Lorentz model can be resolved if these two effects (local field correction and plasma oscillation) are taken into consideration together. In other words, the two mysteries are the solutions to each other: specifically, their effects just cancel each other out. It is important to note that this result of the present section is true for one-, two- and three-dimensional plasmas.

Some monographs and textbooks may identify the plasma oscillation with the Langmuir oscillation, that is, when they talk about the plasma oscillation, they also refer to it as the Langmuir oscillation. Of course, the theoretical mechanisms of both the plasma oscillation and the Langmuir oscillation can be derived from the same set of equations of non-magnetized plasma. If the equations can be solved accurately or numerically without any distinction between unperturbed and perturbed ion and electron number densities, then the plasma oscillation and the Langmuir oscillation can be solved together in a unified form. So you really can no longer distinguish between the plasma oscillation and the Langmuir oscillation. If, however, we decompose the plasma density into a sum of the unperturbed and perturbed parts, in such a case, the plasma oscillation and the Langmuir oscillation have different properties and behaviors.

3. Review of Schemes for Negative Permittivity and Permeability in Quantum-Coherent Atomic Vapor

Here we shall give an overview of some scenarios of negative refractive index with quantum coherence effects. Atomic phase coherence control (or quantum coherence control) in various multi-level atomic systems of atomic vapor (e.g., alkali metal atomic gas) can give rise to a number of novel phenomena in optical wave propagation [56, 57, 58]. One of such phenomena would be the effect of electromagnetically induced transparency (EIT) [56]. The striking effects related to EIT include the ultraslow light wave propagation, superluminal light, optical field storage in three- and four-level atomic vapor as well as atomic ground state cooling [56, 59, 60]. These effects could be useful for developing new techniques in quantum optics and photonics.

We also give a brief review of negative refractive index research. One of the representative media of artificial composite metamaterials is *left-handed media* or *negative refractive index media*, which can exhibit a frequency band of simultaneously negative electric permittivity and magnetic permeability [61, 62, 63]. Such a kind of left-handed media can exhibit various peculiar electromagnetic or optical properties (characteristics). They include anomalous refraction, reversals of both Doppler shift and Cherenkov radiation, evanescent wave amplification [64], modified spontaneous emission rates and unusual effect of photon tunnelling [65]. A combination of two microscopic electromagnetic structures (*array of long metallic wires* and *split ring resonators* [66, 67]) can lead to negative refractive index [63]. Physicists have utilized the classical electromagnetic theory for designing and fabricating a variety of materials having negative refractive index [63, 66, 67, 68, 69]. Apart from the mechanisms within the framework of classical electromagnetic wave theory, some new methods for designing negative refractive index have also been suggested. For example, Oktel *et al.* [70] and us [71] proposed a possibility for realizing simultaneously negative magnetic permeability μ_r and electric permittivity ϵ_r with three-level alkali metal atomic gases.

In this section, we shall review and generalize the three-level quantum coherence work for simultaneously negative magnetic permeability and electric permittivity. We will obtain all the observable optical parameters of a simple three-level atomic vapor and a pumped three-level dressed-state mixed-parity atomic gaseous medium. These are the direct generalizations of the previous studies in negative-index atomic vapors, i.e., based on some mechanisms of a three-level quantum-coherent atomic vapor [70, 71] and a dressed-state assisted negative-index atomic vapor [72, 73].

3.1. An Overview of Negative Optical Parameters in a Three-Level Atomic Vapor

The reason why we need to consider this negative-parameter mechanism based on quantum coherence [70, 71] is that we need to use it to realize the surface electromagnetic wave modes and the so-called parity-time symmetric surface wave modes.

3.1.1. A Three-Level Λ-Type EIT System Involving Magnetic-Transition Contribution

In the following analysis, we shall study the physical conditions for a three-level atomic vapor medium to be a left-handed dielectric. We treat a Λ-configuration three-level atomic system, in which there are two upper energy levels $|a\rangle$, $|c\rangle$ and one lower energy level $|b\rangle$ and (see Fig. 1 for the schematic diagram). This three-level atomic system can be coupled to two optical fields: A coupling laser beam and a probe optical field, which drive a level pair $|a\rangle$-$|c\rangle$ and a level pair $|a\rangle$-$|b\rangle$, respectively. Here, the parity of level $|a\rangle$ is opposite to that of both levels $|b\rangle$ and $|c\rangle$. Then the electric dipole matrix elements (i.e., the transition electric dipole moments) $\vec{\wp}_{ab} = \langle a|e\vec{r}|b\rangle \neq 0$ but $\vec{\wp}_{cb} = \langle c|e\vec{r}|b\rangle = 0$; The magnetic dipole matrix elements (i.e., the transition magnetic dipole moments) $\vec{m}_{cb} = \langle c|(e/2m_{\rm e})(\vec{L} + 2\vec{S})|b\rangle \neq 0$ but $\vec{m}_{ab} = \langle a|(e/2m_{\rm e})(\vec{L} + 2\vec{S})|b\rangle = 0$. Here, $\vec{L}$ and $\vec{S}$ represent respectively the orbital and spinning angular momentum operators. Therefore, when the nearly resonant weak probe light drives the electric-dipole allowed transition between levels $|a\rangle$ and $|b\rangle$ in the present three-level atomic vapor, the transition between levels $|c\rangle$ and $|b\rangle$ (magnetic-dipole) would also be induced. Thus, the electric-dipole allowed transition ($|a\rangle$-$|b\rangle$) and the magnetic-dipole allowed transition ($|c\rangle$-$|b\rangle$) may make their own contribution to the electric polarizability and the magnetic susceptibility, respectively.

It should be noted that the dimensionless ratio $|\vec{m}_{cb}/\left(\vec{\wp}_{ab}c\right)|$ is only the 10^{-2} order of magnitude in an atomic system, and so it is not necessary to take into account the magnetic-dipole allowed transition. But if the level pair ($|a\rangle$-$|c\rangle$) is nearly degenerate, the magnetic-dipole allowed transition needs to be considered. We shall point out that when the population probability at level $|c\rangle$ is much larger than that at the upper level $|a\rangle$ (this can be realized in a three-level EIT condition). In this case, the stronger coupling laser beam could enhance the probability amplitude of level $|c\rangle$ through the quantum coherence (or *atomic-level transition pathway interference*) [6]. Then, the order of magnitude of the off-diagonal density matrix element ρ_{cb} would be larger than the other off-diagonal density matrix element of ρ_{ab}. Thus, we can draw a conclusion: When the weak probe light field is nearly resonant with $|b\rangle$-$|c\rangle$ transition frequency, the magnetic-dipole allowed transition between the levels $|b\rangle$ and $|c\rangle$ will occur, and the transition magnetic dipole moment $2m^{*}_{cb}\rho_{cb}$ would possibly exhibit a same order of magnitude of the transition electric dipole moment $(2c\wp^{*}_{ab}\rho_{ab})$. In this case, it must be necessary to consider the magnetic transition contribution to the three-level atomic vapor. Therefore, a dense EIT atomic vapor medium (with nearly degenerate level pair $|a\rangle$-$|c\rangle$) would exhibit the simultaneously negative permittivity and permeability [70, 71].

The simultaneously negative permittivity and permeability will lead to negative refractive index [61]. We think this can be interpreted easily as follows: We assume a plane electromagnetic wave with phase factor $\exp[-\mathrm{i}(\omega t - \mathbf{k}\cdot\mathbf{r})]$, which agrees with the Maxwell equations: $\mathbf{k}\times\mathbf{E} = \omega\mu_{\mathrm{r}}\mu_0\mathbf{H}$ and $\mathbf{k}\times\mathbf{H} = -\omega\epsilon_{\mathrm{r}}\epsilon_0\mathbf{E}$. In an ordinary electromagnetic dielectric with $\mu_{\mathrm{r}} > 0$ and $\epsilon_{\mathrm{r}} > 0$, $\mathbf{E}$, $\mathbf{H}$ and $\mathbf{k}$ form a right-handed system. But in a medium with simultaneously negative permittivity and permeability, from the aforementioned equations, $\mathbf{E}$, $\mathbf{H}$ and $\mathbf{k}$ form a left-handed system. Now we shall determine the refractive index: We assume this electromagnetic wave is incident (from the vacuum) normally into the medium with simultaneously negative permittivity and permeability (i.e., $\mathbf{E}$ and $\mathbf{H}$ are parallel to the interface). Note that the electromagnetic boundary condition is that the tangential components of $\mathbf{E}$ and $\mathbf{H}$ need to be continuous across the interface between the vacuum and the left-handed dielectric. This means that the tangential components of $\mathbf{E}$ and $\mathbf{H}$ do not change sign at the interface when this electromagnetic wave is incident from the vacuum to the left-handed medium. Since the permittivity ϵ_{r} and permeability μ_{r} change sign, this means that the wave vector $\mathbf{k}$ in the medium of simultaneously negative permittivity and permeability unavoidably changes its direction (opposite to the direction of the wave vector in the vacuum). Thus, the refractive index in the relation $k = n\frac{\omega}{c}$ in the present medium will have a negative sign, i.e., the simultaneously negative permittivity and permeability will cause a negative refractive index [61].

Now we shall obtain the steady solution of the density matrix elements of this three-level atomic system. The off-diagonal density matrix elements, ρ_{ab} and ρ_{cb}, can be written as $\rho_{ab} = \tilde{\rho}_{ab}\exp\left(-\mathrm{i}\omega_{\mathrm{p}}t\right)$ and $\rho_{cb} = \tilde{\rho}_{cb}\exp\left[-\mathrm{i}\left(\omega_{\mathrm{p}} - \omega_{\mathrm{c}}\right)t\right]$. Since the intensity of the probe laser beam is very weak and hence nearly all the atom population probability is in the ground state, *i.e.*, the atomic population probability $\rho_{bb} \simeq 1$ at the atomic level $|b\rangle$. Then, the two off-diagonal density matrix elements $\tilde{\rho}_{ab}$ and $\tilde{\rho}_{cb}$ obey the matrix equation [74]

$$\frac{\partial}{\partial t}\begin{pmatrix}\tilde{\rho}_{ab}\\ \tilde{\rho}_{cb}\end{pmatrix} = \begin{pmatrix}-(\gamma_1 + \mathrm{i}\Delta) & \frac{\mathrm{i}}{2}\Omega_{\mathrm{c}}\\ \frac{\mathrm{i}}{2}\Omega_{\mathrm{c}}^{*} & -\left[\gamma_3 + \mathrm{i}\left(\Delta - \delta\right)\right]\end{pmatrix}\begin{pmatrix}\tilde{\rho}_{ab}\\ \tilde{\rho}_{cb}\end{pmatrix} + \frac{\mathrm{i}}{2\hbar}\begin{pmatrix}\wp_{ab}\mathcal{E}_{\mathrm{p}}\\ m_{cb}\mathcal{B}_{\mathrm{p}}\end{pmatrix}. \tag{3.1}$$

In this 2×2 matrix equation, γ_1 stands for the atomic spontaneous decay rate of level $|a\rangle$; γ_3 is the dephasing rate of level $|c\rangle$; $\mathcal{E}_{\mathrm{p}}$ and $\mathcal{B}_{\mathrm{p}}$ denote the (slowly varying) electric and magnetic field envelopes of the incident probe laser beam; $\wp_{ab}$ and m_{cb} denote the transition electric- and magnetic-dipole

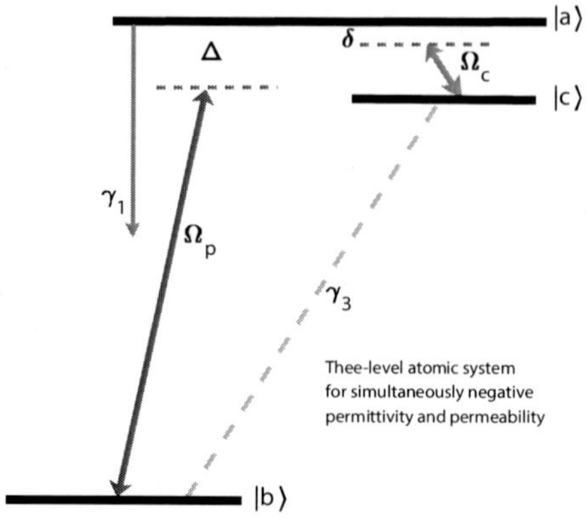

Figure 1. The schematic diagram of a Λ-configuration three-level atomic system for simultaneous electric- and magnetic-dipole allowed transitions. The electric-dipole allowed transition ($|a\rangle$-$|b\rangle$) and the magnetic-dipole allowed transition ($|c\rangle$-$|b\rangle$) could give rise to simultaneously negative electric permittivity and magnetic permeability. The levels $|a\rangle$ and $|c\rangle$ are nearly degenerate, i.e., their energy eigenvalues are almost the same. Such a three-level system could be found in some atoms or semiconductor quantum dots (i.e., nanocrystals or artificial "macroatoms", which also have discrete energy levels inside).

moments, respectively; The Rabi frequency of the coupling laser beam is given by $\Omega_{\rm c} = \wp_{ac}\mathcal{E}_{\rm c}/\hbar$ with $\mathcal{E}_{\rm c}$ the (slowly varying) electric field envelope of the coupling beam; The frequency detunings of probe and coupling beams are $\Delta = \omega_{ab} - \omega_{\rm p}$ and $\delta = \omega_{ac} - \omega_{\rm c}$, in which ω_{ab}, ω_{ac} denote the transition frequencies of $|a\rangle$-$|b\rangle$ and $|a\rangle$-$|c\rangle$; $\omega_{\rm p}$ and $\omega_{\rm c}$ are the angular frequencies of the probe optical field and the coupling optical beam, respectively.

The steady solution of Eq. (3.1) is given by

$$\left\{\begin{array}{l} \tilde{\rho}_{ab} = \frac{1}{2\hbar\mathcal{Z}}\left\{{\rm i}\wp_{ab}\mathcal{E}_{\rm p}\left[\gamma_3 + {\rm i}\left(\Delta - \delta\right)\right] - \frac{m_{cb}\mathcal{B}_{\rm p}\Omega_{\rm c}}{2}\right\}, \\ \tilde{\rho}_{cb} = \frac{1}{2\hbar\mathcal{Z}}\left[{\rm i}m_{cb}\mathcal{B}_{\rm p}\left(\gamma_1 + {\rm i}\Delta\right) - \frac{\wp_{ab}\mathcal{E}_{\rm p}\Omega_{\rm c}^{*}}{2}\right], \end{array}\right. \qquad (3.2)$$

where the parameter $\mathcal{Z}$ is given by

$$\mathcal{Z} = (\gamma_1 + \mathrm{i}\Delta)\left[\gamma_3 + \mathrm{i}\left(\Delta - \delta\right)\right] + \frac{1}{4}\Omega_{\mathrm{c}}^{*}\Omega_{\mathrm{c}}. \tag{3.3}$$

From the steady solution (3.2), it follows that there is a *mediated transition contribution* for $\tilde{\rho}_{ab}$ and $\tilde{\rho}_{cb}$: specifically, apart from the intrinsic electric-dipole allowed transition term $i\wp_{ab}\mathcal{E}_{\mathrm{p}}\left[\gamma_3 + \mathrm{i}(\Delta - \delta)\right]$, an extra magnetic-dipole related term $-m_{cb}\mathcal{B}_{\mathrm{p}}\Omega_{\mathrm{c}}/2$ (mediated through the coupling laser beam Ω_{c}) also makes contribution to the off-diagonal density matrix element $\tilde{\rho}_{ab}$; In the same manner, there is the intrinsic magnetic-transition allowed transition term $im_{cb}\mathcal{B}_{\mathrm{p}}\left(\gamma_1 + \mathrm{i}\Delta\right)$, but an extra electric-dipole related term $-\wp_{ab}\mathcal{E}_{\mathrm{p}}\Omega_{\mathrm{c}}^{*}/2$ (mediated via the coupling beam) could also unavoidably make a contribution to the off-diagonal density matrix element $\tilde{\rho}_{cb}$. In the literature, the cases of strongly driving optical field (i.e., the coupling laser beam), of which the Rabi frequency Ω_{c} ranges from 10^6 to $10^8\ \mathrm{rad}\cdot\mathrm{s}^{-1}$, or $\Omega_{\mathrm{c}} = (0.1 \sim 10) \times \gamma_1$ [75, 76, 77], have been studied. We believe that in some cases under proper conditions, the aforementioned *mediated transition* should also be taken into account.

3.1.2. Microscopic, Macroscopic Parameters and Clausius-Mossotti Relations

Now the relation between the microscopic atomic parameters (i.e., the off-diagonal density matrix elements $\tilde{\rho}_{ab}$ and $\tilde{\rho}_{cb}$) and the macroscopic optical parameters (i.e., the electric permittivity ϵ and the magnetic permeability μ) can be established. Here, the local field effect, which originates from the dipole-dipole electromagnetic interaction between neighboring atoms or dipoles [41, 42] should be taken into account.

i) *The case of electric response:*

The atomic electric polarizability γ_{e} because of the atomic electric-dipole allowed transition between levels $|a\rangle$ and $|b\rangle$ is given by $\gamma_{\mathrm{e}} = 2\wp_{ba}\tilde{\rho}_{ab}/\left(\epsilon_0\mathcal{E}_{\mathrm{p}}\right)$. Substitution of the expression for the off-diagonal density matrix element $\tilde{\rho}_{ab}$ given in Eq. (3.2) into the atomic electric polarizability γ_{e} yields

$$\gamma_{\mathrm{e}} = \frac{1}{\epsilon_0\hbar\mathcal{Z}}\left\{\mathrm{i}|\wp_{ab}|^2\left[\gamma_3 + \mathrm{i}\left(\Delta - \delta\right)\right] - \frac{\wp_{ba}m_{cb}\Omega_{\mathrm{c}}}{2c}\right\}. \tag{3.4}$$

By using the theory of local field correction [41, 42], the relative electric permittivity of the present atomic vapor is $\epsilon = 1 + \frac{N\gamma_{\rm e}}{1-\frac{1}{3}N\gamma_{\rm e}}$, where N denotes the total number of atoms per unit volume. From the equation for the permittivity, the relation between the macroscopic optical parameter (ϵ) and the microscopic atomic parameter ($\gamma_{\rm e}$) is the Clausius-Mossotti relation $\gamma_{\rm e} = \frac{3}{N}\left(\frac{\epsilon-1}{\epsilon+2}\right)$, which describes the electric contribution of neighboring atoms (dipole-dipole interaction) to local electric polarization.

ii) *The case of magnetic response*

Now we shall discuss the local field correction effect in the case of magnetic polarization (i.e, magnetization). The Clausius-Mossotti relation [43] (for the atomic-vapor magnetic susceptibility $\chi_{\rm m}$ and the atomic magnetic polarizability $\gamma_{\rm m}$) in the magnetized vapor medium is given by $\chi_{\rm m} = \frac{N\gamma_{\rm m}}{1-\frac{1}{3}N\gamma_{\rm m}}$, and therefore the relative magnetic permeability μ can be of the form $\mu \equiv \chi_{\rm m} = \frac{1+\frac{2}{3}N\gamma_{\rm m}}{1-\frac{1}{3}N\gamma_{\rm m}}$, and consequently, the relation between the macroscopic optical parameter (μ) and the microscopic atomic parameter ($\gamma_{\rm m}$) is given by $\gamma_{\rm m} = \frac{3}{N}\left(\frac{\mu-1}{\mu+2}\right)$.

Since the atomic (microscopic) magnetic polarizability is given by $\gamma_{\rm m} = 2\mu_0 m_{bc}\rho_{cb}/B_{\rm p}$ [43], where the off-diagonal density matrix element is $\rho_{cb} = \tilde{\rho}_{cb}\exp\left[-{\rm i}\left(\omega_{\rm p}-\omega_{\rm c}\right)t\right]$ and the magnetic induction field is $B_{\rm p} = \mathcal{B}_{\rm p}\exp\left(-{\rm i}\omega_{\rm p}t\right)$. Note that $B_{\rm p}$ and $\mathcal{B}_{\rm p}$ represent the amplitude and envelope, respectively, of the magnetic induction field. Now we can obtain the atomic magnetic polarizability

$$\gamma_{\rm m} = \frac{2\mu_0 m_{bc}\tilde{\rho}_{cb}\exp\left({\rm i}\omega_{\rm c}t\right)}{\mathcal{B}_{\rm p}}. \tag{3.5}$$

In order that a stable effective permeability can be achieved, we should have $\omega_{\rm c} \ll \omega_{\rm p}$ and $\omega_{ac} \ll \omega_{\rm p}$, namely, since $\omega_{ac} = \omega_{ab} - \omega_{cb}$ and ω_{ac} is sufficiently small (nearly degenerate), the transition frequency of the magnetic-dipole allowed transition ω_{cb} needs to be approximately equal to the other transition frequency (electric-dipole allowed transition) ω_{ab}. Now we draw a conclusion: The excited (upper) level $|a\rangle$ and level $|c\rangle$ need to be of near degeneracy in their energy eigenvalues, so that we can possibly realize the significant magnetic-dipole allowed transition that is driven by the magnetic field of the incident weak probe light. Under such a condition, we insert the expression for the off-diagonal density matrix element $\tilde{\rho}_{cb}$ in Eq. (3.2) into the aforementioned formula (3.5) and obtain the atomic microscopic polarizability ($\omega_c \rightarrow 0$)

$$\gamma_m = \frac{\mu_0}{\hbar \mathcal{Z}} \left\{ i|m_{bc}|^2 (\gamma_1 + i\Delta) - \frac{m_{bc}\wp_{ab}c\Omega_c^*}{2} \right\}. \tag{3.6}$$

Some numerical calculations of the present mechanism for achieving simultaneously negative permittivity and permeability in the three-level atomic vapor can be found in our previous paper [71].

3.2. An Overview of Dressed-State Mixed-Parity Transitions for Realizing Negative Permittivity and Permeability

In quantum optics (on atom-field interaction), the dressed states are the linear combinations (or linear superpositions) of two bare energy levels of a two-level atomic system (e.g., alkali metallic atoms). Such dressed states can be realized by using a strong coupling laser beam, which drives the electric- or magnetic-dipole allowed transition in this two-level system. It should be noticed that the dressed states have mixed parities [72], both electric- and magnetic-dipole allowed transitions can simultaneously occur between the dressed states and a third pure-parity bare energy level in this atomic system (see Fig. 2). Such dressed-state mixed-parity transitions in an atomic vapor could possibly lead to simultaneously negative permittivity and permeability (and hence the negative refractive index of the atomic vapor). The advantages of the present system include isotropic negative refractive index and optical-frequency negative optical parameters ($\epsilon_r < 0$ and $\mu_r < 0$ at visible or optical frequencies). In the literature, two-level dressed states for mixed-parity transitions [72] and three-level tunable dressed states [73] for producing simultaneously negative permittivity and permeability in atomic gaseous media have been considered by us. Although such negative refraction index of atomic gases have a bottleneck, i.e., the optical loss is unavoidably large and it would be difficult to be realized or it would find no practical applications in the research field of negative-index materials. But the idea is still instructive, and it would be useful in design of parity-time symmetric optical materials. Here, we shall review the relevant previous work of ours, e.g., refractive index control with a two-level dressed state [72] and a dressed state in a three-level atomic system [73].

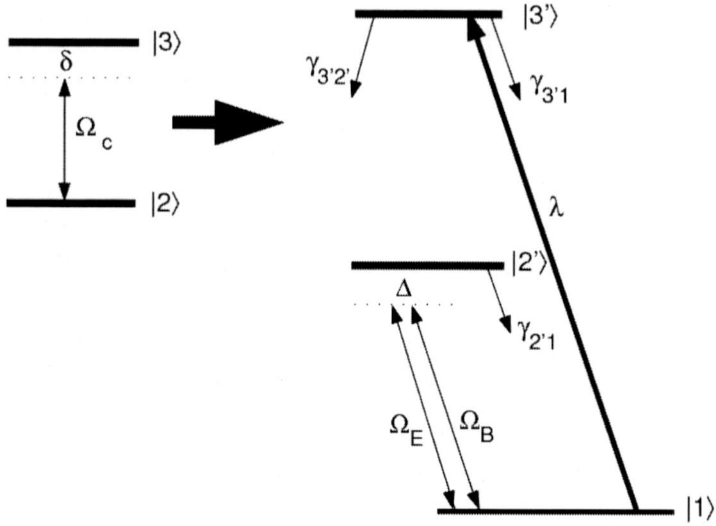

Figure 2. The schematic diagram for the dressed-state mixed-parity transitions. The electric-dipole allowed transition $|3\rangle$-$|2\rangle$ is driven by a strong coupling laser beam, and two dressed states $|3'\rangle$ and $|2'\rangle$ will result from the linear combinations of the two bare levels $|3\rangle$ and $|2\rangle$. The energy level pair $|1\rangle$-$|2'\rangle$ is coupled to the probe electric and magnetic fields. Both electric- and magnetic-dipole allowed transitions between the ground level $|1\rangle$ and the mixed-parity dressed level $|2'\rangle$ will take place. The population of level $|1\rangle$ is pumped to the dressed level $|3'\rangle$ at the pumping rate λ.

For the first, we will give an overview of a scheme of *dressed-state assisted mixed-parity transition* for realizing negative refractive index (a generalized mechanism similar to a previous scenario [73]), in which there is a three-level bare-state atomic system with one ground level $|1\rangle$ and two upper levels $|2\rangle$, $|3\rangle$ (see the schematic diagram in Fig. 2). Here, the two upper bare levels have opposite parities (e.g., energy level $|2\rangle$ possesses an even parity, whereas energy level $|3\rangle$ has an odd parity), and the ground level $|1\rangle$ has an even parity. The alkali metallic atoms can provide such an atomic system. Since the $|2\rangle$-$|3\rangle$ electric-dipole allowed transition can be excited by a strong coupling laser beam, we shall address in detail the dressed states, which involve the interaction between the strong coupling beam and the energy level pair $\{|2\rangle, |3\rangle\}$. The Hamiltonian of the system in the representation of atomic bare-state basis vectors $\{|2\rangle, |3\rangle\}$ is of the form

$$\hat{H} = \begin{pmatrix} \frac{1}{2}\hbar\delta & V \\ V^{*} & -\frac{1}{2}\hbar\delta \end{pmatrix}, \tag{3.7}$$

where the Hamiltonian $\hat{H}$ is written in the interaction picture under the rotating wave approximation [78]. In the Hamiltonian (3.7), $V = -\wp_{32}E_{\mathrm{c}}$ is the electric-dipole allowed transition interaction potential energy; The parameter δ, which is $\delta = \omega_{32} - \omega_{\mathrm{c}}$, is the frequency detuning (ω_{32} is the $|2\rangle$-$|3\rangle$ transition frequency and ω_{c} is the coupling light frequency). Since the applied coupling field intensity E_{c} is quite strong, that is to say, $|V| \gg \hbar\Gamma$, where Γ is the decay rate because of spontaneous emission of atomic upper energy level, we ignore the effect of spontaneous decay in the bare-state system $\{|2\rangle, |3\rangle\}$ [73].

By using the unitary transformation method or algebraic method, the eigen states [73] (i.e., the dressed states) of the above interaction Hamiltonian (3.7) is given by [79, 80]

$$|3'\rangle = \cos\vartheta|3\rangle + \mathrm{e}^{\mathrm{i}\phi}\sin\vartheta|2\rangle, \qquad |2'\rangle = -\sin\vartheta|3\rangle + \mathrm{e}^{\mathrm{i}\phi}\cos\vartheta|2\rangle. \tag{3.8}$$

The relevant parameters $\sin\vartheta$ and $\cos\vartheta$ are of the form

$$\sin^2\vartheta = \frac{1}{2}\left(1 - \frac{1}{\sqrt{4\frac{\Omega_{\mathrm{c}}^2}{\delta^2} + 1}}\right), \qquad \cos^2\vartheta = \frac{1}{2}\left(1 + \frac{1}{\sqrt{4\frac{\Omega_{\mathrm{c}}^2}{\delta^2} + 1}}\right). \tag{3.9}$$

In these relations (3.9), the parameter $\phi = \arg[V]$ (argument). We choose $\phi = 0$. The Rabi frequency of the coupling field is $\Omega_{\mathrm{c}} = \frac{V}{\hbar}$. Obviously, the energy-level frequencies ω_3 and ω_2 of levels $|3\rangle$ and $|2\rangle$ are $\omega_3 = \bar{\omega} + (\omega_{\mathrm{c}} + \delta)/2$ and $\omega_2 = \bar{\omega} - (\omega_{\mathrm{c}} + \delta)/2$. Then we can obtain $\omega_3 + \omega_2 = 2\bar{\omega}$ and $\bar{\omega} = \frac{\omega_3+\omega_2}{2} = \frac{\omega_2}{2} + \frac{\bar{\omega}}{2} + \frac{\omega_{\mathrm{c}}+\delta}{4}$. Thus, we have $\bar{\omega} = \omega_2 + (\omega_{\mathrm{c}} + \delta)/2$. Now we can obtain the transition frequency in the dressed-state $|2'\rangle$-$|3'\rangle$ transition: $\omega_{3'2'} = \omega_{\mathrm{c}} + \sqrt{\delta^2 + \Omega_{\mathrm{c}}^2}$, where there is a modification (compared with the bare-level transition frequency ω_{32}), i.e., $\omega_{32} = \omega_{\mathrm{c}} + \delta$ [73]. Then the energy-level frequencies $\omega_{3'}$ and $\omega_{2'}$ of the dressed-state levels $|3'\rangle$ and $|2'\rangle$ are given by $\omega_{3'} = \bar{\omega} + \left(\omega_{\mathrm{c}} + \sqrt{\delta^2 + \Omega_{\mathrm{c}}^2}\right)/2$ and $\omega_{2'} = \bar{\omega} - \left(\omega_{\mathrm{c}} + \sqrt{\delta^2 + \Omega_{\mathrm{c}}^2}\right)/2$ [73].

Since the atomic bare states have their own pure parities and the dressed states have their mixed parities, the electric field and the magnetic field of the electromagnetic field (weak probe field) can simultaneously excite the (electric- and magnetic-dipole allowed) transitions between the atomic ground level $|1\rangle$ and the atomic mixed-parity level $|2'\rangle$ [73]. Now the interaction Hamiltonian density of the atomic system $\{|2'\rangle, |1\rangle\}$ interacting with the weak probe field is of the form [78]

$$v = -\left\{\phi_{2'}^{*}\left[e\mathbf{r}\cdot\mathbf{E}+\frac{e}{2m}\left(\mathbf{L}+2\mathbf{S}\right)\cdot\mathbf{B}\right]\phi_{1}\right\}\rho_{12'}+\text{H.C.}. \quad (3.10)$$

Here, the $\phi_{2'}$ and ϕ_1 are the wave functions of the two levels $|2'\rangle$ and $|1\rangle$; $\rho_{12'}$ is an off-diagonal density matrix element; $q=-e$ with e the electric charge of an electron; The operator H.C. stands for the Hermitian conjugation term in the Hamiltonian; The operators $\mathbf{L}$ and $\mathbf{S}$ are, respectively, the orbital and spinning angular momenta; $\mathbf{A}$ is the 3-D magnetic vector potential. The field strengths of the electric and magnetic fields of the probe light are given by $\mathbf{E}=\mathbf{e_E}\mathcal{E}\exp(-\mathrm{i}\omega t)$ and $\mathbf{B}=\mathbf{e_B}\mathcal{B}\exp(-\mathrm{i}\omega t)$ [73].

The density matrix element driven by the probe electromagnetic field is given by $\rho_{12'}=\tilde{\rho}_{12'}\exp(\mathrm{i}\omega t)$, where $\mathcal{E}$, $\mathcal{B}$ and $\tilde{\rho}_{12'}$ denote their respective envelopes of such physical quantities. We define the matrix elements of electric and magnetic dipole moments: $\mathbf{p}_{2'1}=\int_V \phi_{2'}^{*}e\mathbf{r}\phi_1\mathrm{d}V$ and $\mathbf{m}_{2'1}=\int_V \phi_{2'}^{*}\frac{e}{2m}\left(\mathbf{L}+2\mathbf{S}\right)\phi_1\mathrm{d}V$ [78]. Here, $\mathrm{d}V$ is a 3-D volume element. It can be shown by using the equation (3.8) that the relations between the electric and magnetic dipole moment elements $\mathbf{p}_{2'1}$, $\mathbf{m}_{2'1}$ in the dressed states and the bare-state pure-parity counterparts $\mathbf{p}_{31}$, $\mathbf{m}_{21}$ turns out to be in the form [73]

$$\begin{aligned}\mathbf{p}_{2'1} &= \langle 2'|e\mathbf{r}|1\rangle = -\mathbf{p}_{31}\sin\vartheta,\\ \mathbf{m}_{2'1} &= \langle 2'|\frac{e}{2m}\left(\mathbf{L}+2\mathbf{S}\right)|1\rangle = \mathbf{m}_{21}\cos\vartheta.\end{aligned} \quad (3.11)$$

This can be obtained as follows:

$$\begin{aligned}\mathbf{p}_{2'1} &= \langle 2'|e\mathbf{r}|1\rangle = \left(-\sin\vartheta\langle 3|+\mathrm{e}^{\mathrm{i}\phi}\cos\vartheta\langle 2|\right)e\mathbf{r}|1\rangle = -\sin\vartheta\langle 3|e\mathbf{r}|1\rangle,\\ \mathbf{m}_{2'1} &= \langle 2'|\frac{e}{2m}\left(\mathbf{L}+2\mathbf{S}\right)|1\rangle\\ &= \left(-\sin\vartheta\langle 3|+\mathrm{e}^{\mathrm{i}\phi}\cos\vartheta\langle 2|\right)\frac{e}{2m}\left(\mathbf{L}+2\mathbf{S}\right)|1\rangle\\ &= \mathrm{e}^{\mathrm{i}\phi}\cos\vartheta\langle 2|\left(\mathbf{L}+2\mathbf{S}\right)|1\rangle,\end{aligned} \quad (3.12)$$

where $\langle 2|e\mathbf{r}|1\rangle = 0$ (because $|1\rangle$ and $|2\rangle$ have the same parity) and $\langle 3|\left(\mathbf{L}+2\mathbf{S}\right)|1\rangle = 0$ (because $|1\rangle$ and $|3\rangle$ have the opposite parity). It follows from the equations (3.11) and (3.12) that both the induced transition electric and dipole moments in the energy pair of $|1\rangle$ and $|2'\rangle$ are no longer zero, and hence the electric- and magnetic-dipole allowed transition can simultaneously occur.

It should be noted that the length scale of the two-level system is much smaller than the wavelength of the electromagnetic wave. Thus the electric- and magnetic-dipole allowed interaction Hamiltonian matrix element of the above system can be written in the form

$$\begin{aligned} V_{\text{int}} &= -\hbar\left(\frac{p_{2'1}\mathcal{E}}{\hbar}+\frac{m_{2'1}\mathcal{B}}{\hbar}\right)\tilde{\rho}_{12'}+\text{H.C.} \\ &= -\hbar\left(\Omega^{\text{E}}+\Omega^{\text{B}}\right)\tilde{\rho}_{12'}+\text{H.C.}, \end{aligned} \tag{3.13}$$

where the optical parameters Ω^{E} and Ω^{B} are defined as $\Omega^{\text{E}} = p_{2'1}\mathcal{E}/\hbar$ and $\Omega^{\text{B}} = m_{2'1}\mathcal{B}/\hbar$, respectively. Here, the $\Omega^{\text{E}}, \Omega^{\text{B}}$ are the electric and magnetic Rabi frequencies, respectively. By following the derivation procedure in Ref. [81], one can obtain the equation of motion of the density matrix elements for the system $\{|2'\rangle, |1\rangle\}$, where the quantum mechanical Schrödinger equation has been utilized. In the present three-level system, the population of level $|1\rangle$ is pumped to the dressed level $|3'\rangle$ at the pumping rate λ. In the rotating wave approximation, the equation of motion of the atomic density matrix can then be written as

$$\dot{\rho}_{11} = -\lambda\rho_{11}+\gamma_{3'1}\rho_{3'3'}+\gamma_{2'1}\rho_{2'2'}-\frac{\text{i}}{2}\left(\Omega\rho_{12'}-\Omega^{*}\rho_{2'1}\right), \tag{3.14a}$$

$$\dot{\rho}_{2'2'} = \gamma_{3'2'}\rho_{3'3'}-\gamma_{2'1}\rho_{2'2'}+\frac{\text{i}}{2}\left(\Omega\rho_{12'}-\Omega^{*}\rho_{2'1}\right), \tag{3.14b}$$

$$\dot{\rho}_{3'3'} = \lambda\rho_{11}-\left(\gamma_{3'1}+\gamma_{3'2'}\right)\rho_{3'3'}, \tag{3.14c}$$

$$\dot{\rho}_{2'1} = -\left(\frac{\gamma_{2'1}}{2}+\text{i}\Delta\right)\rho_{2'1}+\frac{\text{i}}{2}\Omega\left(\rho_{11}-\rho_{2'2'}\right). \tag{3.14d}$$

Here the frequency detuning is $\Delta = \omega_{2'1}-\omega$ and $\Omega = \Omega^{\text{E}}+\Omega^{\text{B}}$. It should be pointed out that this density matrix equation of the optical-gain (pumped) three-level atomic system has been addressed in a previous reference of ours [78], but the exact steady solution has not been considered. Here we shall continue to consider this issue in more details.

We will solve the static density matrix element equation ($\dot{\rho}_{ij} = 0$). From the third and fourth relations in the above equation, we can obtain $\rho_{3'3'}$ and $\rho_{2'1}$:

$$\rho_{3'3'} = \frac{\lambda}{\gamma_{3'1}+\gamma_{3'2'}}\rho_{11}, \qquad \rho_{2'1} = \frac{\frac{\text{i}}{2}\Omega\left(\rho_{11}-\rho_{2'2'}\right)}{\frac{\gamma_{2'1}}{2}+\text{i}\Delta}, \tag{3.15a}$$

and

$$\frac{\mathrm{i}}{2}\left(\Omega\rho_{12'}-\Omega^{*}\rho_{2'1}\right)=\mathcal{K}\left(\rho_{11}-\rho_{2'2'}\right), \tag{3.16}$$

where the parameter $\mathcal{K}$ is given by

$$\mathcal{K}=\frac{\Omega^{*}\Omega}{4}\left(\frac{1}{\frac{\gamma_{2'1}}{2}-\mathrm{i}\Delta}+\frac{1}{\frac{\gamma_{2'1}}{2}+\mathrm{i}\Delta}\right). \tag{3.17}$$

Substitution of $\rho_{3'3'}$ in Eq. (3.15a) and Eq. (3.16) into Eq. (3.14b) yields

$$\rho_{2'2'} = \frac{\gamma_{3'2'}\rho_{3'3'}+\mathcal{K}\rho_{11}}{\gamma_{2'1}+\mathcal{K}}=\frac{1}{\gamma_{2'1}+\mathcal{K}}\left(\frac{\gamma_{3'2'}\lambda}{\gamma_{3'1}+\gamma_{3'2'}}+\mathcal{K}\right)\rho_{11}. \tag{3.18}$$

With the help of $\rho_{11}+\rho_{2'2'}+\rho_{3'3'}=1$ (population probability conservation), we can obtain the population probability ρ_{11} at the ground level:

$$\rho_{11}=\frac{1}{1+\frac{1}{\gamma_{2'1}+\mathcal{K}}\left(\frac{\gamma_{3'2'}\lambda}{\gamma_{3'1}+\gamma_{3'2'}}+\mathcal{K}\right)+\frac{\lambda}{\gamma_{3'1}+\gamma_{3'2'}}} \tag{3.19}$$

and

$$\rho_{11}-\rho_{2'2'}=\frac{1}{\gamma_{2'1}+\mathcal{K}}\left[\frac{\gamma_{2'1}\left(\gamma_{3'1}+\gamma_{3'2'}\right)-\gamma_{3'2'}\lambda}{\gamma_{3'1}+\gamma_{3'2'}}\right]\rho_{11}. \tag{3.20}$$

Then the population inversion $\rho_{11}-\rho_{2'2'}$ in the atom is given as follows:

$$\rho_{11}-\rho_{2'2'} = \frac{\gamma_{2'1}\left(\gamma_{3'1}+\gamma_{3'2'}\right)-\gamma_{3'2'}\lambda}{\gamma_{2'1}\left(\gamma_{3'1}+\gamma_{3'2'}\right)+\lambda\left(\gamma_{3'2'}+\gamma_{2'1}\right)+\left[2\left(\gamma_{3'1}+\gamma_{3'2'}\right)+\lambda\right]\mathcal{K}}. \tag{3.21}$$

The atomic electric polarizability β_{e} (due to $|1\rangle$-$|2'\rangle$ transition) and the relative electric permittivity ϵ_{r} of the present atomic vapor with atomic number density N are expressed by

$$\beta_{\mathrm{e}}=\frac{\left(\Delta+\mathrm{i}\frac{\gamma_{2'1}}{2}\right)\left(\frac{p_{2'1}p_{12'}}{\epsilon_0}+\sqrt{\frac{\mu_0}{\epsilon_0}}m_{2'1}p_{12'}\right)}{\hbar\left[\Delta^2+\left(\frac{\gamma_{2'1}}{2}\right)^2\right]}\left(\rho_{11}-\rho_{2'2'}\right),$$

$$\epsilon_{\mathrm{r}}=1+\frac{N\beta_{\mathrm{e}}}{1-\frac{N\beta_{\mathrm{e}}}{3}}, \tag{3.22}$$

where the local field effect because of the dipole-dipole interaction between the neighboring atoms has been taken into account [the second relation given in Eq. (3.22)] [41].

In the similar fashion, the atomic magnetic polarizability β_{m} (due to $|1\rangle$-$|2'\rangle$ transition) and the relative magnetic permeability μ_{r} of the atomic vapor can be given by

$$\beta_{\mathrm{m}} = \frac{\left(\Delta + \mathrm{i}\frac{\gamma_{2'1}}{2}\right)\left(\mu_0 m_{2'1} m_{12'} + \sqrt{\frac{\mu_0}{\epsilon_0}} p_{2'1} m_{12'}\right)}{\hbar\left[\Delta^2 + \left(\frac{\gamma_{2'1}}{2}\right)^2\right]}\left(\rho_{11} - \rho_{2'2'}\right),$$

$$\mu_{\mathrm{r}} = 1 + \frac{N\beta_{\mathrm{m}}}{1 - \frac{N\beta_{\mathrm{m}}}{3}}. \tag{3.23}$$

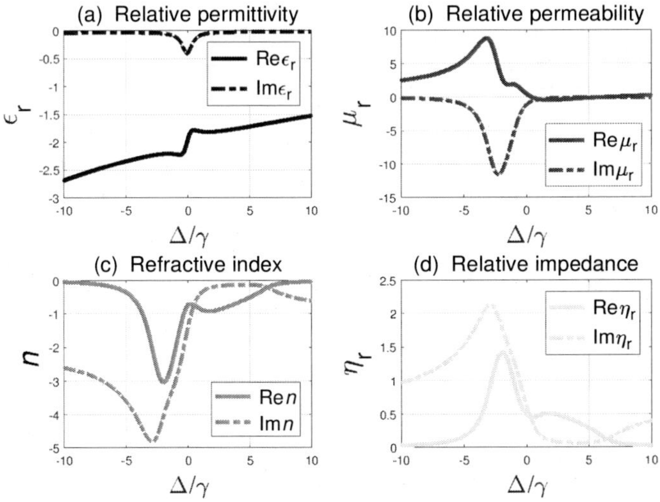

Figure 3. The dispersion behavior of the real and imaginary parts of the relative permittivity ϵ_{r}, the relative permeability μ_{r}, the refractive index n and the relative impedance η_{r} of the pumped three-level dressed-state mixed-parity atomic gaseous medium.

Now we have obtained all the observable optical parameters of the *pumped three-level dressed-state mixed-parity atomic gaseous medium*. In what follows, we present a typical numerical example to show that such a scheme could possibly exhibit a negative refractive index. The typical optical and atomic parameters of the three-level system are chosen as: the transition electric dipole moment is $p_{31} = 3.0 \times 10^{-29}$ C·m, the transition magnetic dipole moment is $m_{21} = 0.45 \times \frac{1}{137} c p_{31} = 3.0 \times 10^{-23}$ C·m^{2}s^{-1}, the frequency detuning of the coupling field is $\delta = 5.0 \times 10^8$ rad · s^{-1} and the Rabi frequency of the coupling field is $\Omega_c = -p_{ab}E_c/\hbar = -2.6 \times 10^7$ rad · s^{-1}. By using the definition (3.11), the magnitudes of $\mathbf{p}_{2'1}$ and $\mathbf{m}_{2'1}$ are given by $p_{2'1} = 1.6 \times 10^{-30}$ C·m and $m_{2'1} = 3.0 \times 10^{-23}$ C·m^{2}s^{-1}, respectively. (Since $(\frac{\Omega_c}{\delta})^2 = 2.7 \times 10^{-3}$ is negligibly small, the effective induced magnetic dipole moment $m_{2'1}$ is close to the bare induced magnetic dipole moment m_{21}.) The number density of atoms is $N = 7.0 \times 10^{22}$ m^{-3}. The decay rates are given by $\gamma_{2'1} = 1.0 \times 10^6$ rad · s^{-1}, $\gamma_{3'1} = 2.0 \times 10^7$ rad·s^{-1} and $\gamma_{3'2'} = 1.0 \times 10^7$ rad·s^{-1}; The pumping rate is $\lambda = 1.0 \times 10^8$ rad·s^{-1}. The Rabi frequency of the probe field driving the $|1\rangle$-$|2'\rangle$ dressed-state mixed transition is $\Omega = 1.0 \times 10^7$ rad · s^{-1}. The frequency parameter that is used to normalize the frequency detuning of the probe light is $\gamma = \gamma_{2'1}$.

From Fig. 3, it can be seen that the permittivity, permeability and the refractive index can have their negative real parts in certain frequency ranges of the probe light. Since the electric- and magnetic-dipole allowed transitions of atoms could be driven by visible and infrared optical fields, we expect that such a mechanism of negative refractive index by taking full advantage of *dressed-state mixed-parity transitions* in atomic vapor would emerge in certain bands of optical frequencies.

4. Surface Plasmon Polaritons and Its SPR in a Prism Coupler

Simple prism coupler systems for exciting surface plasmon polaritons include Otto and Kretschmann's prism couplers [82, 83]. These schemes can be used to detect or measure many dielectric parameters (e.g., permittivity, optical loss, thickness and so on) in metal optics [84]. The excitation of surface plasmon polaritons under the proper conditions (i.e., the phase or momentum matching on two-media planar interface and the Fröhlich condition $\epsilon_{\text{in}} = -2\epsilon_{\text{out}}$ for exciting localized surface plasmons on a metallic nanosphere) can be called "surface plasmon resonance (SPR)" [1, 2]. In this section, we shall discuss the electromagnetic field characteristic equation for surface plasmon polaritons and give an overview on some schemes of "surface plasmon resonance" with multilevel quantum-coherent atomic vapor or quantum-dot materials [15, 16, 21].

4.1. Electromagnetic Field Characteristic Equation and Surface Plasmon Polaritons in a Sandwich Structure

We shall study the characteristic equation of surface electromagnetic wave modes in a sandwich structure consisting of two media forming an interface with a conducting film on the interface.

Let us consider an interface between two optical materials with relative permittivity $\varepsilon_1, \varepsilon_2$. If an interface is coated with a conducting thin film with surface electric polarizability σ_{s}, we shall study theoretically the electromagnetic eigenmodes (confined electromagnetic field state functions) on the interface. We assume the interface is on the y-z plane and the electromagnetic confined eigenstates propagate in the direction of z axis. The media of relative permittivity ε_1 and ε_2 are in the regions of $x > 0$ and $x < 0$, respectively. The amplitude of the surface electromagnetic eigenmodes decay exponentially (i.e., $\mathrm{e}^{-\alpha_1 x}$ and $\mathrm{e}^{+\alpha_2 x}$) into both positive and negative directions of x coordinate. The electromagnetic field is uniform in the y-direction, i.e., $\frac{\partial}{\partial y}(...) = 0$. The magnetic field H_y of the TM mode is given by

$$\begin{aligned} H_{1y} &= H_{1y}^{(0)} \mathrm{e}^{-\alpha_1 x} \mathrm{e}^{\mathrm{i}(\mathrm{q}z-\omega t)} \qquad (x > 0), \\ H_{2y} &= H_{2y}^{(0)} \mathrm{e}^{+\alpha_2 x} \mathrm{e}^{\mathrm{i}(\mathrm{q}z-\omega t)} \qquad (x < 0), \end{aligned} \tag{4.1}$$

where α_1 and α_2 denote the attenuation constants. They are defined through $\alpha_1 = \sqrt{q^2 - \varepsilon_1 \frac{\omega^2}{c^2}}$ and $\alpha_2 = \sqrt{q^2 - \varepsilon_2 \frac{\omega^2}{c^2}}$. According to $\nabla \times \mathbf{H} = \frac{\partial \mathbf{D}}{\partial t}$ (note that there is no current density $\mathbf{J}$ in the equation), where $\frac{\partial \mathbf{D}}{\partial t} = -\mathrm{i}\omega\epsilon_0\varepsilon(E_x, E_y, E_z)$ and

$$\begin{aligned} \nabla \times \mathbf{H_1} &= (-\mathrm{i}qH_{1y}, 0, -\alpha_1 H_{1y}), \\ \nabla \times \mathbf{H_2} &= (-\mathrm{i}qH_{2y}, 0, +\alpha_2 H_{2y}), \end{aligned} \tag{4.2}$$

one can obtain

$$\begin{aligned}
-\mathrm{i}qH_{1y} &= -\mathrm{i}\omega\epsilon_0\varepsilon_1 E_{1x}, \quad 0 = -\mathrm{i}\omega\epsilon_0\varepsilon_1 E_{1y}, \quad -\alpha_1 H_{1y} = -\mathrm{i}\omega\epsilon_0\varepsilon_1 E_{1z}; \\
-\mathrm{i}qH_{2y} &= -\mathrm{i}\omega\epsilon_0\varepsilon_2 E_{2x}, \quad 0 = -\mathrm{i}\omega\epsilon_0\varepsilon_2 E_{2y}, \quad +\alpha_2 H_{2y} = -\mathrm{i}\omega\epsilon_0\varepsilon_2 E_{2z}.
\end{aligned} \tag{4.3}$$

Then the electric field distribution is given by

$$\begin{aligned}
E_{1x} &= \frac{qH_{1y}}{\omega\varepsilon_0\varepsilon_1}, \qquad E_{1y} = 0, \qquad E_{1z} = \frac{-\mathrm{i}\alpha_1 H_{1y}}{\omega\varepsilon_0\varepsilon_1}; \\
E_{2x} &= \frac{qH_{2y}}{\omega\varepsilon_0\varepsilon_2}, \qquad E_{2y} = 0, \qquad E_{2z} = \frac{\mathrm{i}\alpha_2 H_{2y}}{\omega\varepsilon_0\varepsilon_2}.
\end{aligned} \tag{4.4}$$

We employ the boundary condition $\mathbf{e}_n \cdot (\mathbf{D}_1 - \mathbf{D}_2) = \rho_\mathrm{s}$, i.e., $D_{1x}^{(0)} - D_{2x}^{(0)} = \rho_\mathrm{s}$, where $\mathbf{e}_n$ is the normal to the interface ($\mathbf{e}_n = \mathbf{e}_x$), and we obtain $D_{1x}^{(0)} - D_{2x}^{(0)} = \rho_\mathrm{s}$, which can be rewritten as $\varepsilon_1 E_{1x}^{(0)} - \varepsilon_2 E_{2x}^{(0)} = \frac{\rho_\mathrm{s}}{\varepsilon_0}$. By substituting $E_{1x} = \frac{qH_{1y}}{\omega\varepsilon_0\varepsilon_1}$ and $E_{2x} = \frac{qH_{2y}}{\omega\varepsilon_0\varepsilon_2}$ given in Eq. (4.4), we can have

$$\frac{q(H_{1y}^{(0)} - H_{2y}^{(0)})}{\omega} = \rho_\mathrm{s}. \tag{4.5}$$

On the other hand, the magnetic field boundary condition is $\mathbf{e}_n \times (\mathbf{H}_1 - \mathbf{H}_2) = \mathbf{J}_\mathrm{s}$, i.e., $(H_{1y}^{(0)} - H_{2y}^{(0)})\mathbf{e}_z = J_\mathrm{s}\mathbf{e}_z$ and

$$H_{1y}^{(0)} - H_{2y}^{(0)} = J_\mathrm{s}. \tag{4.6}$$

Obviously, from the equations (4.5) and (4.6), the surface electric charge density ρ_s and the surface current density $\mathbf{J}_\mathrm{s} = J_\mathrm{s}\mathbf{e}_z$ agree with the law of charge conservation $\frac{\partial\rho_\mathrm{s}}{\partial t} + \frac{\partial J_\mathrm{s}}{\partial z} = 0$ (i.e., $-\mathrm{i}\omega\rho_\mathrm{s} + \mathrm{i}qJ_\mathrm{s} = 0$). Here, the surface charge density ρ_s and the surface current density J_s are given by

$$\begin{aligned}
-\mathrm{i}\omega\rho_\mathrm{s} &= -\mathrm{i}\omega\frac{q(H_{1y}^{(0)} - H_{2y}^{(0)})}{\omega} = -\mathrm{i}q(H_{1y}^{(0)} - H_{2y}^{(0)}), \\
\mathrm{i}qJ_\mathrm{s} &= \mathrm{i}q(H_{1y}^{(0)} - H_{2y}^{(0)}).
\end{aligned} \tag{4.7}$$

The surface current density is given by $J_\mathrm{s} = \sigma_\mathrm{s} E_{1z}^{(0)}$ or $J_\mathrm{s} = \sigma_\mathrm{s} E_{2z}^{(0)}$, where σ_s is the surface electric conductivity of a conducting thin film, with which the interface is coated. It should be noted that $E_{1z}^{(0)} = E_{2z}^{(0)}$ (boundary condition).

We shall discuss the relation between the surface electric conductivity σ_{s} and the volume electric conductivity σ_{v}: The volume current density is $J_{\mathrm{v}} = \frac{I}{ab}$, or $I = J_{\mathrm{v}}ab$, where ab is the area of the cross section (rectangle) of the wire. If the thickness b of the conducting material is sufficiently small and it cannot be detected temporarily (or it is not necessary to be detected), the physically significant quantity is the surface current density rather $\mathbf{J}_{\mathrm{s}}$ than the volume current density $\mathbf{J}_{\mathrm{v}}$. The surface current density is given by $J_{\mathrm{s}} = \frac{I}{a} = J_{\mathrm{v}}b$. The Ohm microscopic form is $J_{\mathrm{v}} = \sigma_{\mathrm{v}}E$ (ordinary form). Now it can be rewritten as $J_{\mathrm{s}} = J_{\mathrm{v}}b = (\sigma_{\mathrm{v}}E)b = (\sigma_{\mathrm{v}}b)E$, i.e., $J_{\mathrm{s}} = \sigma_{\mathrm{s}}E$ (Ohm microscopic form), where the surface electric conductivity (of the thin layer) is $\sigma_{\mathrm{s}} = \sigma_{\mathrm{v}}b$.

By using $E_{1z}^{(0)} = E_{2z}^{(0)}$ and $J_{\mathrm{s}} = \sigma_{\mathrm{s}}E_{1z}^{(0)}$, from the equation (4.4), one can arrive at $H_{2y}^{(0)} = -\frac{\alpha_1\varepsilon_2}{\alpha_2\varepsilon_1}H_{1y}^{(0)}$ and $J_{\mathrm{s}} = \sigma_{\mathrm{s}} \cdot \left(\frac{-\mathrm{i}\alpha_1 H_{1y}^{(0)}}{\omega\varepsilon_0\varepsilon_1}\right)$; From the equation (4.6), $H_{1y}^{(0)} - H_{2y}^{(0)} = J_{\mathrm{s}}$, we have $(1 + \frac{\alpha_1\varepsilon_2}{\alpha_2\varepsilon_1})H_{1y}^{(0)} = \sigma_{\mathrm{s}} \cdot \left(\frac{-\mathrm{i}\alpha_1 H_{1y}^{(0)}}{\omega\varepsilon_0\varepsilon_1}\right)$. Then we can obtain the characteristic equation

$$1 + \frac{\alpha_1\varepsilon_2}{\alpha_2\varepsilon_1} = \frac{-\mathrm{i}\sigma_{\mathrm{s}}\alpha_1}{\omega\varepsilon_0\varepsilon_1}. \tag{4.8}$$

Alternatively, from the equation (4.4), we can have $H_{1y}^{(0)} = -\frac{\alpha_2\varepsilon_1}{\alpha_1\varepsilon_2}H_{2y}^{(0)}$ and $J_{\mathrm{s}} = \sigma_{\mathrm{s}} \cdot \left(\frac{\mathrm{i}\alpha_2 H_{2y}^{(0)}}{\omega\varepsilon_0\varepsilon_2}\right)$. Then from the equation (4.6), $H_{1y}^{(0)} - H_{2y}^{(0)} = J_{\mathrm{s}}$, we can obtain $-(1 + \frac{\alpha_2\varepsilon_1}{\alpha_1\varepsilon_2})H_{2y}^{(0)} = \sigma_{\mathrm{s}} \cdot \left(\frac{\mathrm{i}\alpha_2 H_{2y}^{(0)}}{\omega\varepsilon_0\varepsilon_2}\right)$, i.e.,

$$-\left(1 + \frac{\alpha_2\varepsilon_1}{\alpha_1\varepsilon_2}\right) = \frac{\mathrm{i}\sigma_{\mathrm{s}}\alpha_2}{\omega\varepsilon_0\varepsilon_2}. \tag{4.9}$$

Clearly, the result (4.9) is equivalent to the other result (4.8), because $E_{1z}^{(0)} = E_{2z}^{(0)}$ and $J_{\mathrm{s}} = \sigma_{\mathrm{s}}E_{1z}^{(0)} = \sigma_{\mathrm{s}}E_{2z}^{(0)}$.

The result (4.8) and (4.9) can be rewritten as

$$\alpha_2\varepsilon_1 + \alpha_1\varepsilon_2 + \frac{\mathrm{i}\sigma_{\mathrm{s}}\alpha_1\alpha_2}{\omega\varepsilon_0} = 0. \tag{4.10}$$

If the surface conductivity $\sigma_{\mathrm{s}} = 0$, it reduces to the form $1 + \frac{\alpha_1\varepsilon_2}{\alpha_2\varepsilon_1} = 0$, where if $\varepsilon_1 > 0$, $\varepsilon_2 < 0$ and $-\varepsilon_2 > \varepsilon_1$, the electromagnetic eigenstates are referred to as the "surface plasmon polaritons" (SPPs) [1, 2].

In what follows, we shall discuss the electromagnetic boundary conditions related to the present surface plasmon polaritons.

In the Maxwell equations, according to the gauge field theoretical viewpoints, $\nabla\cdot\mathbf{D}=\rho_{\mathrm{f}}$ and $\nabla\times\mathbf{H}=\frac{\partial\mathbf{D}}{\partial t}+\mathbf{J}_{\mathrm{f}}$ are the principal equations (corresponding to $\partial_{\mu}F^{\mu\nu}=J^{\nu}$ with $F^{\mu\nu}=\partial^{\mu}A^{\nu}-\partial^{\nu}A^{\mu}$), whereas $\nabla\cdot\mathbf{B}=0$ and $\nabla\times\mathbf{E}=-\frac{\partial\mathbf{B}}{\partial t}$ are the Bianchi identities (corresponding to $\partial_{\mu}\tilde{F}^{\mu\nu}=0$ with dual tensor $\tilde{F}^{\mu\nu}=\frac{1}{2}\epsilon^{\mu\nu}{}_{\alpha\beta}F^{\alpha\beta}$; $\epsilon^{\mu\nu}{}_{\alpha\beta}$ is the Levi-Civita tensor).

In a conductor, by substituting $\mathbf{J}_{\mathrm{f}}=\sigma_{\mathrm{v}}\mathbf{E}$ into $\nabla\times\mathbf{H}=\frac{\partial\mathbf{D}}{\partial t}+\mathbf{J}_{\mathrm{f}}$, the effective relative permittivity is given by $\varepsilon_{\mathrm{eff}}=\bar{\varepsilon}+\frac{\sigma_{\mathrm{v}}}{-\mathrm{i}\omega\varepsilon_{0}}$. On the other hand, according to the conservation law, we have $\nabla\cdot\mathbf{J}_{\mathrm{f}}+\frac{\partial\rho_{\mathrm{f}}}{\partial t}=0$, i.e., $\sigma_{\mathrm{v}}\nabla\cdot\mathbf{E}-\mathrm{i}\omega\rho_{\mathrm{f}}=0$ and $\rho_{\mathrm{f}}=\frac{\sigma_{\mathrm{v}}}{\mathrm{i}\omega}\nabla\cdot\mathbf{E}$. The Gauss law is $\nabla\cdot\mathbf{D}=\rho_{\mathrm{f}}$, which is $\bar{\varepsilon}\varepsilon_{0}\nabla\cdot\mathbf{E}=\frac{\sigma_{\mathrm{v}}}{\mathrm{i}\omega}\nabla\cdot\mathbf{E}$. It can be rearranged as $(\bar{\varepsilon}+\frac{\sigma_{\mathrm{v}}}{-\mathrm{i}\omega\varepsilon_{0}})\varepsilon_{0}\nabla\cdot\mathbf{E}=0$, i.e., $\varepsilon_{\mathrm{eff}}\varepsilon_{0}\nabla\cdot\mathbf{E}=0$ or $\nabla\cdot\mathbf{D}_{\mathrm{eff}}=0$. Here, one can also obtain the effective permittivity $\varepsilon_{\mathrm{eff}}=\bar{\varepsilon}+\frac{\sigma_{\mathrm{v}}}{-\mathrm{i}\omega\varepsilon_{0}}$ that has been obtained in the Ampère circuital law in the above.

According to the Bianchi identities, the electric and magnetic induction fields are given by $\mathbf{E}=-\nabla V-\frac{\partial\mathbf{A}}{\partial t}$ and $\mathbf{B}=\nabla\times\mathbf{A}$. It should be noted that $\mathbf{E}$ and $\mathbf{B}$ are the fundamental quantities, whereas the electric displacement vector $\mathbf{D}$ and the magnetic field $\mathbf{H}$ are the quantities that are not *definite*, which depend on the definitions of permittivity and permeability. For example, there are two definitions of the permittivity (i.e., the original permittivity $\bar{\varepsilon}$ and the effective permittivity $\varepsilon_{\mathrm{eff}}=\bar{\varepsilon}+\frac{\sigma_{\mathrm{v}}}{-\mathrm{i}\omega\varepsilon_{0}}$ that includes the conduction current contribution), and $\bar{\varepsilon}\epsilon_{0}\mathbf{E}$ and $\varepsilon_{\mathrm{eff}}\epsilon_{0}\mathbf{E}$ are different from each other. In other words, $\mathbf{D}$ and $\mathbf{H}$ can be *redefined*, if we adopt the effective dynamical permittivity $\varepsilon_{\mathrm{eff}}$ instead of the original permittivity $\bar{\varepsilon}$.

For the first, we use the effective permittivity, where the contribution of the current density $\mathbf{J}_{\mathrm{f}}$ and the charge density ρ_{f} have been combined into the effective permittivity. In this case, the boundary condition is $D_{1x}^{(0)}-D_{2x}^{(0)}=0$, where $D_{2x}^{(0)}=(\bar{\varepsilon}+\frac{\sigma_{\mathrm{v}}}{-\mathrm{i}\omega\varepsilon_{0}})\varepsilon_{0}E_{2x}^{(0)}$. Thus, we have

$$D_{1x}^{(0)}-\bar{\varepsilon}\varepsilon_{0}E_{2x}^{(0)}=-\frac{\sigma_{\mathrm{v}}}{\mathrm{i}\omega}E_{2x}^{(0)}\quad\Rightarrow\quad D_{1x}^{(0)}-\bar{D}_{2x}^{(0)}=\rho_{\mathrm{s}},\tag{4.11}$$

where $\bar{D}_{2x}^{(0)}=\bar{\varepsilon}\varepsilon_{0}E_{2x}^{(0)}$ and the surface electric charge density is given by

$$\rho_{\mathrm{s}}=-\frac{\sigma_{\mathrm{v}}}{\mathrm{i}\omega}E_{2x}^{(0)}.\tag{4.12}$$

In the similar manner, by using $E_{1x} = \frac{qH_{1y}}{\omega\varepsilon_0\varepsilon_1}$ and $E_{2x} = \frac{qH_{2y}}{\omega\varepsilon_0\varepsilon_{\text{eff}}}$ given in Eq. (4.4), we discuss the boundary condition $0 = H_{1y}^{(0)} - H_{2y}^{(0)}$, where there is no surface current density. The result is given by

$$\begin{aligned} 0 &= H_{1y}^{(0)} - H_{2y}^{(0)} = \frac{\omega\varepsilon_0}{q}(\varepsilon_1 E_{1x}^{(0)}) - \frac{\omega\varepsilon_0}{q}(\varepsilon_{\text{eff}} E_{2x}^{(0)}) \\ &= \frac{\omega\varepsilon_0}{q}(\varepsilon_1 E_{1x}^{(0)}) - \frac{\omega\varepsilon_0}{q}(\bar{\varepsilon} E_{2x}^{(0)}) - \frac{\omega\varepsilon_0}{q}\left(\frac{\sigma_{\text{v}}}{-\mathrm{i}\omega\varepsilon_0} E_{2x}^{(0)}\right). \quad (4.13) \end{aligned}$$

Now we can define a surface electric current density

$$\begin{aligned} J_{\text{s}} &= H_{1y}^{(0)} - \bar{H}_{2y}^{(0)} = \frac{\omega\varepsilon_0}{q}(\varepsilon_1 E_{1x}^{(0)}) - \frac{\omega\varepsilon_0}{q}(\bar{\varepsilon} E_{2x}^{(0)}) \\ &= \frac{\omega\varepsilon_0}{q}\left(\frac{\sigma_{\text{v}}}{-\mathrm{i}\omega\varepsilon_0} E_{2x}^{(0)}\right), \quad (4.14) \end{aligned}$$

i.e.,

$$J_{\text{s}} = \frac{\sigma_{\text{v}}}{-\mathrm{i}q} E_{2x}^{(0)}. \quad (4.15)$$

Let us check whether the law of charge conservation $\frac{\partial \rho_{\text{s}}}{\partial t} + \frac{\partial J_{\text{s}}}{\partial z} = 0$ (i.e., $-\mathrm{i}\omega\rho_{\text{s}} + \mathrm{i}qJ_{\text{s}} = 0$) holds true or not. Here, by substituting the surface electric charge density ρ_{s} given in Eq. (4.12) and the surface electric current density (4.15) into $-\mathrm{i}\omega\rho_{\text{s}} + \mathrm{i}qJ_{\text{s}}$, we have

$$-\mathrm{i}\omega\rho_{\text{s}} + \mathrm{i}qJ_{\text{s}} = -\mathrm{i}\omega\frac{\sigma_{\text{v}}}{-\mathrm{i}\omega}E_{2x}^{(0)} + \mathrm{i}q\frac{\sigma_{\text{v}}}{-\mathrm{i}q}E_{2x}^{(0)} = 0. \quad (4.16)$$

This means that the law of charge conservation $\frac{\partial \rho_{\text{s}}}{\partial t} + \frac{\partial J_{\text{s}}}{\partial z} = 0$ holds true and the above obtained results are self-consistent.

4.2. An Overview of Total Reflection Enhancement and Surface-Plasmon-Like Wave Amplification in an Incoherent-Gain Prism Coupler

In the research history of surface plasmon polaritons and surface plasmon resonance, the effect referred to as ATR (attenuated total reflection) has been found in Otto and Kretschmann's prism couplers [82, 83]. In this working mechanism, a TM-polarized reflected light wave can exhibit a very sharp dip in its reflection spectrum if the angle of incidence on the interface between the prism bottom and a thin metallic film is larger than the total-reflection critical angle [85]. This sudden drawdown in its reflectance is due to the coupling of the incident light wave into the *surface plasmon polariton modes* (surface electromagnetic waves) [84] in the thin metallic layer (located under the prism base or bottom at $x = 0$). Such an ATR technology (and related effects) have become useful tools in determining the characteristics and relevant parameters of metals [84, 85, 86, 87]; Material properties and biological solutions and their concentrations can be measured through the surface plasmon resonance and ATR technology. Clearly, the surface plasmon resonance has been a very fundamental working principle for designing various optical devices, including optical sensors and modulators [88, 89] and biomolecular and biological solution sensors [90].

Since the surface plasmon excitation (resonance) has found many applications in electromagnetic optics (metal optics), nanophotonics and the relevant technologies [91, 92, 93], e.g., it can be utilized for designing new various micro-and nano-sized photonic devices [91, 92, 93], we shall suggest and study a scheme (incoherent-gain-assisted prism coupler) that might be experimentally feasible (i.e., the relevant main characteristics of this scheme have been suggested in a previous reference of ours [15]). Here, an incoherent-gain three-level atomic medium (e.g., alkali-metal atomic vapor or semiconductor-quantum-dot solid) plays the role of a substrate of the base of the prism coupler (see Fig. 4 for its schematic diagram). The simplest optical SPR couplers include Otto and Kretschmann type prism couplers [82, 83]. Here we adopt a Kretschmann type prism coupler, of which the theoretical description of optical wave incidence, reflection and surface plasmon resonance can be found in some monographs, e.g., Ref. [94]. In this scheme, the spectrum of the energy of an incident weak probe light wave that can be coupled into the so-called "surface-plasmon-like electromagnetic confined state modes" could be influenced by the $|1\rangle$-$|3\rangle$ pumping action. Here, this pumping action may give rise to the transfer of population probability from the atomic ground state $|1\rangle$ to the atomic upper state $|3\rangle$ [see Fig. 4(b)]. Then the so-called "*surface-plasmon-like optical field amplification*" and "enhanced total reflection" can be manipulated by such an incoherently pumped transition $|1\rangle$-$|3\rangle$ (electric-dipole allowed) of the present three-level atomic system. Note that an effect of total reflection already means the reflectance of $R = 1$.

We should interpret the physical meanings of "total reflection enhancement" or "enhanced total reflection": When the reflectance $R > 1$, we refer to this effect as "total reflection enhancement", which arises when the incidence angle of the weak probe light makes the transmitted optical field (i.e., the incident weak probe light driving the $|2\rangle$-$|3\rangle$ transition) decay exponentially as the function of distance from the interface between the quantum-coherent dielectric ($0 < x < d$) and the vacuum ($x < 0$) in Fig. 20(b) [15].

4.2.1. The Solution of the Density Matrix Equation

For the first, let us consider a three-level energy level system $\{|1\rangle, |2\rangle, |3\rangle\}$ [see the schematic diagram in Fig. 4(b)]. The two upper levels $|2\rangle, |3\rangle$ are chosen to have their opposite parities, which will lead to the electric-dipole allowed transition between them; The parity of the ground-state energy level $|1\rangle$ in the present three-level system (quantum dots or atoms) is even. This system could be found in some neutral alkali-metal atoms (such as lithium, potassium, sodium and rubidium) and in some "*artificial big atoms*" (such as semiconductor quantum dots [95]). As an illustrative example, we can choose the system $\{|1\rangle, |2\rangle, |3\rangle\}$ as $\{3^2\mathrm{S}_{\frac{1}{2}}, 3^2\mathrm{P}_{\frac{1}{2}}, 4^2\mathrm{S}_{\frac{1}{2}}\}$ in the neutral sodium atom or $\{5^2\mathrm{S}_{\frac{1}{2}}, 5^2\mathrm{P}_{\frac{1}{2}}, 4^2\mathrm{D}_{\frac{3}{2}}\}$ in the neutral rubidium atom, which can be employed in our scenario.

An optical pumping process giving rise to *incoherent population transfer* is introduced. This *incoherent population transfer* is to pump the population probability from the atomic or quantum-dot ground level $|1\rangle$ to its upper level $|3\rangle$. This would enable the alkali-metal atomic vapor or the semiconductor quantum-dot solid to exhibit the permittivity with a negative imaginary part. This is the physical meaning of the incoherent-gain amplification of the incident probe light, which excites the $|2\rangle$-$|3\rangle$ electric-dipole allowed transition in the present three-level system.

With the help of the quantum mechanical Schrödinger equation for the energy-level density matrix, the equation of the three-level system that is

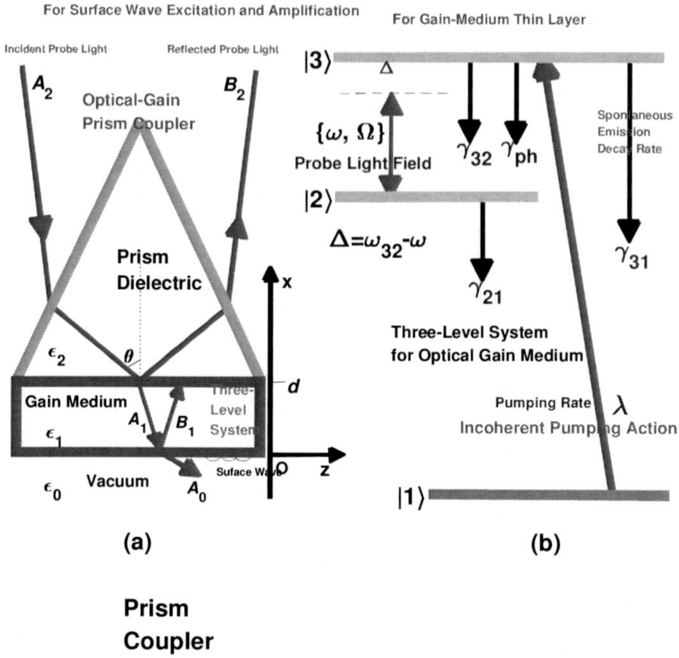

Figure 4. The schematic diagrams of an incoherent-gain assisted prism coupler (a) and a three-level incoherently pumped system (b). In (a), a weak probe light wave is incident and it is then reflected on the prism base (the interface between the prism dielectric and the three-level quantum coherent medium). A thin layer of incoherent-gain medium (e.g., a quantum-dot solid film or a three-level atomic vapor cell) is assumed to be adjacent to the prism base. The substrate (i.e., the medium 0) in the region of the coordinate $x < 0$ is the vacuum ($x < 0$) and the candidate of the incoherent-gain layer (in $0 < x < d$) can be chosen as a three-level semiconductor-quantum-dot film or an alkali metallic atomic vapor.

depicted in Fig. 4(b) can be written as

$$\begin{aligned}
\dot{\rho}_{11} &= -\lambda\rho_{11} + \gamma_{31}\rho_{33} + \gamma_{21}\rho_{22}, \\
\dot{\rho}_{22} &= -\gamma_{21}\rho_{22} + \gamma_{32}\rho_{33} - \frac{\mathrm{i}}{2}\Omega\rho_{23} + \frac{\mathrm{i}}{2}\Omega^{*}\rho_{32}, \\
\dot{\rho}_{33} &= \lambda\rho_{11} - \left(\gamma_{31} + \gamma_{32}\right)\rho_{33} + \frac{\mathrm{i}}{2}\Omega\rho_{23} - \frac{\mathrm{i}}{2}\Omega^{*}\rho_{32}, \\
\dot{\rho}_{32} &= -\left(\frac{\gamma_{31} + \gamma_{32} + \gamma_{21} + \gamma_{\mathrm{ph}}}{2} + \mathrm{i}\Delta\right)\rho_{32} + \frac{\mathrm{i}}{2}\Omega\left(\rho_{22} - \rho_{33}\right), \\
\dot{\rho}_{23} &= -\left(\frac{\gamma_{31} + \gamma_{32} + \gamma_{21} + \gamma_{\mathrm{ph}}}{2} - \mathrm{i}\Delta\right)\rho_{23} - \frac{\mathrm{i}}{2}\Omega^{*}\left(\rho_{22} - \rho_{33}\right)
\end{aligned} \tag{4.17}$$

In the above equations, the diagonal elements ρ_{11}, ρ_{22} and ρ_{33} of the density matrix obey the probability conservation, i.e., $\rho_{11} + \rho_{22} + \rho_{33} = 1$. The optical-field Rabi frequency $\Omega = \wp_{32}\mathcal{E}_{\mathrm{p}}/\hbar$, where $\wp_{32}$ and $\mathcal{E}_{\mathrm{p}}$ denote the transition electric dipole moment and the slowly-varying amplitudes of the incident optical field (the weak probe light), respectively. The decay parameters γ_{ij} $(i, j = 1, 2, 3)$ denote the energy-level spontaneous emission decay rates; γ_{ph} denotes the collisional dephasing rate (determined by the temperature or the atom number density, i..e, the total number of atoms per unit volume). The parameter λ is the pumping rate of the population probability from the atomic or quantum-dot ground level $|1\rangle$ to its upper level $|3\rangle$. The frequency detuning Δ of the incident probe light is defined as $\Delta = \omega_{32} - \omega$, where ω_{32} denotes the transition frequency between the level $|3\rangle$ and the level $|2\rangle$, and ω is the angular frequency of the incident probe optical field. Apart from the equations given in Eq. (4.17), two dynamical equations of ρ_{21} and ρ_{31} can characterize the phase coherence [7] or the quantum coherence between the level transition pathways [6]:

$$\begin{aligned} \dot{\rho}_{21} &= -\left(\frac{\gamma_{21} + \lambda + \gamma_{\mathrm{ph}}^{(21)}}{2}\right)\rho_{21}, \\ \dot{\rho}_{31} &= -\left(\frac{\gamma_{31} + \gamma_{32} + \lambda + \gamma_{\mathrm{ph}}^{(31)}}{2}\right)\rho_{31}. \end{aligned} \tag{4.18}$$

Both ρ_{21} and ρ_{31} can exponentially decay to zero value. Therefore, we need to address only the (atomic or quantum-dot system) density matrix elements ρ_{11}, ρ_{22}, ρ_{33} and ρ_{32}. With the help of the last two relations in Eq. (4.17), it can be obtained

$$\rho_{32} = \frac{\frac{\mathrm{i}}{2}\Omega\left(\rho_{22} - \rho_{33}\right)}{\gamma + \mathrm{i}\Delta}, \qquad \rho_{23} = \frac{-\frac{\mathrm{i}}{2}\Omega^*\left(\rho_{22} - \rho_{33}\right)}{\gamma - \mathrm{i}\Delta}. \tag{4.19}$$

Here, $\gamma \equiv \frac{\gamma_{31}+\gamma_{32}+\gamma_{21}+\gamma_{\mathrm{ph}}}{2}$. We substitute the results in Eq. (4.19) into the second relation in Eq. (4.17) and obtain

$$\begin{aligned} &\left[\frac{1}{4}\Omega^*\Omega\left(\frac{1}{\gamma + \mathrm{i}\Delta} + \frac{1}{\gamma - \mathrm{i}\Delta}\right) + \gamma_{21}\right]\rho_{22} \\ = &\left[\frac{1}{4}\Omega^*\Omega\left(\frac{1}{\gamma + \mathrm{i}\Delta} + \frac{1}{\gamma - \mathrm{i}\Delta}\right) + \gamma_{32}\right]\rho_{33}. \end{aligned} \tag{4.20}$$

By inserting the equation (4.19) and the relation $\rho_{11} = 1 - \rho_{22} - \rho_{33}$ into the third formula in Eq. (4.17), we can have

$$-\Big[\frac{1}{4}\Omega^*\Omega\Big(\frac{1}{\gamma+\mathrm{i}\Delta}+\frac{1}{\gamma-\mathrm{i}\Delta}\Big)+\Big(\lambda+\gamma_{31}+\gamma_{32}\Big)\Big]\rho_{33}+\lambda$$
$$+\left[\frac{1}{4}\Omega^*\Omega\Big(\frac{1}{\gamma+\mathrm{i}\Delta}+\frac{1}{\gamma-\mathrm{i}\Delta}\Big)-\lambda\right]\rho_{22}=0. \tag{4.21}$$

For convenience, these two relations (4.20) and (4.21) can be written in the form

$$\mathcal{A}\rho_{22}=\mathcal{B}\rho_{33}, \qquad \mathcal{C}\rho_{22}+\mathcal{D}\rho_{33}+\lambda=0. \tag{4.22}$$

The parameters $\mathcal{A}$, $\mathcal{B}$, $\mathcal{C}$ and $\mathcal{D}$ are defined by

$$\begin{aligned}
\mathcal{A} &= \frac{1}{4}\Omega^*\Omega\left(\frac{1}{\gamma+\mathrm{i}\Delta}+\frac{1}{\gamma-\mathrm{i}\Delta}\right)+\gamma_{21},\\
\mathcal{B} &= \frac{1}{4}\Omega^*\Omega\left(\frac{1}{\gamma+\mathrm{i}\Delta}+\frac{1}{\gamma-\mathrm{i}\Delta}\right)+\gamma_{32},\\
\mathcal{C} &= \frac{1}{4}\Omega^*\Omega\left(\frac{1}{\gamma+\mathrm{i}\Delta}+\frac{1}{\gamma-\mathrm{i}\Delta}\right)-\lambda,\\
\mathcal{D} &= -\left[\frac{1}{4}\Omega^*\Omega\left(\frac{1}{\gamma+\mathrm{i}\Delta}+\frac{1}{\gamma-\mathrm{i}\Delta}\right)+(\lambda+\gamma_{31}+\gamma_{32})\right]
\end{aligned} \tag{4.23}$$

The solution to the equation (4.22) can be written explicitly as

$$\rho_{33}=-\frac{\lambda\mathcal{A}}{\mathcal{C}\mathcal{B}+\mathcal{A}\mathcal{D}}, \qquad \rho_{22}=-\frac{\lambda\mathcal{B}}{\mathcal{C}\mathcal{B}+\mathcal{A}\mathcal{D}}. \tag{4.24}$$

The $\rho_{33} - \rho_{22}$ can describe the population inversion in the present system. The explicit expression of $\rho_{33} - \rho_{22}$ can be of the form

$$\rho_{33}-\rho_{22}=\frac{\lambda\left(\gamma_{21}-\gamma_{32}\right)}{\tilde{\Omega}^2+\lambda\left(\gamma_{32}+\gamma_{21}\right)+\gamma_{21}\left(\gamma_{31}+\gamma_{32}\right)}.$$
$$\tilde{\Omega}^2=\frac{1}{4}\Omega^*\Omega\left(\frac{1}{\gamma+\mathrm{i}\Delta}+\frac{1}{\gamma-\mathrm{i}\Delta}\right)\left(2\lambda+\gamma_{31}+\gamma_{21}\right). \tag{4.25}$$

When the incident optical field (driving the $|1\rangle$-$|2\rangle$ electric dipole allowed transition) is very weak, e.g., $\Omega^*\Omega$ is negligibly small, the "population inversion" $\rho_{33} - \rho_{22}$ reduces to the form

$$\rho_{33}-\rho_{22}\simeq\frac{\lambda\left(\gamma_{21}-\gamma_{32}\right)}{\lambda\left(\gamma_{32}+\gamma_{21}\right)+\gamma_{21}\left(\gamma_{31}+\gamma_{32}\right)}. \tag{4.26}$$

From this result (4.26), in order to achieve the population inversion between $|2\rangle$ and $|3\rangle$, the decay rate γ_{21} must be larger than the other decay rate γ_{32}, that is to say, the lifetime of the (atomic or quantum-dot) energy level $|3\rangle$ needs to be longer than the lifetime of the other energy level $|2\rangle$.

4.2.2. The Optical Response of the Gain-Assisted Prism Coupler System

The electric-dipole allowed $|2\rangle$-$|3\rangle$ transition is driven by the electric field of the incident probe field. The steady density matrix element ρ_{32} can describe the electric-dipole allowed transition $|2\rangle$-$|3\rangle$. The expression for the microscopic electric polarizability of the present three-level system is $\beta_e = 2\wp_{23}\rho_{32}/(\epsilon_0 \mathcal{E}_p)$. According to the quantum optics references on atom-light interaction [74], this can be rewritten as the form

$$\beta_e = \frac{2|\wp_{32}|^2}{\epsilon_0 \hbar \Omega}\rho_{32}. \tag{4.27}$$

Here, the vacuum permittivity is ϵ_0. For convenience of readers, we present a proof here:

The atomic microscopic electric polarizability is defined as $\beta_e = \frac{\mathbf{p}}{\epsilon_0 \mathbf{E}}$, where $\mathbf{p}$ is the induced electric dipole moment of a single atom and $\mathbf{E}$ is the electric field exciting the atomic level transition. The induced electric dipole moment of a single atom is $\mathbf{p} = \langle\psi|e\mathbf{r}|\psi\rangle$ with the atom energy-level state vector $|\psi\rangle = c_2|2\rangle + c_3|3\rangle$. It can be expanded as

$$\begin{aligned} \mathbf{p} &= \left(c_2^*\langle 2| + c_3^*\langle 3|\right)e\mathbf{r}\left(c_2|2\rangle + c_3|3\rangle\right) \\ &= c_2 c_3^*\langle 3|e\mathbf{r}|2\rangle + c_3 c_2^*\langle 2|e\mathbf{r}|3\rangle \\ &= \rho_{23}\vec{\wp}_{32} + \rho_{32}\vec{\wp}_{23}, \end{aligned} \tag{4.28}$$

where $\vec{\wp}_{32}$ and $\vec{\wp}_{23}$ are the transition electric dipole moments (or the electric dipole matrix elements) and ρ_{23} and ρ_{32} are the density matrix elements of the atom. In general, we calculate the atomic density matrix elements in the interaction picture rather than in the Schrödinger picture, and ρ_{23} and ρ_{32} are the slowly varying amplitudes (envelopes) of the density matrix elements. In the Schrödinger picture, the induced electric dipole moment of the single atom and the electric field strength are given by

$$\begin{aligned}\mathbf{p}(t) &= \rho_{23}\mathrm{e}^{\mathrm{i}\omega t}\vec{\wp}_{32}+\rho_{32}\mathrm{e}^{-\mathrm{i}\omega t}\vec{\wp}_{23}=\mathbf{p}_{-}(t)+\mathbf{p}_{+}(t),\\ \mathbf{E}(t) &= \mathbf{E}_0\cos(\omega t)=\frac{1}{2}\mathbf{E}_0(\mathrm{e}^{\mathrm{i}\omega t}+\mathrm{e}^{-\mathrm{i}\omega t})=\mathbf{E}_{-}(t)+\mathbf{E}_{+}(t)\end{aligned} \tag{4.29}$$

Note that $\mathbf{p}_{-}(t)$ and $\mathbf{p}_{+}(t)$ are the induced electric dipole moments corresponding to the negative and positive frequencies, respectively. Thus, the microscopic electric polarizability $\beta_{\mathrm{e}}=\frac{\mathbf{p}}{\epsilon_0\mathbf{E}}$ can be rearranged as $\beta_{\mathrm{e}}=\frac{\mathbf{p}_{+}(t)}{\epsilon_0\mathbf{E}_{+}(t)}$, of which the explicit expression is given by

$$\beta_{\mathrm{e}}=\frac{\rho_{32}\mathrm{e}^{-\mathrm{i}\omega t}\vec{\wp}_{23}}{\epsilon_0\frac{1}{2}\mathbf{E}_0\mathrm{e}^{-\mathrm{i}\omega t}}=\frac{2\rho_{32}\wp_{23}}{\epsilon_0 E_0}=\frac{2|\wp_{32}|^2}{\epsilon_0\hbar\Omega}\rho_{32}, \tag{4.30}$$

where we have used the Rabi frequency $\Omega=\frac{\wp_{32}E_0}{\hbar}$ and $E_0=\frac{\hbar\Omega}{\wp_{32}}$. In the equation (4.30), the electric dipole vector $\vec{\wp}_{23}$ is assumed to be parallel to the electric field vector $\mathbf{E}_0$ that drives the atomic level transition.

As is known, in order that the negative permittivity can be achieved, the three-level atomic vapor or quantum-dot solid should be chosen as a dense medium (the number density N of atoms or quantum dots is large), i.e., we need to take account of the local field effect [41, 43]. In this effect, we must distinguish between the applied (incident) macroscopic field and the microscopic local field acting on the atoms or quantum dots [41] when we treat how the properties of the energy-level transitions are related to the three-level electric susceptibility [41, 43]. The Clausius-Mossotti relation that can describe the connection between the macroscopic quantity (e.g., the electric permittivity ε_{r}) and the microscopic polarizability (i.e., β_{e}) is given by [41, 43]

$$\varepsilon_{\mathrm{r}}=\frac{1+\frac{2}{3}N\beta_{\mathrm{e}}}{1-\frac{1}{3}N\beta_{\mathrm{e}}}, \tag{4.31}$$

where N denotes the total number of atoms or quantum dots per unit volume.

The real part of the electric permittivity, ε_1, of the present three-level incoherent-gain medium (i.e., atomic vapor or semiconductor quantum-dot solid) could be either positive or negative. This depends on the frequency detuning Δ, which is shown in Fig. 10. Here, the frequency detuning of the incident probe light (driving the $|2\rangle$-$|3\rangle$ transition) is defined as $\Delta=$

$\omega_{32} - \omega$, where ω_{32} denotes the energy-level transition frequency and ω is the angular frequency of the incident weak probe light.

In the following numerical example, the dispersion characteristics of the reflection spectrum will be presented. The unusual and interesting effect of "*total reflection enhancement*" of the incoherent-gain assisted prism coupler system will also be addressed in the following numerical example.

4.2.3. The Reflection Spectrum of the Gain-Assisted Prism Coupler

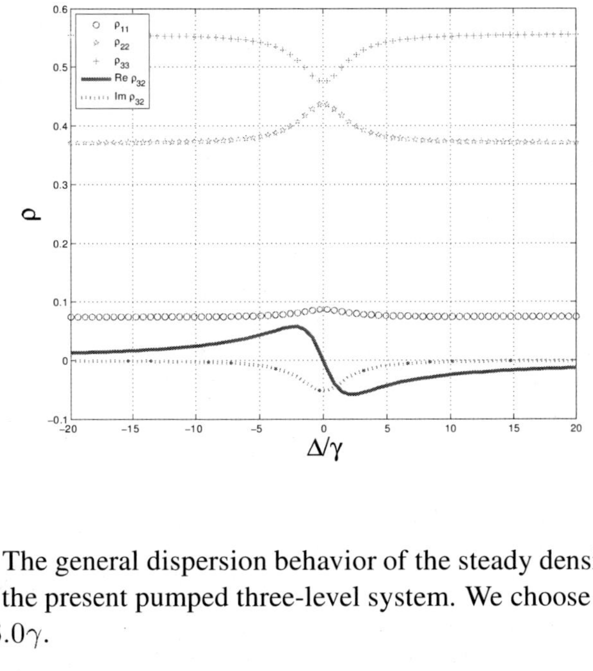

Figure 5. The general dispersion behavior of the steady density matrix elements of the present pumped three-level system. We choose the pumping rate $\lambda = 6.0\gamma$.

In the present gain-prism coupler, the thickness of the incoherent-gain medium layer (or atomic vapor cell) is chosen as $d = 50$ nm, the electric permittivity of the dielectric of prism is given by $\varepsilon_2 = 3.24$, and the wavelength of the incident light wave in vacuum is taken to be $\lambda_0 = 632.8$ nm [86, 94]. As an example, the transition electric-dipole moment of the three-level system is $\wp_{32} = 3.0 \times 10^{-29}$ C·m. Because the level $|3\rangle$ and the level $|1\rangle$ have the same parity, the induced spontaneous emission decay rate (induced by the electric-dipole allowed transi-

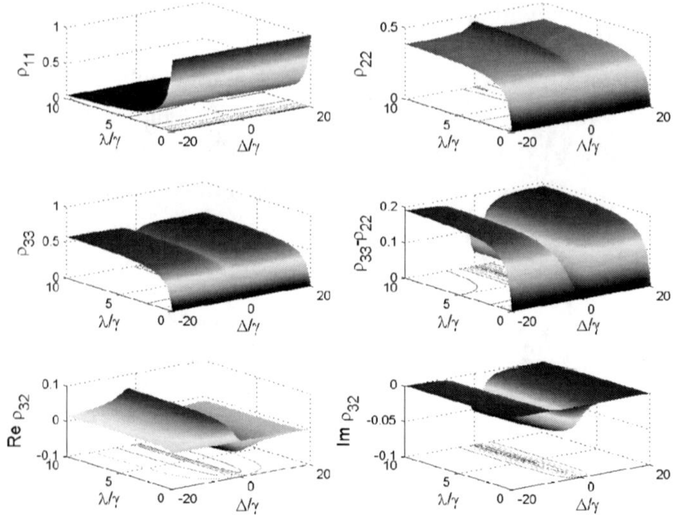

Figure 6. The general dispersion behavior of the steady density matrix elements of the present pumped three-level system versus the frequency detuning $\frac{\Delta}{\gamma}$ and the pumping rate $\frac{\lambda}{\gamma}$. The parameters are chosen the same as those used in Fig. 5.

tion) vanishes, the decay rates (due to energy-level spontaneous emission) are given by $\gamma_{31} = 0$. (The magnetic-dipole allowed transition induced spontaneous emission decay rate is negligibly small.) The other decay rates are chosen as $\gamma_{32} = 2.0 \times 10^7\ \mathrm{s}^{-1}$, $\gamma_{21} = 3.0 \times 10^7\ \mathrm{s}^{-1}$, and the collisional dephasing rate $\gamma_{\mathrm{ph}} = 2.0 \times 10^5\ \mathrm{s}^{-1}$. The γ, which is defined as $\gamma \equiv (\gamma_{31} + \gamma_{32} + \gamma_{21} + \gamma_{\mathrm{ph}})/2$, serves as a normalized parameter. In Figs. 8-10, and 12-15, γ is $2.5 \times 10^7\ \mathrm{s}^{-1}$. The pumping rate for the population transfer from the ground level $|1\rangle$ to the upper level $|3\rangle$ is $\lambda = 6.0\gamma$. The number density (concentration) of the atoms or quantum dots is $N = 2.0 \times 10^{21}$ Atoms m^{-3}. Note that all the *normalized* wave numbers (k_{0x}, k_{1x}, k_{2x} and β) demonstrated in Figs. 11-13 and Fig. 15 are normalized by the vacuum wave number ω/c ($2\pi/\lambda_0$), i.e., the vacuum wave number of the incident light wave (probe field) driving the $|2\rangle$-

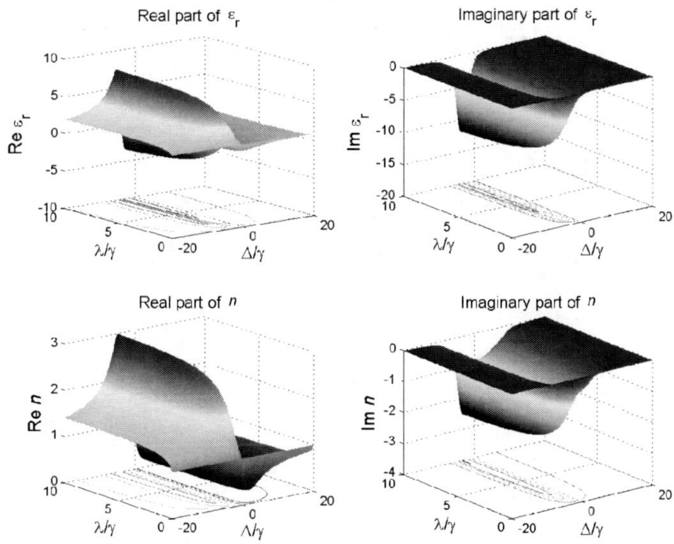

Figure 7. The general dispersion behavior of the electric permittivity and the refractive index of the present pumped three-level medium versus the frequency detuning $\frac{\Delta}{\gamma}$ and the pumping rate $\frac{\lambda}{\gamma}$. The dipole number density is $N = 2.0 \times 10^{21}\ \mathrm{m}^{-3}$. Other parameters are chosen the same as those used in Fig. 5.

$|3\rangle$ transition is the denominator in these dimensionless normalized wave numbers (after normalization, k_{1x}, k_{2x} and β are written as $\frac{k_{1x}}{\frac{\omega}{c}}$, $\frac{k_{2x}}{\frac{\omega}{c}}$ and $\frac{\beta}{\frac{\omega}{c}}$, respectively).

We solve the density matrix equations of the present pumped three-level system and obtain the behavior of the steady density matrix elements versus the frequency detuning $\frac{\Delta}{\gamma}$ and the pumping rate $\frac{\lambda}{\gamma}$. These results are presented in Figs. 5 and 6. The 3-D behavior of the electric permittivity and the refractive index of the present pumped three-level medium versus the frequency detuning $\frac{\Delta}{\gamma}$ and the pumping rate $\frac{\lambda}{\gamma}$ is shown in Fig. 7. The 2-D dispersion characteristics of the three-level permittivity ε_1 and the corresponding refractive index n of the incoherent-gain medium are given in

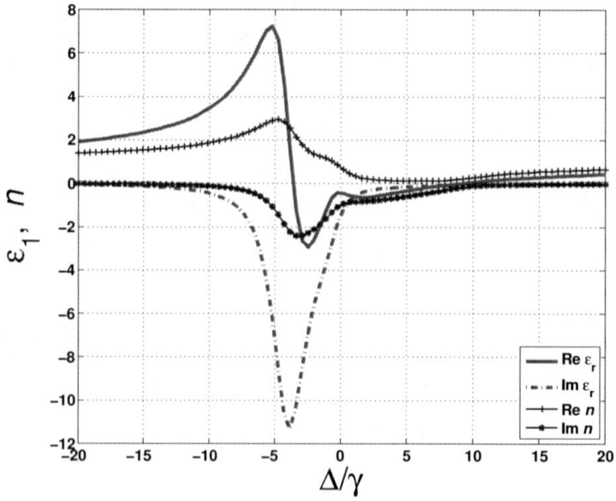

Figure 8. The typical dispersion behavior or characteristics of the electric permittivity ε_1 and the relevant refractive index n of the present incoherent-gain three-level medium if the electric Rabi frequency of the incident probe field (driving the $|2\rangle$-$|3\rangle$ transition) is chosen as $\Omega = 7.0 \times 10^7$ s^{-1}.

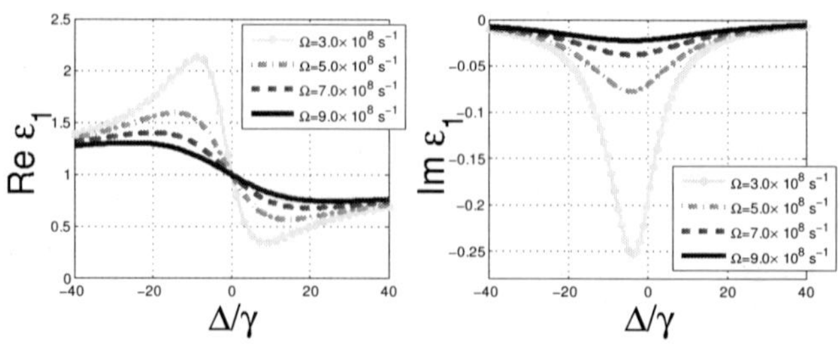

Figure 9. The dispersion characteristics of the permittivity ε_1 of the incoherent-gain medium when the electric Rabi frequency of the incident field becomes large.

Figs. 8 and 9. The electric-dipole-transition Rabi frequency of the incident optical field is $\Omega = 7.0 \times 10^7$ s^{-1} in Fig. 8. It can be found that the imaginary parts of both the permittivity and the refractive index are quite negative, e.g., the imaginary part $\mathrm{Im}n = -2.4$ at the frequency detuning $\Delta = -3.4\gamma$. But it should be pointed out that such a negative permittivity with a negative imaginary part is unstable. When the electric Rabi frequency of the incident field increases, the permittivity will be shifted into positive ranges, and the imaginary parts of both the permittivity and the refractive index will be small. For example, we plot in Fig. 9 the dispersion curves of the permittivity when the electric Rabi frequency of the incident field is chosen as $\Omega = 3.0 \times 10^8$, 5.0×10^8, 7.0×10^8, and 9.0×10^8 s^{-1}, where the imaginary parts of both the permittivity and the refractive index will become relatively small (e.g., $[-0.25, 0.0]$).

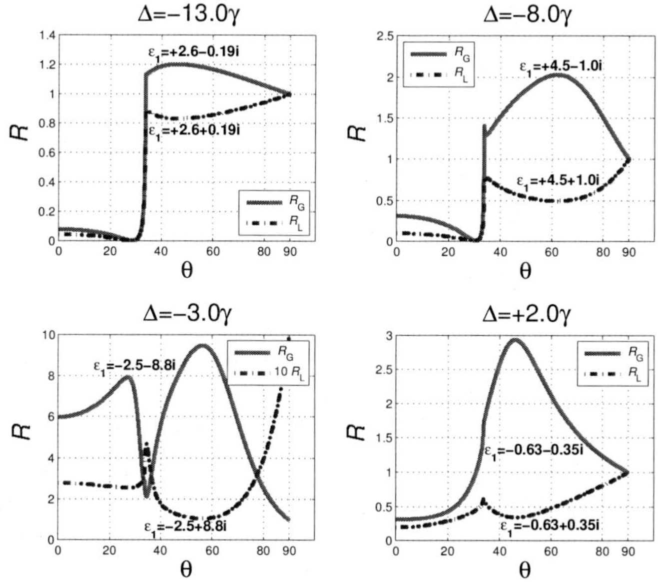

Figure 10. The tunable spectrum of reflection of the incident probe light wave in the incoherent-gain assisted prism coupler corresponding to various probe-light frequency detuning values Δ. The two reflectances (R_{G} and R_{L}), which correspond to a *gain* dielectric layer and a *lossy* dielectric layer, respectively, are given also in this figure.

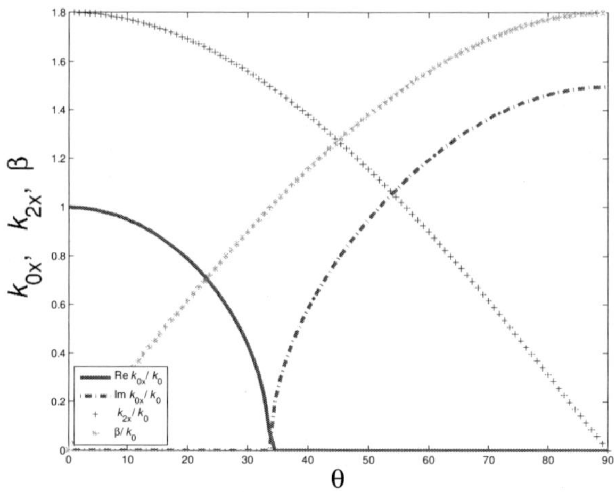

Figure 11. The wave vector components k_{0x}, k_{2x} (perpendicular to the thin incoherent-gain layer surface, i.e., the medium-substrate interface) and the phase constant β (with its direction parallel to the interface) of the incident light in the prism coupler.

The reflectance of the incident electromagnetic wave, e.g., a TM wave, is given by $R \equiv |B_2/A_2|^2$ (see Fig. 4(a)). The tunable reflection spectrum in the incoherent-gain prism coupler corresponding to various frequency detuning values Δ is presented in Fig. 10. It should be noted that the imaginary part of the relative permittivity, ε_1, of the incoherent-gain atomic medium layer is negative because of the population inversion between levels $|2\rangle$ and $|3\rangle$, i.e., $\varepsilon_1 = +2.6 - 0.19\text{i}$, $+4.5 - 1.0\text{i}$, $-2.5 - 8.8\text{i}$, and $-0.63 - 0.35\text{i}$ corresponding to the probe frequency detuning $\Delta = -13.0\gamma$, -8.0γ, -3.0γ, and $+2.0\gamma$, respectively. We also present some hypothetical but conventional cases of layer-dielectric electric permittivity ε_1 whose imaginary parts are positive for comparison in Fig. 10. It can be seen that, if the layer dielectric is a gain medium, the reflectance $R_{\text{G}} > 1$ on the prism base when the angle of incidence θ is larger than the critical angle of total reflection ($\theta_{\text{crit}} \simeq 34-36$ degrees). Whereas, when the layer dielectric is a lossy medium, the reflectance $R_{\text{L}} < 1$ on the prism base. In the cases of $\Delta = -13.0\gamma$, -8.0γ and $+2.0\gamma$, the re-

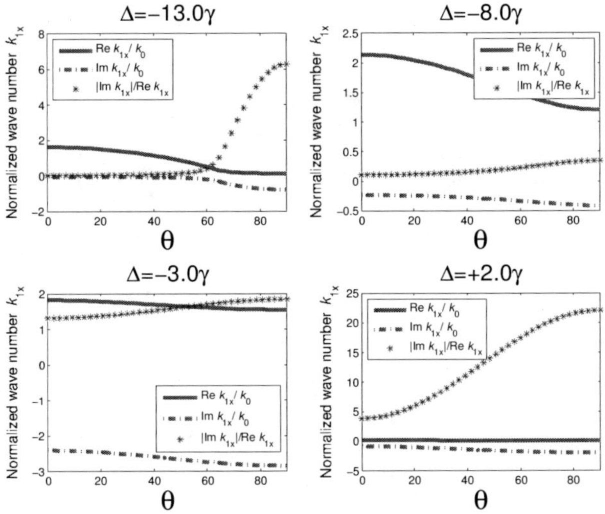

Figure 12. The wave number k_{1x} of the surface-plasmon-like modes corresponding to various frequency detuning values Δ.

flectance R_{G} increases drastically at $\theta = 34 - 36$ degrees because k_{0x} (the wave vector component in vacuum in the direction perpendicular to the interface at position $x = 0$) becomes an imaginary number (i.e., the wave number k_{0x} shown in Fig. 11 is a real number when $\theta < 34$ degrees. Whereas, it becomes an imaginary number when θ is larger than about 34 degrees), and hence the transmitted wave (through the incoherent-gain medium layer into the substrate) decreases or its power is coupled into the evanescent surface-plasmon-like wave because of surface wave excitation (at the layer-substrate interface). Then there is an effect of "resonant amplification" (after multiple internal reflection or back-and-forth beam/wave passages) in the incoherent-gain thin layer, and the amplified wave is reflected back to the prism. Therefore, the reflectance is $R_{\mathrm{G}} > 1$ on the prism base in these cases. In the case of $\Delta = -3.0\gamma$ (in Fig. 10), the imaginary part of the permittivity, ε_1, of the incoherent-gain layer is quite negative ($\mathrm{Im}\varepsilon_1 = -8.8$), and even in the normal incidence and small-angle incidence (e.g., $\theta < 30$ degrees) the reflectance R_{G} on the prism base is

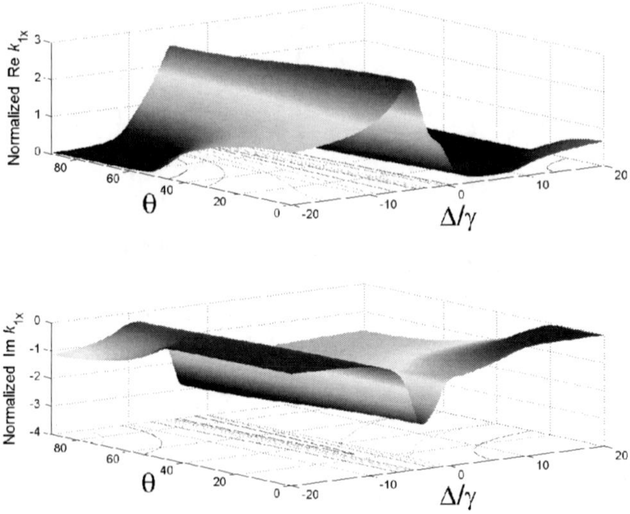

Figure 13. The wave number k_{1x} of the surface-plasma-like modes corresponding to various frequency detuning values Δ.

larger than 1. When $\theta > 30$ degrees, however, the reflectance R_G dramatically drops from 6 to 2 (still high-gain optical amplification) due to the excitation and amplification of surface-plasmon-like wave at the interface (i.e., $x = 0$) between the substrate (vacuum in the region $x < 0$) and the incoherent-gain layer (the region of $[0, d]$ in the x coordinate). When $34 < \theta < 55$ degrees, the reflectance increases again caused by the resonant enhancement in the incoherent-gain thin layer.

The imaginary part of the wave number k_{1x} in the incoherent-gain medium (in the direction perpendicular to the interface) in the case of $\Delta = -3.0\gamma$ is quite negative ($-3 < \mathrm{Im}k_{1x}/k_0 < -2$), while in the cases of $\Delta = -13.0\gamma$, -8.0γ and $+2.0\gamma$, it is relatively small, i.e., close to zero ($-1 < \mathrm{Im}k_{1x}/k_0 < 0$), as shown in Fig. 12. Therefore, the resonant enhancement (via multiple internal reflections) in the thin atomic medium layer is quite large at $\Delta = -3.0\gamma$ (this leads to large reflectance ($R_G = 6 \sim 8$) as shown in Fig. 10). The dispersion characteristics of the

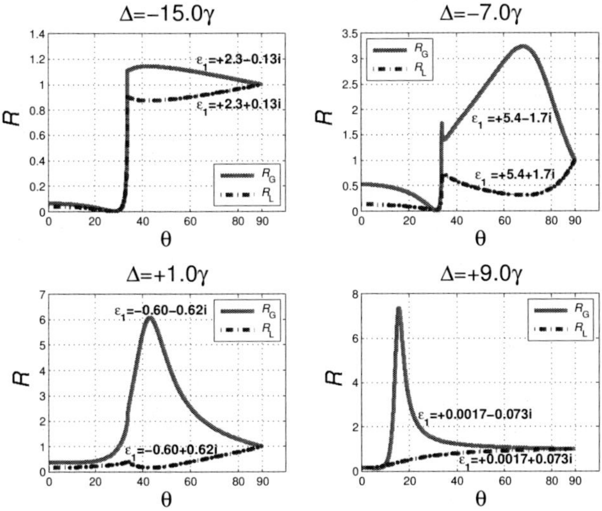

Figure 14. The spectrum of tunable reflection in this incoherent gain assisted prism coupler versus the frequency detuning values Δ. Here, the reflectance R_{G} and R_{L} correspond respectively to an optical *gain* dielectric layer and a *lossy* dielectric layer (i.e., the switchable three-level medium).

real and imaginary parts of the wave number k_{1x} of the surface-plasmon-like modes (depending upon the angle of incidence θ and the frequency detuning Δ) are shown in Fig. 13. Since the imaginary part of the wave number k_{1x} is always negative, the reflectance on the prism base will always be larger than 1 when the angle of incidence is close to the critical angle of total reflection.

In Figs. 14 and 15, we also present the reflectance and the wave number (depending upon the incidence angle) in the prism coupler corresponding to some other probe frequency detunings, e.g., $\Delta = -15.0\gamma$, -7.0γ, $+1.0\gamma$, and $+9.0\gamma$. There is also reflected wave amplification in these cases. Such a reflection enhancement effect (on the prism base) can be utilized as a fundamental mechanism of designing photonic transistors, which might have an application in some new topics in the fields such as photonic microcircuits (or integrated optical circuits).

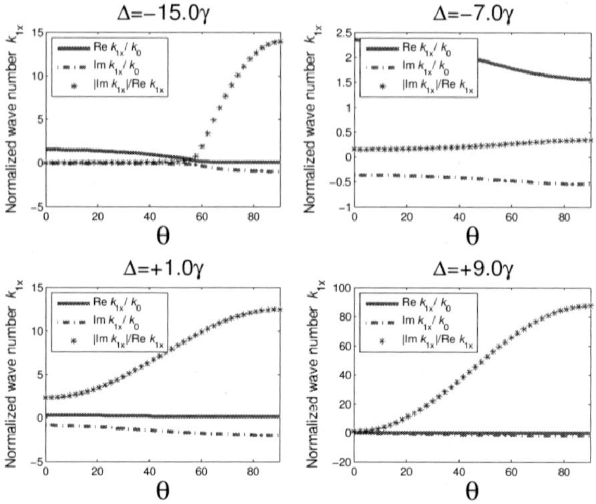

Figure 15. The wave number k_{1x} of the surface-plasmon-like modes corresponding to various frequency detuning values Δ.

4.2.4. The Spatial Field Profile in the Gain-Assisted Prism Coupler

The spatial field profile of the magnetic field is plotted in Figs. 16 and 17. There is a transmitted wave in the substrate (at $x < 0$) when $\theta = 33$ degrees. But the transmitted wave will disappear when $\theta > 34$ degrees, and the evanescent surface-plasmon-like wave will appear when $\theta > 34$ degrees. In Fig. 19, for example, we plot the simulation result of the excitation of the surface-plasmon-like wave at the interface between substrate (medium 0, i.e., vacuum) and the incoherent-gain layer when the angle of incidence θ changes. The relative permittivity of the incoherent-gain medium is $\varepsilon_1 = -2.5 - 8.8\mathrm{i}$ at the frequency detuning $\Delta = -3.0\gamma$. In order to manifest the excited modes more clear, the amplified reflected wave in the prism (in the region of $x > d$) has been removed, so that there is only the incident wave (illuminating light) present in the region of $x > d$. It can be found that the amplitude excited at the interface is larger than that of the illuminating light. This can be considered to be a visual criterion (i.e., for intuitionistic proof) of surface wave excitation. The present effect can then be utilized to enhance the interaction of evanescent surface waves with atoms, since the surface wave amplitude A_0 at the layer-substrate interface (the plane of $x = 0$) can be dramatically enhanced (as shown in Fig. 18), where the gain rate $|A_0/A_2|^2$ is about $2.0 \sim 4.1$ when the angle of incidence θ is in the range $[30, 38]$ degrees.

It should be pointed out that the dispersion in the reflectance and the surface wave parameters (e.g., k_x) is quite large (in Figs. 10-15), e.g., 10^7 times that in conventional metal-substrate prism couplers. If, for example, the frequency detuning, Δ ($\sim 10^8$ s^{-1}), changes at the level of one part in 10^7 in the frequency ($\sim 10^{15}$ s^{-1}) of the incident light (the typical value of the frequency of the applied incident light for surface wave excitation and atomic electric-dipole allowed transition has the order of magnitude of 10^{15} s^{-1}), the reflectance will change from, e.g., 2 to 8 (for an incoherent-gain medium as a thin layer substrate), i.e., it exhibits a frequency-sensitive optical characteristic (response). The physical mechanism (working principle) involved in the present dispersion-sensitive effect can be easily interpreted as follows: The conventional surface plasmon waves were generated (excited by surface plasmon resonance) at the interface between a positive-permittivity dielectric and a negative-permittivity metal [94]. The relative permittivity, $\epsilon_{\rm r} = 1 - \omega_{\rm p}^2/\omega^2$ (the optical loss is ignored), of a metal is not quite sensitive to the probe-field frequency, because the variable (frequency) appears as $1/\omega^2$ in the metal permittivity. But the atomic system or quantum dot in such a medium shown in the present work of surface plasmon resonance is governed by quantum mechanics, in which the Schrödinger equation is a first order differential equation of time t, the frequency will appear as the function $1/\Delta$, where the frequency detuning is defined as $\Delta = \omega_{32} - \omega$ appearing in the atomic electric polarizability and in the atomic vapor permittivity. Therefore, the material dispersion in the electric permittivity of the atomic medium is unusually large in certain frequency bands. As an illustrative example, we can evaluate the dispersion. For a two-level atomic vapor, the electric permittivity is $\epsilon_{\rm atom} = 1 + \frac{N\wp^2}{\epsilon_0\hbar}\frac{\rm i}{\Gamma + {\rm i}\Delta} \sim 1 + \frac{N\wp^2}{\epsilon_0\hbar}\frac{1}{\Delta}$, where, for convenience, Γ is ignored. It can be found that the atomic vapor dispersion $|\frac{1}{\epsilon_{\rm atom}}\frac{\partial\epsilon_{\rm atom}}{\partial\omega}|$ is much larger than the metal dispersion $|\frac{1}{\epsilon_{\rm r}}\frac{\partial\epsilon_{\rm r}}{\partial\omega}|$ by many orders of magnitude [15].

It should be pointed out that the results shown in Figs. 17-19 are approximate because the amplification factor of the probe light that trav-

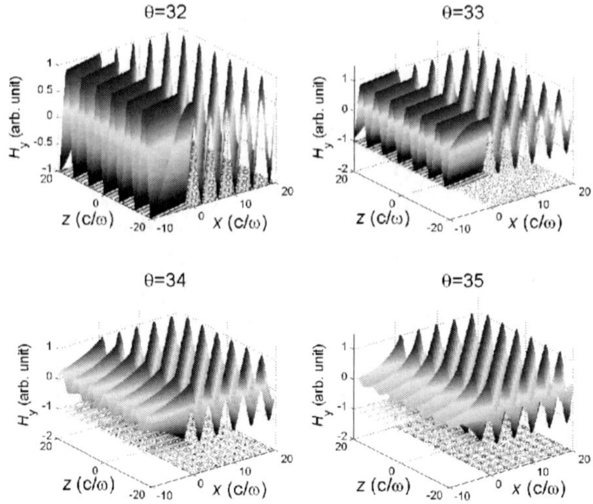

Figure 16. The surface-plasmon-like wave magnetic field profile in the present prism coupler when the angle of incidence θ of the probe light changes. The relative permittivity of the incoherent-gain three-level medium is given by $\varepsilon_1 = 2.6 - 0.19\mathrm{i}$ when the frequency detuning is chosen as $\Delta = -13.0\gamma$.

els across the thickness of the gain-medium layer $\varepsilon_1 = -2.5 - 8.8\mathrm{i}$ is $\exp\left(2\pi|\mathrm{Im}\sqrt{\varepsilon_1}|\frac{d}{\lambda_0}\right) = 3.3$, and then the permittivity of the gain-medium layer depends upon the $\hat{x}$ coordinate (in the direction of thickness of the quantum-coherent film in the region of $0 < x < d$). Since when the field intensity increases, the real part of the permittivity becomes positive (shown in Fig. 9), the effective thickness that supports the electromagnetic surface wave is less than 50 nm.

Let us conclude with some remarks on the present mechanism of "total reflection enhancement and surface-plasmon-like wave amplification in an incoherent-gain prism coupler". The incoherent-gain prism coupler, where a thin layer (just below the bottom of the prism) is a three-level medium, can exhibit three characteristic: i) surface-plasmon-like amplification of optical field and total reflection enhancement through the mechanism of

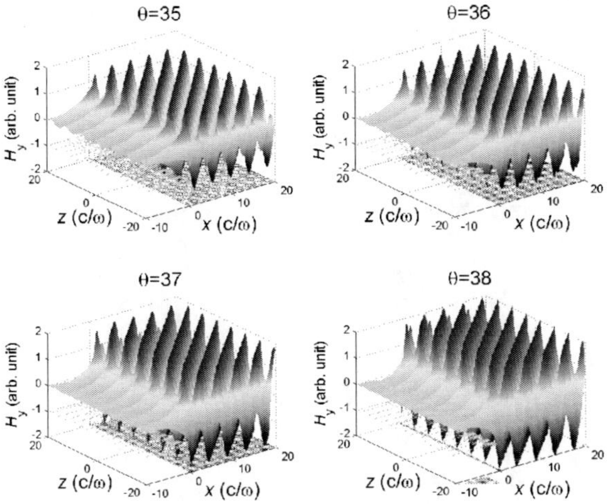

Figure 17. The magnetic field profile of the surface-plasmon-like modes (depending on the angle θ of incidence of the weak probe light) in the prism coupler. The relative permittivity of the three-level medium is chosen as $\varepsilon_1 = -2.5 - 8.8\mathrm{i}$ at the weak probe light frequency detuning $\Delta = -3.0\gamma$.

population transfer in the three-level system, ii) controllable or tunable surface electromagnetic eigen modes via the optical incoherent pumping action, iii) the surface electromagnetic wave that is very sensitive to the dispersion in this three-level quantum incoherent medium.

The tunable reflectance of the incident weak probe light on the prism base is determined by the pumping rate λ and the frequency detuning δ, and it would be easy to design and fabricate a tunable reflector, of which the reflectance of the incident optical field can be amplified (and manipulated controllably) in such an incoherent-gain assisted prism coupler. Here, the changeable frequency detuning δ can be realized with the Stark and Zeeman effects.

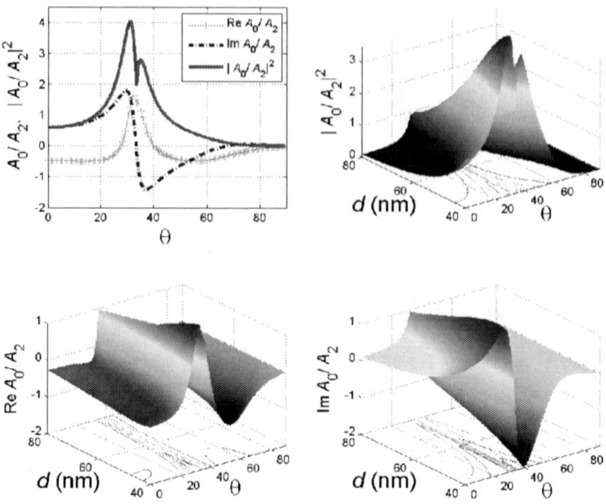

Figure 18. The effect of surface-plasmon-like optical field amplification. The surface wave amplitude A_0 in the prism coupler can be amplified when the electric permittivity of the incoherent-gain three-level medium is $\varepsilon_1 = -2.5 - 8.8\mathrm{i}$. Here, the frequency detuning of the incident optical field is chosen as $\Delta = -3.0\gamma$.

4.3. An Overview of Tunable Quantum-Interference Assisted EIT-Prism Coupler for Surface Plasmon Resonance

Over the past four decades, artificial (composite) electromagnetic or optical materials, which can manifest various novel optical properties and effects have attracted considerable or intensive attention in various scientific and technological research fields [96]. For example, with the development in applied electromagnetics, photonic technology and new quantum optical working mechanisms, scientists in a number of different new areas suggested many new ideas to manipulate lightwave propagations with artificial composite electromagnetic metamaterials [97] (and references therein). One of the technologies by using light to coherently control light is the *phase coherence control* [7] (quantum coherence) in multilevel systems (i.e., atomic vapor and semiconductor quantum dots) [56, 98, 99]. This in-

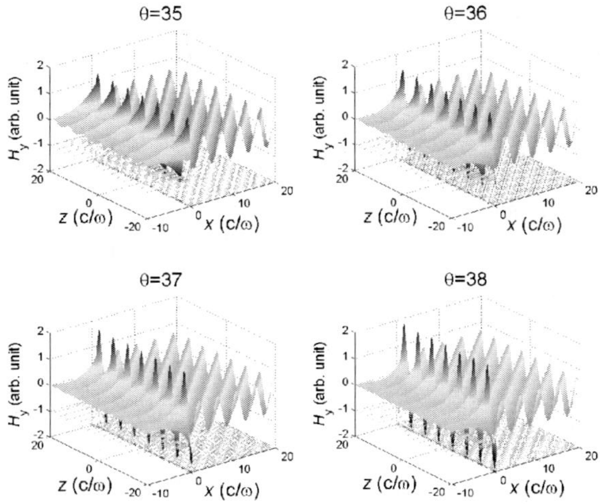

Figure 19. The surface-plasmon-like optical field excitation in the prism coupler. Here, the angle of incidence θ changes. The relative permittivity of the incoherent-gain three-level medium is $\varepsilon_1 = -2.5 - 8.8\mathrm{i}$ when the incident light frequency detuning is given by $\Delta = -3.0\gamma$. In order that the excited mode can be exhibited clearly, we remove the light wave (in the region of $x > d$) that is reflected from the prism base.

cludes many interesting effects such as electromagnetically induced transparency (EIT) [56], spontaneous emission cancellation [57], inversion-less optical field amplification [100], multi-photon population trapping [101] and quantum coherence assisted negative refractive index [73]. It should be pointed out that most of such effects can be found in three-level systems [56]. Here we shall study a four-level atomic system because when it is driven by three optical fields (i.e., two control beams and one probe optical field), it can exhibit interesting quantum *destructive* and *constructive* interference [102, 103, 104, 105] among the atomic level transition pathways [6]. An experimentally feasible configuration for EIT-prism coupler system will be considered. In this scenario, a so-called *quantum-interference assisted tunable reflection spectrum* can be coherently manipulated by external control fields when an incident probe light wave is coupled into surface electromagnetic wave modes below the bottom of the prism (see Fig. 20 for the EIT-prism schematic diagram). In this process, the quantum destructive interference arises in the atomic level transition pathways [6], which are driven by the two control fields ($\Omega_{\rm c}$ and $\Omega_{\rm c'}$).

Surface plasmon polaritons (SPPs) have captured extensive interest and attention in electromagnetics, optics, photonics, material science and biological technologies, since this electromagnetic effect can exhibit a strong coupling of fields with dielectric structures and can be utilized in designing a variety of micro- and nano-scale devices [91, 92, 93]. In the present EIT-prism coupling system (see Fig. 20 for the schematic diagram), a four-level atomic vapor would have a large imaginary part in its refractive index when the atomic transitions (i.e., $|c\rangle$-$|a\rangle$ and $|c'\rangle$-$|a\rangle$) experiences quantum destructive interference [16, 21]. In this process, an effect of surface-plasmon-like excitation at the interface ($x = 0$ plane) between the prism coupler bottom dielectric ($x < 0$) and the four-level atomic vapor cell (in the region of $0 < x < d$) will occur under certain proper conditions. Then by taking advantage of the present effect of *coherent switching control* or *atomic phase coherence control* [7], some new electromagnetic or photonic devices, including logic and functional gates and sensitively switchable devices, could be designed [16, 21]. Such sensitively quantum-interference assisted switchable devices might serve as certain fundamental building blocks in some new technologies, e.g., photonic microcircuits on silicon, where the high-speed-response light beam can replace the conventional electron motion [22].

4.3.1. The Tunable Quantum Interference between the Energy-Level Transitions in a Four-Level System

In order to indicate the tunable reflection spectrum due to *destructive and constructive quantum interference* in an EIT-prism coupling system [16, 21], here we shall for the first treat the optical properties or characteristics of a four-level alkali-metal atomic vapor, where the atomic system can be found in Fig. 20(a). These properties include the two-level and three-level dark states (superposition of atomic bare states) that lead to $|c\rangle$-$|a\rangle$ and $|c'\rangle$-$|a\rangle$ atomic-transition destructive and constructive quantum interference. These atomic transitions $|c\rangle$-$|a\rangle$ and $|c'\rangle$-$|a\rangle$ are driven by two control fields (characterized by the Rabi frequencies $\Omega_{\rm c}$ and $\Omega_{\rm c'}$).

The dispersion characteristics of the optical response in the present four-level atomic vapor will also be addressed. Based on these studies, the tunable reflection spectrum (due to the $|c\rangle$-$|a\rangle$ and $|c'\rangle$-$|a\rangle$ atomic-transition destructive and constructive quantum interference) in this EIT-prism coupling system will be presented as an illustrative example. The behavior of the reflectance depending on the angle of incidence θ, the probe frequency detuning $\delta_{\rm p}$ as well as the Rabi frequencies $\Omega_{\rm c}$ and $\Omega_{\rm c'}$ of the two control beam fields will be considered in a numerical example with emphasis on an effect of ATR (attenuated total reflection [94]) induced by the surface-plasmon-like resonance [see the schematic diagram in Fig. 20(b)]. Such a surface-plasmon-like resonance can couple the incident (probe) light wave (i.e., TM mode) into the EIT-prism interface [16, 21]. The mathematical theoretical mechanism of ATR used here is adopted from the previous reference [94].

Now we shall address the effects relevant to the quantum coherence, where the intriguing dispersion characteristics in the present double-control atomic system in Fig. 20(a) are involved [in the region of $0 < x < d$ in Fig. 20(b)]. The equation of motion of the atomic-level density matrix of the present double-control tripod-configuration atomic system [104, 105] (see Fig. 20(a) for the tripod-configuration atomic energy level configuration) will be presented below [21]. A simplified theoretical description of density matrix equation of the four-level system has been given in our previous reference [78]. Here we shall discuss this problem in more details. We assume that this four-level atomic system that can be found in alkali metallic atoms (e.g., Li, Na, K and Rb), can interact with three applied optical beams (i.e., the two control fields $\Omega_{\rm c}$, $\Omega_{\rm c'}$ and one probe field $\Omega_{\rm p}$, which drives the electric-dipole allowed transitions of level pairs $|c\rangle$-$|a\rangle$, $|c'\rangle$-$|a\rangle$ and $|b\rangle$-$|a\rangle$, respectively). The Rabi frequencies of the three optical fields are given by $\Omega_{\rm c} = \wp_{cb}E_{\rm c}/\hbar$, $\Omega_{\rm c'} = \wp_{c'b}E_{\rm c'}/\hbar$, and $\Omega_{\rm p} = \wp_{ab}E_{\rm p}/\hbar$, where $\wp_{cb}$, $\wp_{c'b}$ and $\wp_{ab}$ denote the transition electric dipole moments for $|c\rangle$-$|b\rangle$, $|c'\rangle$-$|b\rangle$ and $|a\rangle$-$|b\rangle$; $E_{\rm c}$, $E_{\rm c'}$ and $E_{\rm p}$ are the envelopes of electric fields, which are the slowly varying amplitudes of the two strong control fields ($\Omega_{\rm c}$, $\Omega_{\rm c'}$) and the weak probe light ($\Omega_{\rm p}$), respectively. The frequency detunings $\delta_{\rm c}$, $\delta_{\rm c'}$ and $\delta_{\rm p}$ of these three optical beams are given by $\delta_{\rm c} = \omega_{\rm c} - \omega_{ac}$, $\delta_{\rm c'} = \omega_{\rm c'} - \omega_{ac'}$, and $\delta_{\rm p} = \omega_{\rm p} - \omega_{ab}$. Here, ω_{ac}, $\omega_{ac'}$ and ω_{ab} represent the transition frequencies of the corresponding atomic energy-level pairs; $\omega_{\rm c}$, $\omega_{\rm c'}$, $\omega_{\rm p}$ denote the angular frequencies of the

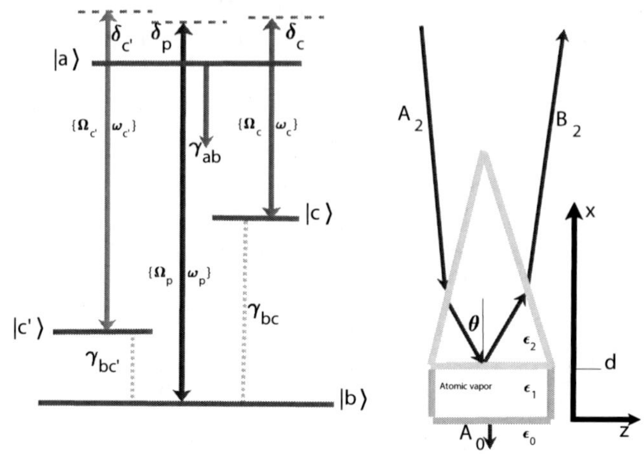

(a) Four-level atomic system (b) Kretschmann prism coupler

Figure 20. The schematic diagrams of a four-level atomic system (a) and an EIT-prism coupler (b). The four-level atomic system is driven by two strong control fields $\Omega_{\rm c}$, $\Omega_{\rm c'}$ and one weak probe light field $\Omega_{\rm p}$. In (a), the $|c\rangle$-$|a\rangle$ and $|c'\rangle$-$|a\rangle$ electric-dipole allowed transitions are excited by two control fields. The $|b\rangle$-$|a\rangle$ transition, which is driven by the incident probe optical field, can be adjusted through the quantum destructive and constructive interference between the electric-dipole allowed transitions (such as the $|c\rangle$-$|a\rangle$ and $|c'\rangle$-$|a\rangle$ atomic level transitions). If the quantum destructive interference between the transition pathways $|c\rangle$-$|a\rangle$ and $|c'\rangle$-$|a\rangle$ arises, it seems that the two atomic energy levels $|c\rangle$ and $|c'\rangle$ were not present. In this case, the atomic vapor medium would be opaque to the weak probe optical field. In (b), the probe wave with the amplitude A_2 is incident into the prism and then strikes the bottom (i.e., at $x = d$) of the prism, i.e., the interface between the dielectric of prism and the four-level vapor cell, and then reflected (with the amplitude B_2) from the bottom of the prism.

two control beams and the probe light, respectively. By taking advantage of the Schrödinger equation for the atom-light interaction, the equations of motion of the diagonal density matrix elements of the four-level system can be written as

$$\begin{aligned}
\mathrm{i}\frac{\partial \rho_{aa}}{\partial t} &= -\mathrm{i}\left(\gamma_{ab}^{(\mathrm{sp})}+\gamma_{ac}^{(\mathrm{sp})}+\gamma_{ac'}^{(\mathrm{sp})}\right)\rho_{aa} \\
&\quad +\left(\frac{\Omega_{\mathrm{p}}^{*}}{2}\rho_{ab}+\frac{\Omega_{\mathrm{c}}^{*}}{2}\rho_{ac}+\frac{\Omega_{\mathrm{c}'}^{*}}{2}\rho_{ac'}\right)-\left(\frac{\Omega_{\mathrm{p}}}{2}\rho_{ba}+\frac{\Omega_{\mathrm{c}}}{2}\rho_{ca}+\frac{\Omega_{\mathrm{c}'}}{2}\rho_{c'a}\right), \\
\mathrm{i}\frac{\partial \rho_{bb}}{\partial t} &= \mathrm{i}\left(\gamma_{ab}^{(\mathrm{sp})}\rho_{aa}+\gamma_{cb}^{(\mathrm{sp})}\rho_{cc}+\gamma_{c'b}^{(\mathrm{sp})}\rho_{c'c'}\right)+\frac{\Omega_{\mathrm{p}}}{2}\rho_{ba}-\frac{\Omega_{\mathrm{p}}^{*}}{2}\rho_{ab}, \\
\mathrm{i}\frac{\partial \rho_{cc}}{\partial t} &= -\mathrm{i}\left(\gamma_{cb}^{(\mathrm{sp})}\rho_{cc}-\gamma_{ac}^{(\mathrm{sp})}\rho_{aa}-\gamma_{c'c}^{(\mathrm{sp})}\rho_{c'c'}\right)+\frac{\Omega_{\mathrm{c}}}{2}\rho_{ca}-\frac{\Omega_{\mathrm{c}}^{*}}{2}\rho_{ac}, \\
\mathrm{i}\frac{\partial \rho_{c'c'}}{\partial t} &= -\mathrm{i}\left(\gamma_{c'b}^{(\mathrm{sp})}\rho_{c'c'}+\gamma_{c'c}^{(\mathrm{sp})}\rho_{c'c'}-\gamma_{ac'}^{(\mathrm{sp})}\rho_{aa}\right)+\frac{\Omega_{\mathrm{c}'}}{2}\rho_{c'a}-\frac{\Omega_{\mathrm{c}'}^{*}}{2}\rho_{ac'}, \qquad (4.32)
\end{aligned}$$

and the equations of the off-diagonal density matrix elements are

$$\begin{aligned}
\mathrm{i}\frac{\partial \rho_{ab}}{\partial t} &= -\mathrm{i}\left(\gamma_{ab}-\mathrm{i}\delta_{\mathrm{p}}\right)\rho_{ab}+\frac{\Omega_{\mathrm{p}}}{2}\rho_{aa}-\left(\frac{\Omega_{\mathrm{p}}}{2}\rho_{bb}+\frac{\Omega_{\mathrm{c}}}{2}\rho_{cb}+\frac{\Omega_{\mathrm{c}'}}{2}\rho_{c'b}\right), \\
\mathrm{i}\frac{\partial \rho_{ac}}{\partial t} &= -\mathrm{i}\left(\gamma_{ac}-\mathrm{i}\delta_{\mathrm{c}}\right)\rho_{ac}+\frac{\Omega_{\mathrm{c}}}{2}\rho_{aa}-\left(\frac{\Omega_{\mathrm{p}}}{2}\rho_{bc}+\frac{\Omega_{\mathrm{c}}}{2}\rho_{cc}+\frac{\Omega_{\mathrm{c}'}}{2}\rho_{c'c}\right), \\
\mathrm{i}\frac{\partial \rho_{ac'}}{\partial t} &= -\mathrm{i}\left(\gamma_{ac'}-\mathrm{i}\delta_{\mathrm{c}'}\right)\rho_{ac'}+\frac{\Omega_{\mathrm{c}'}}{2}\rho_{aa}-\left(\frac{\Omega_{\mathrm{p}}}{2}\rho_{bc'}+\frac{\Omega_{\mathrm{c}}}{2}\rho_{cc'}+\frac{\Omega_{\mathrm{c}'}}{2}\rho_{c'c'}\right), \\
\mathrm{i}\frac{\partial \rho_{bc}}{\partial t} &= -\mathrm{i}\left[\gamma_{bc}+\mathrm{i}\left(\delta_{\mathrm{p}}-\delta_{\mathrm{c}}\right)\right]\rho_{bc}+\frac{\Omega_{\mathrm{c}}}{2}\rho_{ba}-\frac{\Omega_{\mathrm{p}}^{*}}{2}\rho_{ac}, \\
\mathrm{i}\frac{\partial \rho_{bc'}}{\partial t} &= -\mathrm{i}\left[\gamma_{bc'}+\mathrm{i}\left(\delta_{\mathrm{p}}-\delta_{\mathrm{c}'}\right)\right]\rho_{bc'}+\frac{\Omega_{\mathrm{c}'}}{2}\rho_{ba}-\frac{\Omega_{\mathrm{p}}^{*}}{2}\rho_{ac'}, \\
\mathrm{i}\frac{\partial \rho_{cc'}}{\partial t} &= -\mathrm{i}\gamma_{cc'}\rho_{cc'}+\frac{\Omega_{\mathrm{c}'}}{2}\rho_{ca}-\frac{\Omega_{\mathrm{c}}^{*}}{2}\rho_{ac'}. \qquad (4.33)
\end{aligned}$$

In the above equations, the atomic-level decay rate parameters $\gamma_{ij}^{(\mathrm{sp})}$, where i, j represent the atomic energy levels $|a\rangle$, $|b\rangle$, $|c\rangle$ and $|c'\rangle$, are mostly the atomic-level population spontaneous emission decay rates, which are caused by the atomic electric- or magnetic-dipole allowed transitions (such transitions are induced by the quantum-vacuum zero-point fluctuation modes). Some other atomic-level population decay rates γ_{ij} in Eq. (1) can be written as

$$\begin{aligned}
\gamma_{ab} &= \frac{\gamma_{ab}^{(\mathrm{sp})}+\gamma_{ac}^{(\mathrm{sp})}+\gamma_{ac'}^{(\mathrm{sp})}}{2}, \qquad \gamma_{ac} = \frac{\gamma_{ab}^{(\mathrm{sp})}+\gamma_{ac}^{(\mathrm{sp})}+\gamma_{ac'}^{(\mathrm{sp})}+\gamma_{cb}^{(\mathrm{sp})}}{2}, \\
\gamma_{ac'} &= \frac{\gamma_{ab}^{(\mathrm{sp})}+\gamma_{ac}^{(\mathrm{sp})}+\gamma_{ac'}^{(\mathrm{sp})}+\gamma_{c'b}^{(\mathrm{sp})}+\gamma_{c'c}^{(\mathrm{sp})}}{2}, \\
\gamma_{cc'} &= \frac{\gamma_{c'b}^{(\mathrm{sp})}+\gamma_{c'c}^{(\mathrm{sp})}+\gamma_{cb}^{(\mathrm{sp})}}{2}+\gamma_{cc'}^{(\mathrm{ph})}, \\
\gamma_{bc} &= \frac{\gamma_{cb}^{(\mathrm{sp})}}{2}+\gamma_{bc}^{(\mathrm{ph})}, \qquad \gamma_{bc'} = \frac{\gamma_{c'b}^{(\mathrm{sp})}+\gamma_{c'c}^{(\mathrm{sp})}}{2}+\gamma_{bc'}^{(\mathrm{ph})}. \qquad (4.34)
\end{aligned}$$

In the above, $\gamma_{bc}^{(\mathrm{ph})}$, $\gamma_{bc'}^{(\mathrm{ph})}$ and $\gamma_{cc'}^{(\mathrm{ph})}$ denote the collisional dephasing rates (proportional to the atomic number density and the gas temperature). It should be noted that the atomic level population decay rates in Eq. (4.34) are the general expressions. As the contribution of the magnetic-dipole allowed transitions can be negligibly small, we focus only on the electric-dipole allowed transitions, i.e., we can have $\gamma_{c'c}^{(\mathrm{sp})} = \gamma_{c'b}^{(\mathrm{sp})} = \gamma_{cb}^{(\mathrm{sp})} = 0$, where the superscript "sp" stands for "spontaneous emission decay rates". Then, the decay rates γ_{bc}, $\gamma_{bc'}$ and $\gamma_{cc'}$ of the present atomic system are determined by the dephasing rates (that is to say, $\gamma_{bc} = \gamma_{bc}^{(\mathrm{ph})}$, $\gamma_{bc'} = \gamma_{bc'}^{(\mathrm{ph})}$, and $\gamma_{cc'} = \gamma_{cc'}^{(\mathrm{ph})}$).

According to the above equation of motion of the atomic-level density matrix, the off-diagonal elements $\rho_{cb}, \rho_{c'b}, \rho_{ab}$, which are related to the atomic ground state $|b\rangle$, of this atomic system appears to be of the form

$$\begin{aligned}
\dot{\rho}_{cb} &= -\left[\gamma_{bc} - \mathrm{i}\left(\delta_{\mathrm{p}} - \delta_{\mathrm{c}}\right)\right]\rho_{cb} + \frac{\mathrm{i}}{2}\Omega_{\mathrm{c}}^{*}\rho_{ab}, \\
\dot{\rho}_{c'b} &= -\left[\gamma_{bc'} - \mathrm{i}\left(\delta_{\mathrm{p}} - \delta_{\mathrm{c'}}\right)\right]\rho_{c'b} + \frac{\mathrm{i}}{2}\Omega_{\mathrm{c'}}^{*}\rho_{ab}, \\
\dot{\rho}_{ab} &= -\left(\gamma_{ab} - \mathrm{i}\delta_{\mathrm{p}}\right)\rho_{ab} + \frac{\mathrm{i}}{2}\Omega_{\mathrm{c}}\rho_{cb} + \frac{\mathrm{i}}{2}\Omega_{\mathrm{c'}}\rho_{c'b} + \frac{\mathrm{i}}{2}\Omega_{\mathrm{p}}\rho_{bb}.
\end{aligned} \tag{4.35}$$

As the other off-diagonal density matrix elements (e.g., $\rho_{ac}, \rho_{ac'}, \rho_{cc'}$) can be negligibly small compared with $\rho_{cb}, \rho_{c'b}, \rho_{ab}$ (related intimately to the atomic ground level $|b\rangle$), their evolutional equations will not be taken into account. On the other hand, compared with the parameters (with dimension of frequency, e.g., $\gamma_{ab}, \Omega_{\mathrm{c}}$ and $\Omega_{\mathrm{c'}}$), the Rabi frequency Ω_{p} for the interaction between $|b\rangle$-$|a\rangle$ and the probe field is sufficiently small, and this means that the population probability ρ_{aa} at the atomic level $|a\rangle$ can almost vanish [74]. By making use of the third relation given in Eq. (4.35), we can obtain the steady form of ρ_{ab}:

$$\rho_{ab} = \frac{\mathrm{i}}{2}\left(\frac{\Omega_{\mathrm{c}}\rho_{cb} + \Omega_{\mathrm{c'}}\rho_{c'b} + \Omega_{\mathrm{p}}\rho_{bb}}{\gamma_{ab} - \mathrm{i}\delta_{\mathrm{p}}}\right). \tag{4.36}$$

From the first and second equations in Eq. (4.35), it follows that the following steady relations can be derived:

$$\begin{aligned}\rho_{cb} &= \frac{\mathrm{i}}{2}\left[\frac{\Omega_{\mathrm{c}}^{*}}{\gamma_{bc}-\mathrm{i}\left(\delta_{\mathrm{p}}-\delta_{\mathrm{c}}\right)}\right]\rho_{ab}, \\ \rho_{c'b} &= \frac{\mathrm{i}}{2}\left[\frac{\Omega_{\mathrm{c}'}^{*}}{\gamma_{bc'}-\mathrm{i}\left(\delta_{\mathrm{p}}-\delta_{\mathrm{c}'}\right)}\right]\rho_{ab}.\end{aligned} \tag{4.37}$$

We substitute the equation (4.37) into a *double-control* atomic-transition-pathway quantum destructive interference condition $\Omega_{\mathrm{c}}\rho_{cb}+\Omega_{\mathrm{c}'}\rho_{c'b}=0$ (between the two transitions $|c\rangle$-$|a\rangle$ and $|c'\rangle$-$|a\rangle$ driven by the two control fields Ω_{c}, $\Omega_{\mathrm{c}'}$) and obtain [78]

$$\begin{aligned}\Omega_{\mathrm{c}}\rho_{cb}+\Omega_{\mathrm{c}'}\rho_{c'b} &= \frac{\mathrm{i}}{2}\left[\frac{\Omega_{\mathrm{c}}^{*}\Omega_{\mathrm{c}}}{\gamma_{bc}-\mathrm{i}\left(\delta_{\mathrm{p}}-\delta_{\mathrm{c}}\right)}+\frac{\Omega_{\mathrm{c}'}^{*}\Omega_{\mathrm{c}'}}{\gamma_{bc'}-\mathrm{i}\left(\delta_{\mathrm{p}}-\delta_{\mathrm{c}'}\right)}\right]\rho_{ab} \\ &\simeq \frac{\mathrm{i}}{2}\left(\frac{\Omega_{\mathrm{c}}^{*}\Omega_{\mathrm{c}}}{\delta_{\mathrm{p}}-\delta_{\mathrm{c}}}+\frac{\Omega_{\mathrm{c}'}^{*}\Omega_{\mathrm{c}'}}{\delta_{\mathrm{p}}-\delta_{\mathrm{c}'}}\right)\rho_{ab}.\end{aligned} \tag{4.38}$$

As is well known, in a dilute atomic gas at room temperature, the dephasing rates $\gamma_{bc}^{(\mathrm{ph})}$ and $\gamma_{bc'}^{(\mathrm{ph})}$ can be ignored (negligibly small) because they are much less than the other decay rates (due to atomic spontaneous emission) and the Rabi frequencies, Ω_{c} and $\Omega_{\mathrm{c}'}$, of the control fields. Then if we need to maintain (respect) the condition (4.38) of the quantum destructive interference, it is necessary to impose the following constraint on the present atom-field interacting system:

$$\frac{\Omega_{\mathrm{c}}^{*}\Omega_{\mathrm{c}}}{\delta_{\mathrm{p}}-\delta_{\mathrm{c}}}+\frac{\Omega_{\mathrm{c}'}^{*}\Omega_{\mathrm{c}'}}{\delta_{\mathrm{p}}-\delta_{\mathrm{c}'}}=0. \tag{4.39}$$

Now since the so-called condition of double-control quantum destructive interference ($\Omega_{\mathrm{c}}\rho_{cb}+\Omega_{\mathrm{c}'}\rho_{c'b}=0$) has been satisfied, the off-diagonal density matrix element ρ_{ab} (driven by the probe field Ω_{p}) reduces to the well-known steady solution similar to a two-level system $\{|a\rangle, |b\rangle\}$:

$$\rho_{ab}=\frac{\mathrm{i}}{2}\left(\frac{\Omega_{\mathrm{p}}}{\gamma_{ab}-\mathrm{i}\delta_{\mathrm{p}}}\right)\rho_{bb}. \tag{4.40}$$

It can now be found that the $|c\rangle$-$|a\rangle$ and $|c'\rangle$-$|a\rangle$ destructive interference exactly eliminate the population evolution from the lower levels $|c\rangle$ and $|c'\rangle$ to the upper atomic level $|a\rangle$, that is to say, a *two-level dark state* consisting of the atomic levels $|c\rangle$ and $|c'\rangle$ forms. Clearly, such a $|c\rangle$-$|c'\rangle$ *two-level dark state* gives rise to nearly-resonant optical absorption caused by the two-level excitation pathway $|b\rangle$-$|a\rangle$. This can be shown in Eq. (4.40). This formula has exactly the same mathematical structure as that in a well-known electric-dipole allowed transition of a two-level atomic system. If, however, the atomic levels $|c\rangle$, $|c'\rangle$ and $|b\rangle$ constitute a three-level dark state, that is to say, the quantum interference condition $\Omega_{\mathrm{c}}\rho_{cb} + \Omega_{\mathrm{c'}}\rho_{c'b} + \Omega_{\mathrm{p}}\rho_{bb} = 0$ is fulfilled, from Eq. (4.36) it follows that the steady density matrix element is given by $\rho_{ab} = 0$. Under this condition, the quantum destructive interference in the following three transition pathways $|c\rangle$-$|a\rangle$, $|c'\rangle$-$|a\rangle$ and $|b\rangle$-$|a\rangle$ are exactly cancelled out, and the population probability pumped from the three lower levels $|b\rangle, |c\rangle, |c'\rangle$ to the upper level $|a\rangle$ vanishes. From the relation $\Omega_{\mathrm{c}}\rho_{cb} + \Omega_{\mathrm{c'}}\rho_{c'b} + \Omega_{\mathrm{p}}\rho_{bb} = 0$, it follows that the relation for quantum interference in the present atomic system is of the form

$$\frac{\Omega_{\mathrm{c}}\rho_{cb} + \Omega_{\mathrm{c'}}\rho_{c'b}}{\Omega_{\mathrm{p}}\rho_{bb}} = -1. \tag{4.41}$$

Therefore, the quantum constructive interference between the two atomic level transition pathways ($|c\rangle$-$|a\rangle$ and $|c'\rangle$-$|a\rangle$ excited by two control fields) occurs. The steady and transient optical responses of the four-level system have been studied in the literature [102, 103, 104, 105] (and references therein).

We shall point out that the quantum destructive interference in the three energy levels $|c\rangle$, $|c'\rangle$ and $|b\rangle$ can occur, and simultaneously there is also an effect of quantum constructive interference between $|c\rangle$-$|a\rangle$ and $|c'\rangle$-$|a\rangle$ [21]. In this case, the population probability in the upper energy level $|a\rangle$ is almost zero, and this could lead to the so-called "*double-control*" electromagnetically induced transparency (EIT) [102, 103, 104, 105]. Such an effect has been shown theoretically and reviewed in the preceding subsection.

4.3.2. General Optical Responses of the Four-Level Atomic Vapor

The general optical response behaviour of the present atom-field interacting system will be presented here. We substitute the equation (4.37) into

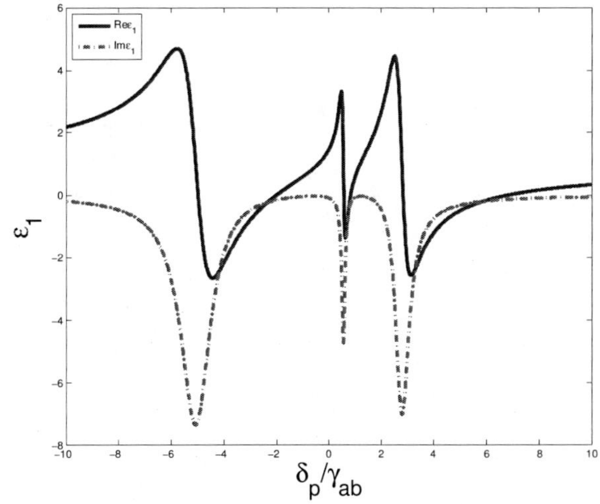

Figure 21. The dispersion property of the real part ($\mathrm{Re}\varepsilon_1$) and the imaginary part ($\mathrm{Im}\varepsilon_1$) of the four-level atomic vapor electric permittivity (in the convention of engineers). It can be found that a sensitive dispersion behavior giving rise to novel optical characteristics caused by the four-level (double-control) quantum interference among the various atomic level transition pathways can be exhibited.

the equation (4.36) and the steady solution of Eq. (4.35) can be given by

$$\begin{aligned}
\rho_{cb} &= -\frac{1}{4\mathcal{D}}\Omega_{\mathrm{p}}\Omega_{\mathrm{c}}^{*}\left[\gamma_{bc'} - \mathrm{i}\left(\delta_{\mathrm{p}} - \delta_{\mathrm{c}'}\right)\right]\rho_{bb}, \\
\rho_{c'b} &= -\frac{1}{4\mathcal{D}}\Omega_{\mathrm{p}}\Omega_{\mathrm{c}'}^{*}\left[\gamma_{bc} - \mathrm{i}\left(\delta_{\mathrm{p}} - \delta_{\mathrm{c}}\right)\right]\rho_{bb}, \\
\rho_{ab} &= \frac{\mathrm{i}}{2\mathcal{D}}\Omega_{\mathrm{p}}\left[\gamma_{bc} - \mathrm{i}\left(\delta_{\mathrm{p}} - \delta_{\mathrm{c}}\right)\right]\left[\gamma_{bc'} - \mathrm{i}\left(\delta_{\mathrm{p}} - \delta_{\mathrm{c}'}\right)\right]\rho_{bb}. \quad (4.42)
\end{aligned}$$

In the above density matrix elements, the denominator $\mathcal{D}$ is written explicitly as

$$\begin{aligned}
\mathcal{D} &= \left(\gamma_{ab} - \mathrm{i}\delta_{\mathrm{p}}\right)\left[\gamma_{bc} - \mathrm{i}\left(\delta_{\mathrm{p}} - \delta_{\mathrm{c}}\right)\right]\left[\gamma_{bc'} - \mathrm{i}\left(\delta_{\mathrm{p}} - \delta_{\mathrm{c}'}\right)\right] \\
&\quad + \frac{1}{4}\Omega_{\mathrm{c}'}^{*}\Omega_{\mathrm{c}'}\left[\gamma_{bc} - \mathrm{i}\left(\delta_{\mathrm{p}} - \delta_{\mathrm{c}}\right)\right] + \frac{1}{4}\Omega_{\mathrm{c}}^{*}\Omega_{\mathrm{c}}\left[\gamma_{bc'} - \mathrm{i}\left(\delta_{\mathrm{p}} - \delta_{\mathrm{c}'}\right)\right] \quad (4.43)
\end{aligned}$$

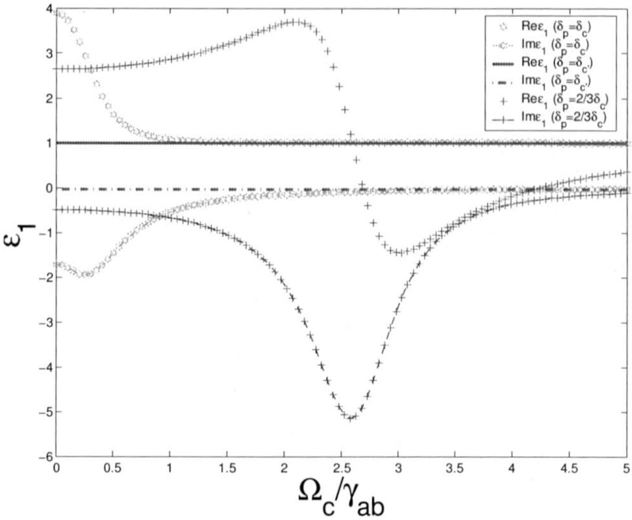

Figure 22. The dependence of the electric permittivity of the four-level EIT atomic vapor on the Rabi frequency $\Omega_{\rm c}$ of the control field. The Rabi frequency of the other control field is $\Omega_{\rm c'} = 6.0\gamma_{ab}$.

The explicit formula of the electric susceptibility (corresponding to the $|b\rangle$-$|a\rangle$ transition) of the four-level atomic vapor will be given. The electric polarizability of the $|b\rangle$-$|a\rangle$ electric-dipole allowed transition driven by the probe field is $\varrho = 2\wp_{ba}\rho_{ab}/(\epsilon_0 E_{\rm p}) = 2|\wp_{ba}|^2\rho_{ab}/(\epsilon_0\hbar\Omega_{\rm p})$. By using this definition, the quantity $N\varrho$ of the atom-field interacting system is given by [78]

$$
\begin{aligned}
N\varrho &= -\frac{N|\wp_{ab}|^2}{\epsilon_0\hbar}\frac{(\delta_{\rm p}-\delta_{\rm c}+{\rm i}\gamma_{bc})(\delta_{\rm p}-\delta_{\rm c'}+{\rm i}\gamma_{bc'})}{\mathcal{F}+\mathcal{G}}, \\
\mathcal{F} &= (\delta_{\rm p}-\delta_{\rm c}+{\rm i}\gamma_{bc})(\delta_{\rm p}-\delta_{\rm c'}+{\rm i}\gamma_{bc'})(\delta_{\rm p}+{\rm i}\gamma_{ab}), \\
\mathcal{G} &= -\frac{1}{4}\Omega_{\rm c}^*\Omega_{\rm c}(\delta_{\rm p}-\delta_{\rm c'}+{\rm i}\gamma_{bc'})-\frac{1}{4}\Omega_{\rm c'}^*\Omega_{\rm c'}(\delta_{\rm p}-\delta_{\rm c}+{\rm i}\gamma_{bc}), \quad (4.44)
\end{aligned}
$$

The level population probability in the atomic ground state $|b\rangle$ can be considered to be unity ($\rho_{bb}\simeq 1$) [74]. In the equation (4.44), N denotes the total atom number per unit volume (atomic number density) in the present atomic vapour.

If the frequency detuning $\delta_{\mathrm{p}} \rightarrow \delta_{\mathrm{c}}$, the four-level electromagnetically induced transparency effect [102, 103, 104, 105] would reduce to a three-level one $\{|a\rangle, |b\rangle, |c\rangle\}$ (i.e., the control field $\Omega_{c'}$ seems to be absent). This can be described by the relation:

$$N\varrho \rightarrow -\frac{N|\wp_{ab}|^2}{\epsilon_0 \hbar} \frac{\delta_{\mathrm{p}} - \delta_{\mathrm{c}} + \mathrm{i}\gamma_{bc}}{(\delta_{\mathrm{p}} - \delta_{\mathrm{c}} + \mathrm{i}\gamma_{bc})(\delta_{\mathrm{p}} + \mathrm{i}\gamma_{ab}) - \frac{1}{4}\Omega_{\mathrm{c}}^*\Omega_{\mathrm{c}}}. \tag{4.45}$$

However, when the condition (4.39) of the double-control quantum destructive interference between $|c\rangle$-$|a\rangle$ and $|c'\rangle$-$|a\rangle$ transition pathways is fulfilled, the quantity $N\varrho$ will reduce to the form describing the well-known two-level nearly resonant transition:

$$N\varrho \rightarrow -\frac{N|\wp_{ab}|^2}{\epsilon_0 \hbar} \frac{1}{\delta_{\mathrm{p}} + \mathrm{i}\gamma_{ab}}. \tag{4.46}$$

Therefore, under this condition, both the two atomic levels $|c\rangle$, $|c'\rangle$ and the two control fields with Rabi frequencies Ω_{c}, $\Omega_{\mathrm{c'}}$ seem to be absent.

We should point out that the Clausius-Mossotti relation, which accounts for the contribution of neighboring atoms to the local electric polarization [41], should be taken into account. Then the relative electric permittivity of the four-level atomic vapour is of the form (equivalent to the Clausius-Mossotti relation [45])

$$\varepsilon_1 = 1 + \frac{N\varrho}{1 - \frac{N\varrho}{3}}. \tag{4.47}$$

The dispersion behavior of the relative permittivity ε_1 of the four-level atomic vapor is shown in Fig. 21. The typical parameters (atomic and optical) for this system are given as follows: The atomic-level decay rate (due to the spontaneous emission) is $\gamma_{ab} = 1.0 \times 10^7 \mathrm{s}^{-1}$, the dephasing rates (depending on the thermal collision of atoms) are $\gamma_{bc} = \gamma_{ab}/50$ and $\gamma_{bc'} = \gamma_{ab}/25$, the transition electric dipole moment (electric dipole matrix element) $\wp_{ab} = 1.0 \times 10^{-29}\mathrm{C} \cdot \mathrm{m}$, the Rabi frequency $\Omega_{\mathrm{c'}} = 6.0\gamma_{ab}$, the detuning frequencies of the transitions are $\delta_{\mathrm{c}} = 1.2\gamma_{ab}$ and $\delta_{\mathrm{c'}} = -0.4\gamma_{ab}$, and the number density of atoms is given by $N = 7.0 \times 10^{20}\mathrm{At}/\mathrm{m}^3$ (total number of atoms per unit volume). In Fig. 21, the Rabi frequency Ω_{c} of another control field is $\Omega_{\mathrm{c}} = 4.0\gamma_{ab}$.

As is well known, in the physicist convention for the oscillation factor of a time harmonic wave, we choose $\exp(-\mathrm{i}\omega t)$ as its time-dependent phase factor (the positive and negative imaginary parts in the electric permittivity represents the optical loss and gain, respectively). In the above, we have chosen such a physicist convention. But in the engineer convention, the time oscillation factor is chosen as $\exp(\mathrm{j}\omega t)$ for a time harmonic wave [106], where the positive and negative imaginary parts in the electric permittivity represents the optical gain and loss, respectively). Then the negative and positive imaginary parts in the electric permittivity represent the optical loss and gain, respectively. In what follows, we shall adopt the convention of engineers [106].

Since there are more peaks and valleys in the curve of dispersion of a four-level atomic vapor electric susceptibility than in that of an ordinary three-level atomic vapor susceptibility, it can be found that the dispersion behavior in the permittivity of a four-level atomic vapor would be much more sensitive (to the probe field frequency) than that in an ordinary three-level atomic vapor [21]. Thus, a four-level atomic vapor would possibly find more interesting applications in classical and quantum optics, including photonics. The tunable permittivity (determined by the control-field Rabi frequency Ω_{c}) [21] in this four-level atomic vapor is presented given in Fig. 22, in which three cases are addressed: the frequency detuning of the probe field is $\delta_{\mathrm{p}} = \delta_{\mathrm{c}}$, $\delta_{\mathrm{p}} = \delta_{\mathrm{c}'}$, and $\delta_{\mathrm{p}} = 2/3\delta_{\mathrm{c}}$. From Fig. 22, it can be found that the four-level vapor exhibits an effect of "*electromagnetically induced transparency*" [56], which can be characterized by $\mathrm{Re}\varepsilon_1 = 1$ and $\mathrm{Im}\varepsilon_1 = 0$. Here, the frequency detuning of the probe light is $\delta_{\mathrm{p}} = \delta_{\mathrm{c}'}$. The optical or electromagnetic transparency (to the probe beam optical field) would also take place in the present prism coupler when the Rabi frequency Ω_{c} (representing the atom-light coupling) of the control field is chosen as a magnitude larger than $1.5\gamma_{ab}$. But the quantum destructive interference [21] between the two excitation pathways $|c\rangle$-$|a\rangle$ and $|c'\rangle$-$|a\rangle$ (driven by the two control fields Ω_{c} and $\Omega_{\mathrm{c}'}$, respectively) occurs if the Rabi frequency, Ω_{c}, of one of the two control fields is smaller than $1.5\gamma_{ab}$, and therefore the imaginary part $|\mathrm{Im}\varepsilon_1|$ of the four-level atomic vapor permittivity has a large magnitude. This can be referred to as "*switchable quantum interference*" [21], which could be utilized in order to achieve the effect that is similar to the well-known surface plasmon resonance in metal optics [94]. But we should point out that there is no metal in the *surface-plasmon-resonance prism coupler*, i.e., the four-level atomic vapor [104, 105] (or four-level quantum-dot media) is employed instead of metals. Thus, we can refer to the above effect as "surface-plasmon-like resonance" on the EIT-prism interface [16].

4.3.3. Tunable Reflection Spectrum in a System of EIT-Prism Coupling

Now here we are in a position to consider the mechanism of EIT-prism coupling. As has been pointed out, simple prism coupler systems for surface plasmon polariton resonance include Otto and Kretschmann's prism couplers [82, 83]. Here, we choose Kretschmann's prism coupler, where the metallic film (located under the prism and close to the bottom of the prism seen in Fig. 20) is replaced with the four-level quantum medium (atomic vapor or quantum-dot medium). With the help of the Maxwell electromagnetic field equations and the electromagnetic boundary conditions, the distribution (field profile) of the magnetic field **H** in the three regions of the present EIT-prism coupler can be expressed as [94]

$$H_y(x) = \begin{cases} A_2 e^{\alpha_2(x-d)} + B_2 e^{-\alpha_2(x-d)}, & x > d \\ A_1 e^{\alpha_1 x} + B_1 e^{-\alpha_1 x}, & 0 < x < d \\ A_0 e^{\alpha_0 x}, & x < 0 \end{cases} \tag{4.48}$$

where the attenuation coefficients are $\alpha_j = (\beta^2 - k_0^2\varepsilon_j)^{1/2}$ $(j = 0, 1, 2)$ and the phase constant (in the z direction) is $\beta = k_0\sqrt{\varepsilon_2}\sin\theta$. By using the boundary conditions on the interfaces at $x = 0$ and $x = d$ in Fig. 20(b), the reflectance R of an incident probe field (i.e., the TM mode), is defined as $R \equiv |B_2/A_2|^2$. This reflectance R can be explicitly written as [94]

$$R = \left| \frac{g_{12}\exp(+\alpha_1 d) + g_{01}\exp(-\alpha_1 d)}{\exp(+\alpha_1 d) + g_{12}g_{01}\exp(-\alpha_1 d)} \right|^2 \tag{4.49}$$

with $g_{01} = (\varepsilon_0\alpha_1 - \varepsilon_1\alpha_0)/(\varepsilon_0\alpha_1 + \varepsilon_1\alpha_0)$ and $g_{12} = (\varepsilon_1\alpha_2 - \varepsilon_2\alpha_1)/(\varepsilon_1\alpha_2 + \varepsilon_2\alpha_1)$. In the EIT-prism coupler system (See Fig. 20(a) for the schematic diagrams of the system of EIT-prism coupling), the thickness of the quantum-coherent thin film (i.e., the EIT vapor cell) is $d = 50$ nm, the relative electric permittivity of the prism dielectric is $\varepsilon_2 = 3.24$, and the wavelength of the incident light (i.e., the probe light that drives the $|b\rangle$-$|a\rangle$ electric dipole allowed transition) is $\lambda = 632.8$ nm [94].

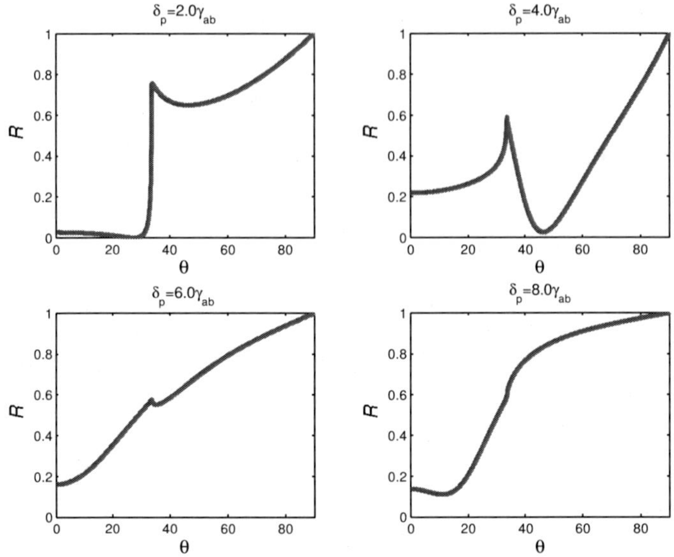

Figure 23. The reflection spectrum depending upon the angle of incidence, θ, on the EIT-prism interface for the probe frequency detuning $\delta_{\mathrm{p}} = 2.0\gamma_{ab}$, $4.0\gamma_{ab}$, $6.0\gamma_{ab}$, and $8.0\gamma_{ab}$. When $\delta_{\mathrm{p}} = 2.0\gamma_{ab}$, the incident light is highly transmitted into EIT, and the reflectance R is small at $\theta < 30$ degrees. Then when $\theta > 30$ degrees, R increases because θ will approach the angle of total reflection. When the probe frequency detuning is $\delta_{\mathrm{p}} = 4.0\gamma_{ab}$, the permittivity of the atomic vapor has a large imaginary part, and the incident light can be highly reflected when the angle of incidence θ is in the range $[0, 35]$ degrees (e.g., $R \rightarrow 0.6$ at $\theta = 35$ degrees). It will, however, drops to almost zero at $\theta = 45$ degrees. This is an effect of ATR because the surface-plasmon-like resonance on the EIT-prism interface occurs, and the incident light has been coupled into the four-level quantum-coherent thin film with a subwavelength thickness d.

The tunable reflection spectrum corresponding to various probe frequency detunings are presented in Fig. 23. In the case of $\delta_{\mathrm{p}} = 2.0\gamma_{ab}$, the reflectance is quite small ($R < 0.03$) when the angle of incidence is in the range $[0, 30]$ degrees. This is because δ_{p} is close to the one of the two-photon resonant frequencies ($\delta_{\mathrm{p}} \simeq \delta_{\mathrm{c}}$), and thus the atomic vapor exhibits

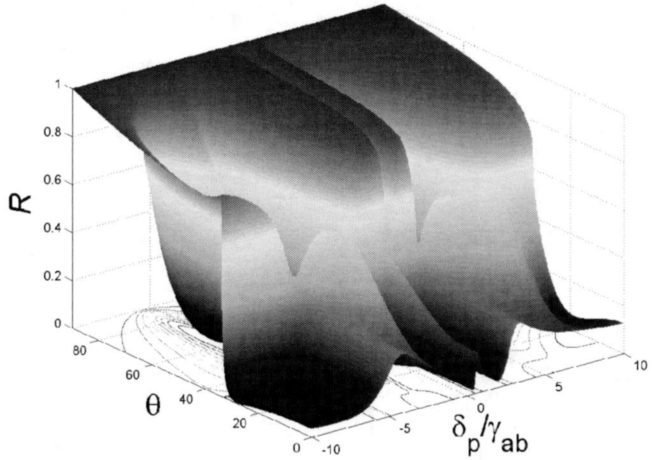

Figure 24. The general dispersion characteristic of reflection spectrum depending upon the angle of incidence on the EIT-prism interface. There are some regions when $\theta > 40$ degrees, in which the probe light reflectance R is small because of surface-plasmon-like resonance that couples the incident light into the atomic vapor.

EIT effect. As the angle of incidence on the EIT-prism interface is largen than 30 degrees, the reflectance increases rapidly because of the mechanism relevant to total reflection. In the other three cases ($\delta_{\rm p} = 4.0\gamma_{ab}$, $6.0\gamma_{ab}$, and $8.0\gamma_{ab}$), no EIT effect would occur since the probe frequency detuning $\delta_{\rm p}$ is far away from the two-photon resonant frequencies, and hence the reflectance increases at small angle of incidence ($[0, 30]$ degrees) is large (0.2-0.6). In these cases, the permittivity of the four-level atomic vapor has a large negative imaginary part, and has exhibited a high reflectance. It should, however, be pointed out that the reflectance then drops to almost zero at the angle of incidence $\theta \simeq 45$ degrees. This is an effect that is related intimately to so-called ATR (Attenuated Total Reflection [94]). Such an ATR effect has been observed in Otto and Kretschmann's configuration for coupling of incident waves into surface plasmon polariton modes [82, 83]. In the EIT-prism coupling system, the wave vector of the incident probe field in the z-direction (along the EIT-prism interface) can be equal to the wave vector of surface plasma wave if we tune the

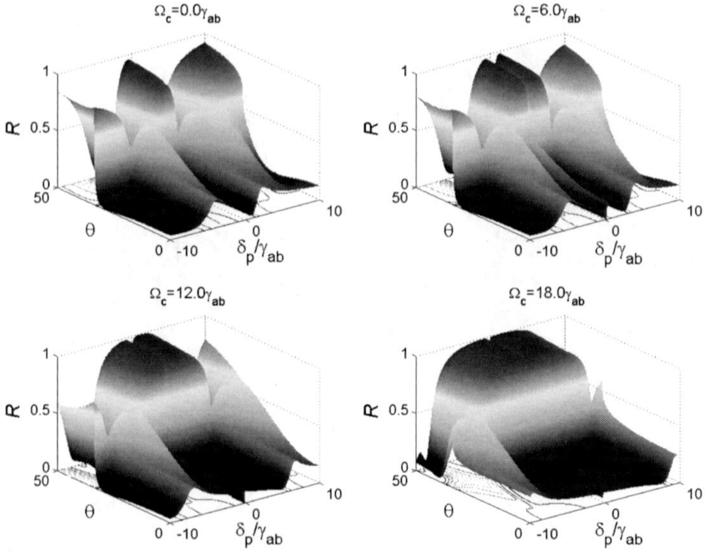

Figure 25. The tunable reflection spectrum as the Rabi frequency $\Omega_{\rm c}$ changes. The Rabi frequency of the other control field (driving the $|c'\rangle$-$|a\rangle$ transition) is $\Omega_{\rm c'} = 6.0\gamma_{ab}$.

angle of incidence on the EIT-prism interface. Such a surface plasmonic wave can be excited at the interface if we choose a proper thickness d of the EIT vapor cell, and this means that the EIT-prism system couples the incident probe wave into the surface plasmon polariton modes, i.e., the reflectance on the EIT-prism interface decreases dramatically (see the case of $\delta_{\rm p} = 4.0\gamma_{ab}$ in Fig. 23). A general behavior corresponding to the reflection spectrum shown in Fig. 23 is given in Fig. 24, where the incidence angle θ and the frequency detuning $\delta_{\rm p}$ are the tunable variables of the reflectance R. The tunable reflection spectrum as the Rabi frequency $\Omega_{\rm c}$ changes is plotted in Fig. 25. The effect of double-control interference (determined by the parameters $\Omega_{\rm c}$ and $\delta_{\rm p}$) on the surface-plasmon-like excitation (and hence on the reflectance R) on the EIT-prism interface, where the total reflection should have already taken place when $\theta > 30$ degrees but it has actually been replaced by the attenuated total reflection (ATR), is shown in Figs. 24 and 25.

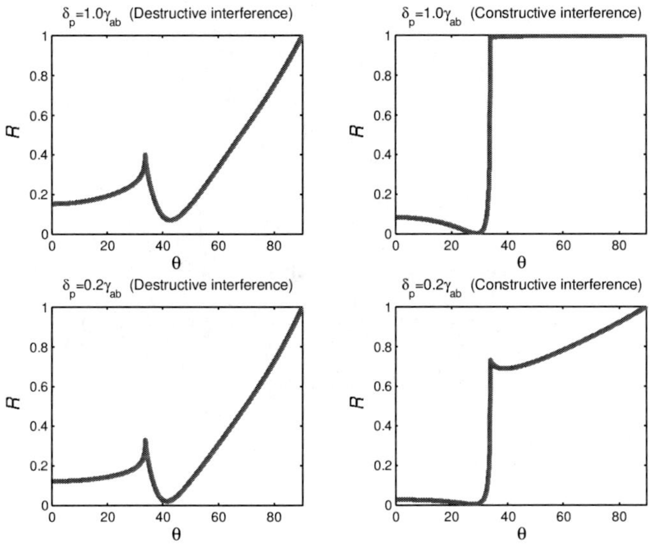

Figure 26. The quantum-interference switchable reflection spectrum as the Rabi frequency Ω_{c} changes. The Rabi frequency of the other control field is $\Omega_{\mathrm{c}'} = 6.0\gamma_{ab}$. In the case of $\delta_{\mathrm{p}} = 1.0\gamma_{ab}$, the Rabi frequency is $\Omega_{\mathrm{c}} = \Omega_{\mathrm{c}'}/\sqrt{7} = 2.3\gamma_{ab}$ for the destructive quantum interference, and $\Omega_{\mathrm{c}} = \sqrt{7}\Omega_{\mathrm{c}'} = 16\gamma_{ab}$ for the constructive quantum interference between the $|c\rangle$-$|a\rangle$ and $|c'\rangle$-$|a\rangle$ transitions. In the case of $\delta_{\mathrm{p}} = 0.2\gamma_{ab}$, the Rabi frequency $\Omega_{\mathrm{c}} = \sqrt{5/3}\Omega_{\mathrm{c}'} = 7.7\gamma_{ab}$ for the destructive quantum interference, and $\Omega_{\mathrm{c}} = \Omega_{\mathrm{c}'}/\sqrt{5/3} = 4.6\gamma_{ab}$ for the constructive quantum interference between the $|c\rangle$-$|a\rangle$ and $|c'\rangle$-$|a\rangle$ transitions.

We shall address the quantum-interference switchable reflection spectrum when the Rabi frequency Ω_{c} changes. Here, as an illustrative example, we discuss two cases of probe frequency detuning ($\delta_{\mathrm{p}} = 1.0\gamma_{ab}$ and $\delta_{\mathrm{p}} = 0.2\gamma_{ab}$), where both destructive and constructive quantum interference have been involved. In the destructive quantum interference between the $|c\rangle$-$|a\rangle$ and $|c'\rangle$-$|a\rangle$ transitions, the four-level atomic vapor has a quite large negative imaginary part of the permittivity, and then the reflectance R is large (0.2-0.4) even the angle of incidence is small (e.g., θ is less than 30 degree). But the reflectance becomes small at $\theta \simeq 40$ degree caused

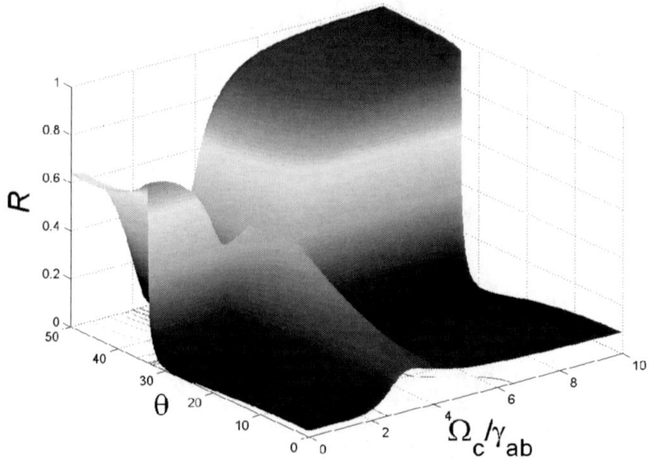

Figure 27. The quantum-interference switchable reflection spectrum as the Rabi frequency $\Omega_{\rm c}$ changes. The Rabi frequency of the other control field is $\Omega_{\rm c'} = 6.0\gamma_{ab}$. The probe frequency detuning is $\delta_{\rm p} = 0.8\gamma_{ab}$.

by ATR effect (related to the large negative imaginary part of the refractive index of the four-level atomic vapor), in which the surface plasmon resonance occurs at the interfaces among the prism dielectric ($x > d$), substrate ($x < 0$) and the EIT vapor cell ($0 < x < d$). In the constructive quantum interference between the $|c\rangle$-$|a\rangle$ and $|c'\rangle$-$|a\rangle$ transitions, however, the four-level atomic vapor can exhibit an EIT effect [102, 103, 104, 105]. The transmittance is large, and hence the reflectance is quite small (0.0–0.1). In particular, R decreases to zero at $\theta = 30$ degrees, and then increases to unity because of total reflection (see the case of constructive quantum interference when $\delta_{\rm p} = 1.0\gamma_{ab}$ in Fig. 26).

In Fig. 27, we present the general behavior of the dependence of the reflection spectrum on the Rabi frequency $\Omega_{\rm c}$ (and hence on the quantum interference between the $|c\rangle$-$|a\rangle$ and $|c'\rangle$-$|a\rangle$ transitions) [16, 21]. Since here $\delta_{\rm p} = 0.8\gamma_{ab}$ is close to one of the two-photon resonant frequency $\delta_{\rm p} \simeq \delta_{\rm c}$, the atomic vapor is transparent to the probe field. As the property of reflection at small angles of incidence is of particular interest, we shall again consider the case of $\theta < 30$ degrees. It can be found that the reflection is very small when the Rabi frequency $\Omega_{\rm c}$ is taken to be in the range

$[0, 2\gamma_{ab}]$ and in the range $> 5\gamma_{ab}$. In the range $[2\gamma_{ab}, 5\gamma_{ab}]$ of Ω_c, however, the quantum interference between the $|c\rangle$-$|a\rangle$ and $|c'\rangle$-$|a\rangle$ transitions occurs, and then the atomic vapor exhibits a high reflectance [16, 21].

To close this work, we conclude this section with some remarks on *quantum-coherence assisted surface plasmon resonance*. We have shown that a four-level atomic system can exhibit an intriguing effect of *destructive* and *constructive* quantum interference [104, 105] that could manipulate the optical response of an atomic vapor [21]. Here, by taking full advantage of quantum coherent control induced by external control fields [16], the EIT-prism coupling system (i.e., Kretschmann's prism coupler for surface plasmon resonance [83, 84], where the metal film located under the prism and close to the bottom of the prism is replaced by the four-level quantum-coherent medium) can give rise to quantum-interference assisted reflection of the incident probe wave. If the atomic system experience a double-control constructive quantum interference [21], the atomic vapor would exhibit EIT [104, 105], and the reflectance on the EIT-prism interface (at $x = 0$) is small at small angles of incidence. Whereas, the reflectance R of the probe light on the EIT-prism interface would become large if a double-control destructive quantum interference occurs [104, 105] when we choose proper intensity ratio (characterized by $\Omega_c^*\Omega_c/\Omega_{c'}^*\Omega_{c'}$) for the control fields [16, 21]. But the reflectance R would drop to almost zero when the angle of incidence of the probe field increases to a certain value, where the critical condition of "surface plasmon resonance" is fulfilled (i.e., the wave vector, $\sqrt{\epsilon_2}\frac{\omega}{c}\sin\theta$, of the incident probe wave along the prism-EIT interface in the z direction equals the wave vector of surface plasma wave β) [15, 16, 21].

When the present quantum-interference assisted switchable (tunable) reflection spectrum of light based on the so-called *surface-plasmon-like resonance* of the four-level atomic vapor [16, 21] in the prism coupler is realized experimentally, we expect that such a quantum coherence mechanism can be utilized in techniques of photonics, e.g., all-optical switches on silicon on chips [22, 79], in which a first optical field can be controlled with a second optical field, and optical logic gates. The photonic logic gates, photonic transistors and optical switching based on the scenario of *quantum coherence manipulation* mechanism [21] may be needed in integrated optical circuits [22].

5. Field Quantization of Localized Surface Plasmons

Surface plasmonic fields include two big categories: one is the "*surface plasmon polaritons*" that are sustained on metal-dielectric planar interfaces [82, 83] (with an infinite curvature radius), the other is the "*localized surface plasmons*" supported on subwavelength metallic nanospheres [54, 55] (with a sufficiently small curvature radius). In the above, we have considered the quantum-coherence assisted surface plasmon resonance of the planar-interface "*surface plasmon polaritons*" based on the theoretical frameworks of certain monographes such as waveguide optics [94]. Now we shall consider the metallic-nanosphere "*localized surface plasmons*" [55]. Here, the topics of "*localized surface plasmons*" include quantum surface plasmons in a metal nanosphere, plasmonic vacuum-vacuum interaction and quantum levitation of a metallic nanosphere, and localized surface plasmon resonance on subatomic particles.

5.1. Quantum Surface Plasmons in a Metal Nanosphere

We consider a model, in which an incident optical field is illuminating a metallic nanosphere and the surface plasmons are excited on the nanosphere [55]. The Helmholtz equation $\nabla^2\Phi + k^2\Phi = 0$ for an electromagnetic wave (at wavelength longer than, e.g., 300 nm) inside or in the vicinity of a nanosphere can reduce to the simple Laplace equation $\nabla^2\Phi = 0$ of an electrostatic field [54], since the magnitude of the term $k^2\Phi$ is much less than the Laplace operator term $\nabla^2\Phi$. If, for example, $\Phi(r) = \sum_n C_n r^n$, where we consider the component $\varphi_n = C_n r^n$, the term $\nabla^2\Phi$ contains $\frac{\partial^2}{\partial r^2}\varphi_n = \frac{\partial^2}{\partial r^2}(C_n r^n) = \frac{n(n-1)}{r^2}\varphi_n$. Since $|\frac{n(n-1)}{r^2}|$ with $r \simeq 50$ nm is larger than $k^2 = \frac{4\pi^2}{\lambda^2}$ (with $\lambda \simeq 500$ nm) by a few orders of magnitude, the latter (related to the k^2 term) can be ignored. In this issue, the magnetic field can also be ignored. In other words, when the electric field excitation fulfills the surface plasmon resonance condition $\epsilon_{\text{in}} + 2\epsilon_{\text{out}} = 0$, the magnetic surface plasmon resonance condition $\mu_{\text{in}} + 2\mu_{\text{out}} = 0$ can in general not be satisfied, because the frequency dispersion in permittivity and permeability of subwavelength sphere and surrounding medium cannot have the same mathematical structure.

A metallic nanosphere of radius a with the relative permittivity $\epsilon_{\text{m}} < 0$

interacts with a uniform driving field, namely, the present nanosphere seems to be placed in an initially uniform electric field E_0, which (at large distances from the sphere) is directed along the z axis (or parallel to the z axis). From the axial symmetry of the geometry as well as the boundary condition at $r = a$ (i.e., the electric scalar potential Φ should be continuous across the nanosphere surface and the electric displacement vector $\mathbf{D}$ perpendicular to the nanosphere surface should also be continuous at $r = a$), we can have the following field distribution inside and outside the metal nanosphere [55]

$$\begin{aligned}\Phi_{\mathrm{in}} &= -\left(\frac{3\epsilon_{\mathrm{d}}}{\epsilon_{\mathrm{m}}+2\epsilon_{\mathrm{d}}}\right) E_0 r\cos\theta & (r<a),\\ \Phi_{\mathrm{out}} &= \left(\frac{\epsilon_{\mathrm{m}}-\epsilon_{\mathrm{d}}}{\epsilon_{\mathrm{m}}+2\epsilon_{\mathrm{d}}}\right) E_0 \frac{a^3}{r^2}\cos\theta - E_0 r\cos\theta & (r\geq a),\end{aligned} \tag{5.1}$$

where $-E_0 r\cos\theta$ is the incident external electric scalar potential at infinity (i.e., $r \to \infty$) away from the nanosphere. In the solution (5.1) of the Laplace equation, ϵ_{m} and ϵ_{d} denote the permittivities of *the metallic nanosphere* and the *ambient dielectric* surrounding the nanosphere, respectively. For convenience, the time-harmonic factor such as $\begin{matrix}\cos\\ \sin\end{matrix}\,\omega t$ has been omitted. The electric field strength on the metallic nanosphere is given by [54] $\mathbf{E}_{\mathrm{in}} = \left(\frac{3\epsilon_{\mathrm{d}}}{\epsilon_{\mathrm{m}}+2\epsilon_{\mathrm{d}}}\right) E_0\mathbf{e}_z$ $(r < a)$ and $\mathbf{E}_{\mathrm{out}} = \left(\frac{\epsilon_{\mathrm{m}}-\epsilon_{\mathrm{d}}}{\epsilon_{\mathrm{m}}+2\epsilon_{\mathrm{d}}}\right) E_0\frac{a^3}{r^3}\left(2\cos\theta\mathbf{e}_r+\sin\theta\mathbf{e}_\theta\right)+E_0\mathbf{e}_z$ $(r \geq a)$ [55], where $\mathbf{E}_{\mathrm{in}}$ and $\mathbf{E}_{\mathrm{out}}$ are the electric fields inside and outside the present subwavelength sphere, respectively. Now let us consider the surface plasmon quantization:

The quantized factor is given by $\xi_{\mathrm{vac}}\frac{\hat{a}e^{-i\omega t}+\hat{a}^{\dagger}e^{i\omega t}}{\sqrt{2}}$ and now the quantized electric field (outside and inside the nanosphere) is

$$\begin{aligned}\hat{\mathbf{E}}_{\mathrm{in}} &= \left(\frac{3\epsilon_{\mathrm{d}}}{\epsilon_{\mathrm{m}}+2\epsilon_{\mathrm{d}}}\right) E_0\xi_{\mathrm{vac}}\hat{\mathcal{F}}\mathbf{e}_z,\\ \hat{\mathbf{E}}_{\mathrm{out}} &= \left(\frac{\epsilon_{\mathrm{m}}-\epsilon_{\mathrm{d}}}{\epsilon_{\mathrm{m}}+2\epsilon_{\mathrm{d}}}\right) E_0\frac{a^3}{r^3}\left(2\cos\theta\mathbf{e}_r+\sin\theta\mathbf{e}_\theta\right)\xi_{\mathrm{vac}}\hat{\mathcal{F}}+E_0\mathbf{e}_z,\end{aligned} \tag{5.2}$$

where the quantum operator $\hat{\mathcal{F}}$ (containing the creation operator $\hat{a}^{\dagger}$ and the annihilation operator $\hat{a}$ of nanosphere-sustained surface plasmons) is given by

$$\hat{\mathcal{F}} = \frac{\hat{a}\mathrm{e}^{-\mathrm{i}\omega t} + \hat{a}^{\dagger}\mathrm{e}^{\mathrm{i}\omega t}}{\sqrt{2}}. \tag{5.3}$$

When the incident (exciting) optical field is in resonance with the nanosphere (i.e., $\epsilon_{\mathrm{m}} + 2\epsilon_{\mathrm{d}} \to 0$), in $\hat{\mathbf{E}}_{\mathrm{out}}$, the incident electric field strength E_0 is much less than the field strength on the metallic nanosphere. When we calculate the total energy, the term $E_0\mathbf{e}_z$ can be ignored [107]. In a dispersive medium, the electric field energy density $\hat{\mathcal{E}}_{\mathrm{in}} = \frac{1}{2}\epsilon_0\frac{\partial(\omega\epsilon_{\mathrm{m}})}{\partial\omega}\hat{\mathbf{E}}_{\mathrm{in}}^2 \cdot \frac{4\pi}{3}a^3$ with the metal permittivity $\epsilon_{\mathrm{m}} = \epsilon_{\infty} - \frac{\omega_{\mathrm{p}}^2}{\omega^2}$. It should be noted that the electric field energy density in a dispersive medium is $\frac{1}{2}\epsilon_0\frac{\partial(\omega\epsilon_{\mathrm{m}})}{\partial\omega}\mathbf{E}^2$ [108] rather than $\frac{1}{2}\epsilon_0\epsilon_{\mathrm{m}}\mathbf{E}^2$, the relevant interpretation of which can be found in some references, e.g., Prade and Vinet's paper [108]. In general, the positive ion background permittivity ϵ_{∞} is less than $\frac{\omega_{\mathrm{p}}^2}{\omega^2}$ at the resonance frequency (this makes $\epsilon_{\mathrm{m}} + 2\epsilon_{\mathrm{d}} \to 0$). Thus, in this case, for simplicity, we choose the relative permittivity of plasma, i.e, $\epsilon_{\mathrm{m}} = 1 - \frac{\omega_{\mathrm{p}}^2}{\omega^2}$. The energy density of the quantized electric field inside the nanosphere is

$$\hat{\mathcal{E}}_{\mathrm{in}} = \frac{1}{2}\epsilon_0\left(1 + \frac{\omega_{\mathrm{p}}^2}{\omega^2}\right)\left(\frac{3\epsilon_{\mathrm{d}}}{\epsilon_{\mathrm{m}} + 2\epsilon_{\mathrm{d}}}\right)^2 E_0^2\xi_{\mathrm{vac}}^2\hat{\mathcal{F}}^2 \cdot \frac{4\pi}{3}a^3. \tag{5.4}$$

The energy density of the quantized electric field outside the nanosphere is

$$\begin{aligned}\hat{\mathcal{E}}_{\mathrm{out}} &= \int_{r\geq a}\frac{1}{2}\epsilon_0\epsilon_{\mathrm{d}}\left(\frac{\epsilon_{\mathrm{m}} - \epsilon_{\mathrm{d}}}{\epsilon_{\mathrm{m}} + 2\epsilon_{\mathrm{d}}}\right)^2 E_0^2\frac{a^6}{r^6}\left(3\cos^2\theta + 1\right)\xi_{\mathrm{vac}}^2\hat{\mathcal{F}}^2\mathrm{d}\tau \\ &= \frac{1}{2}\epsilon_0\epsilon_{\mathrm{d}}\left(\frac{\epsilon_{\mathrm{m}} - \epsilon_{\mathrm{d}}}{\epsilon_{\mathrm{m}} + 2\epsilon_{\mathrm{d}}}\right)^2 E_0^2\xi_{\mathrm{vac}}^2\hat{\mathcal{F}}^2 \cdot \frac{8\pi}{3}a^3,\end{aligned} \tag{5.5}$$

where the volume element is $\mathrm{d}\tau = r^2\sin\theta\mathrm{d}r\mathrm{d}\theta\mathrm{d}\varphi$.

By using the integral

$$\begin{aligned}&\int_{r\geq a}\frac{a^6}{r^6}\left(3\cos^2\theta + 1\right)\mathrm{d}\tau = 2\pi\int_a^{+\infty}\mathrm{d}r\int_0^{\pi}\mathrm{d}\theta\frac{a^6}{r^4}\left(3\cos^2\theta + 1\right)\sin\theta \\ &= 2\pi a^6\left(-\frac{1}{3}\frac{1}{r^3}\right)\Big|_a^{+\infty}\int_0^{\pi}\left[-\left(3\cos^2\theta + 1\right)\right]\mathrm{d}\cos\theta = \frac{8\pi}{3}a^3,\end{aligned} \tag{5.6}$$

the total energy of the quantized surface plasmon on the nanosphere is $\hat{\mathcal{E}}_{\mathrm{sp}} = \hat{\mathcal{E}}_{\mathrm{in}} + \hat{\mathcal{E}}_{\mathrm{out}}$, which can be explicitly expressed as

$$\hat{\mathcal{E}}_{\mathrm{sp}} = \frac{1}{2}\epsilon_0 E_0^2\xi_{\mathrm{vac}}^2\hat{\mathcal{F}}^2\mathcal{W} \cdot \frac{4\pi}{3}a^3. \tag{5.7}$$

Here, $\hat{\mathcal{F}}^2 = \frac{1}{2}\left(\hat{a}^2 \mathrm{e}^{-\mathrm{i}2\omega t} + \hat{a}^{\dagger 2}\mathrm{e}^{\mathrm{i}2\omega t} + \hat{a}\hat{a}^\dagger + \hat{a}^\dagger\hat{a}\right)$ and

$$\mathcal{W} = \left(1 + \frac{\omega_\mathrm{p}^2}{\omega^2}\right)\left(\frac{3\epsilon_\mathrm{d}}{\epsilon_\mathrm{m} + 2\epsilon_\mathrm{d}}\right)^2 + 2\epsilon_\mathrm{d}\left(\frac{\epsilon_\mathrm{m} - \epsilon_\mathrm{d}}{\epsilon_\mathrm{m} + 2\epsilon_\mathrm{d}}\right)^2. \tag{5.8}$$

In this dispersive optical material structure, the quantized energy of the surface plasmons on the nanosphere is $\mathcal{E}_\mathrm{sp} = \langle n|\hat{\mathcal{E}}_\mathrm{sp}|n\rangle$ would have the order of magnitude of $\left(n + \frac{1}{2}\right)\hbar\omega$, or $\mathcal{E}_\mathrm{sp} = \langle n|\hat{\mathcal{E}}_\mathrm{sp}|n\rangle \simeq \left(n + \frac{1}{2}\right)\hbar\omega$ (if there is no dispersion in a material structure, the approximately equal sign "$\simeq$" can be replaced by the equal sign "$=$"). Now we have

$$\frac{1}{2}\epsilon_0 E_0^2 \xi_\mathrm{vac}^2 \left(n + \frac{1}{2}\right)\mathcal{W}\cdot\frac{4\pi}{3}a^3 \simeq \left(n + \frac{1}{2}\right)\hbar\omega. \tag{5.9}$$

Thus, we can obtain

$$\xi_\mathrm{vac} \simeq \sqrt{\frac{\hbar\omega}{\frac{2\pi}{3}\epsilon_0 E_0^2 a^3 \mathcal{W}}}. \tag{5.10}$$

On the nanosphere surface with $r = a$, the magnitude of the quantized surface plasmons is $|\mathbf{E}|^2_\mathrm{max}(r = a) = 4\left(n + \frac{1}{2}\right)\left(\frac{\epsilon_\mathrm{m} - \epsilon_\mathrm{d}}{\epsilon_\mathrm{m} + 2\epsilon_\mathrm{d}}\right)^2 E_0^2\xi_\mathrm{vac}^2$. Thus, the order of the surface plasmon electric field at quantum-vacuum level can be evaluated.

In the above, we assume that the quantum surface plasmons are in the Fock state. In what follows, we shall address the case of quantum surface plasmons in a coherent state. As is well known, in quantum optics (in the coherent-state representation), the coherent state is given by $|\alpha\rangle = \exp\left(-\frac{1}{2}\alpha^*\alpha\right)\sum_{n=0}^{\infty}\frac{\alpha^n}{\sqrt{n!}}|n\rangle$ and $\hat{a}|\alpha\rangle = \alpha|\alpha\rangle$. The energy of the surface plasmons is $\mathcal{E}_\mathrm{in} = \langle\alpha|\hat{\mathcal{E}}_\mathrm{in}|\alpha\rangle$ and $\mathcal{E}_\mathrm{out} = \langle\alpha|\hat{\mathcal{E}}_\mathrm{out}|\alpha\rangle$. The mean value of the number of quantum surface plasmons is $\bar{n} = \langle\alpha|a^\dagger a|\alpha\rangle = \alpha^*\alpha$.

In the equation (5.2), we have $\langle\alpha|\xi_\mathrm{vac}\hat{\mathcal{F}}|\alpha\rangle = \begin{matrix}\cos\\ \sin\end{matrix}\,\omega t$, where $\langle\alpha|\xi_\mathrm{vac}\hat{\mathcal{F}}|\alpha\rangle = \frac{1}{\sqrt{2}}\xi_\mathrm{vac}\left[(\alpha + \alpha^*)\cos\omega t - \mathrm{i}\,(\alpha - \alpha^*)\sin\omega t\right]$. For simplicity, we assume $\alpha^* = \alpha$. Thus, this requires $\sqrt{2}\xi_\mathrm{vac}\alpha = 1$, i.e., $\sqrt{2}\xi_\mathrm{vac}\sqrt{\bar{n}} = 1$. By substituting the result (5.10) into $\bar{n} = \frac{1}{2\xi_\mathrm{vac}^2}$, we have

$$\begin{aligned}\bar{n} &= \frac{\pi\epsilon_0 E_0^2 a^3 \mathcal{W}}{3\hbar\omega}\\ &= \frac{\pi\epsilon_0 E_0^2 a^3}{3\hbar\omega}\left[\left(1 + \frac{\omega_\mathrm{p}^2}{\omega^2}\right)\left(\frac{3\epsilon_\mathrm{d}}{\epsilon_\mathrm{m} + 2\epsilon_\mathrm{d}}\right)^2 + 2\epsilon_\mathrm{d}\left(\frac{\epsilon_\mathrm{m} - \epsilon_\mathrm{d}}{\epsilon_\mathrm{m} + 2\epsilon_\mathrm{d}}\right)^2\right].\end{aligned} \tag{5.11}$$

Thus, the total number $\bar{n}$ of the quantized surface plasmons on the nanosphere is determined by the applied external exciting field E_0 (classical field), metal nanosphere radius a, background dielectric permittivity $\epsilon_{\rm d}$, and the frequency ω. By choosing the proper values of these parameters $E_0, a, \epsilon_{\rm d}, \omega$, the quantum surface plasmon number $\bar{n}$ can be manipulated. The quantum vacuum effect and single or few surface plasmon effects on the nanospheres deserve consideration.

To close this issue, we shall suggest a concept of "*localized chromodynamic surface plasmons*" for quantum chromodynamics. In electromagnetics, localized surface plasmons can be drastically excited on metal nanospheres with negative dielectric coefficient of plasma when the nanospheres are irradiated by external electromagnetic waves at suitable frequencies (i.e., $\epsilon_{\rm in} + 2\epsilon_{\rm out} = 0$ and the fields $\mathbf{E}_{\rm in}$ and $\mathbf{E}_{\rm out}$ become very large). This surface plasmon resonance mechanism can be used in hadrons, such as protons and neutrons, which contain valence quarks, sea quarks and gluon fields. The chromodynamic plasma (or quark-gluon plasma) would also produce negative chromodynamic permittivity. Therefore, by the irradiation of external chromo-electric field, the *chromodynamic localized surface plasmons* may also be excited in and on protons and neutrons, which could change the rest masses of proton and neutron (our viewpoint). High energy physics experiments have produced resonance particle states of hadrons, which are highly excited and high-spin bound states of quarks confined in their linear confining potential. We believe that the chromodynamic localized surface plasmons can inevitably be excited on such resonance particles. The protons and neutrons in a nucleus can be considered to be in a chromo-electric field of neighboring protons and neutrons. The protons and neutrons in the nucleus differ from those free protons and neutrons in that the chromodynamic surface plasmons may be sustained on the former. The plasma frequency in protons and neutrons is six or seven orders of magnitude higher than the ordinary plasma frequency in metals. Therefore, the energy of *quantized chromodynamic surface plasmons* can be of the order of MeV. In addition to the chromodynamic surface plasmons, the ordinary (electric) surface plasmons can also be excited in such hadrons. In order that the localized surface plasmons can be excited, the excitation frequency should be as high as the order of 10^{22} s^{-1}, i.e., we may believe that the γ photons could excite such localized surface plasmons on protons and neutrons.

5.2. Plasmonic Vacuum-Vacuum Interaction and Quantum Levitation of Metallic Nanosphere in a Quantum Cavity

5.2.1. Surface Plasmons on a Subwavelength Metallic Nanosphere

We shall investigate the effect of nanosphere's quantum levitation, where the plasmon-cavity interaction is involved. In this issue, a metallic nanosphere with a scale of one tenth of light wavelength [55] is placed into a quantum cavity with a scale of the same order of magnitude as the wavelength of light. Since a nanosphere with the excited surface plasmons sustained on it can be identified as an electric dipole [55], it can interact with the quantum electromagnetic modes in the quantum cavity. Such an interaction potential can lead to an induced force exerted on the metallic nanosphere. If this quantum optical-tweezers force is larger than the gravity of metallic nanosphere, the nanosphere will be found suspended by such a quantum optically induced force.

Now we shall suggest and study the relevant theoretical mechanism of this quantum induced force (including the quantum-vacuum induced force acting on the metallic nanosphere). The following mathematical formalism of localized surface plasmon modes at classical level was prepared based on certain monographs of plasmonics, e.g., Maier's reference [55], and then we develop a quantum treatment of surface plasmons sustained on a plasmonic nanosphere.

On a subwavelength plasmonic nanosphere, when an external optical field with $E_{\rm ext}$ is illuminating a nanosphere, the electric scalar potential (by solving the Laplace equation $\nabla^2\Phi = 0$) is given by [55] $\Phi_{\rm in} = -\frac{3\epsilon_{\rm out}}{\epsilon_{\rm in}+2\epsilon_{\rm out}}E_{\rm ext}r\cos\theta$ $(r \leq R)$ and $\Phi_{\rm out} = -E_{\rm ext}r\cos\theta + \frac{\epsilon_{\rm in}-\epsilon_{\rm out}}{\epsilon_{\rm in}+2\epsilon_{\rm out}}E_{\rm ext}R^3\frac{\cos\theta}{r^2}$ $(r > R)$, where R denotes the radius of the subwavelength plasmonic nanosphere; $\epsilon_{\rm in}$ is the electric permittivity of the nanosphere and $\epsilon_{\rm out}$ is the dielectric coefficient of the medium surrounding the nanosphere. By using the gradient of the electric scalar potential $\nabla\Phi = \frac{\partial\Phi}{\partial r}\mathbf{e}_r + \frac{1}{r}\frac{\partial\Phi}{\partial\theta}\mathbf{e}_\theta + \frac{1}{r\sin\theta}\frac{\partial\Phi}{\partial\phi}\mathbf{e}_\phi$ in the spherical polar coordinate, we have the electric field inside the plasmonic nanosphere [54, 55]: $E^r_{\rm in} = -\frac{\partial\Phi_{\rm in}}{\partial r} = \frac{3\epsilon_{\rm out}}{\epsilon_{\rm in}+2\epsilon_{\rm out}}E_{\rm ext}\cos\theta$, $E^\theta_{\rm in} = -\frac{1}{r}\frac{\partial\Phi_{\rm in}}{\partial\theta} = -\frac{3\epsilon_{\rm out}}{\epsilon_{\rm in}+2\epsilon_{\rm out}}E_{\rm ext}\sin\theta$, and $E^\phi_{\rm in} = 0$. Therefore, $\mathbf{E}_{\rm in} = \frac{3\epsilon_{\rm out}}{\epsilon_{\rm in}+2\epsilon_{\rm out}}E_{\rm ext}\left(\cos\theta\mathbf{e}_r - \sin\theta\mathbf{e}_\theta\right) = \frac{3\epsilon_{\rm out}}{\epsilon_{\rm in}+2\epsilon_{\rm out}}\mathbf{E}_{\rm ext}$ [55], where $\cos\theta\mathbf{e}_r - \sin\theta\mathbf{e}_\theta = \mathbf{e}_z$ and $\mathbf{E}_{\rm ext} = E_{\rm ext}\mathbf{e}_z$.

Since the electric field outside the nanosphere is given by [54, 55]

$E^r_{\rm out} = -\frac{\partial \Phi_{\rm out}}{\partial r} = E_{\rm ext}\cos\theta + 2\left(\frac{\epsilon_{\rm in}-\epsilon_{\rm out}}{\epsilon_{\rm in}+2\epsilon_{\rm out}}E_{\rm ext}\right)\frac{R^3}{r^3}\cos\theta$, $E^\theta_{\rm out} = -\frac{1}{r}\frac{\partial \Phi_{\rm out}}{\partial \theta} = -E_{\rm ext}\sin\theta + \frac{\epsilon_{\rm in}-\epsilon_{\rm out}}{\epsilon_{\rm in}+2\epsilon_{\rm out}}E_{\rm ext}\frac{R^3}{r^3}\sin\theta$, and $E^\phi_{\rm out} = 0$, it can be rewritten as $\mathbf{E}_{\rm out} = E_{\rm ext}\left(\cos\theta\mathbf{e}_r - \sin\theta\mathbf{e}_\theta\right) + \frac{\epsilon_{\rm in}-\epsilon_{\rm out}}{\epsilon_{\rm in}+2\epsilon_{\rm out}}E_{\rm ext}\frac{R^3}{r^3}\left(2\cos\theta\mathbf{e}_r + \sin\theta\mathbf{e}_\theta\right)$, or $\mathbf{E}_{\rm out} = \mathbf{E}_{\rm ext} + \frac{\epsilon_{\rm in}-\epsilon_{\rm out}}{\epsilon_{\rm in}+2\epsilon_{\rm out}}\frac{R^3}{r^3}E_{\rm ext}\left[3(\mathbf{n}\cdot\mathbf{e}_z)\mathbf{n} - \mathbf{e}_z\right]$, where $3(\mathbf{n}\cdot\mathbf{e}_z)\mathbf{n} - \mathbf{e}_z = 2\cos\theta\mathbf{e}_r + \sin\theta\mathbf{e}_\theta$. Here, $\mathbf{n}$ denotes the *unit vector* of the 3-D coordinate vector $\mathbf{r}$ (i.e., $\mathbf{n} = \frac{\mathbf{r}}{r}$).

Now we have $\mathbf{E}_{\rm in} = \frac{3\epsilon_{\rm out}}{\epsilon_{\rm in}+2\epsilon_{\rm out}}\mathbf{E}_{\rm ext}$ and $\mathbf{E}_{\rm out} = \mathbf{E}_{\rm ext} + \left(\frac{\epsilon_{\rm in}-\epsilon_{\rm out}}{\epsilon_{\rm in}+2\epsilon_{\rm out}}\right)\frac{R^3}{r^3}\left[\frac{3(\mathbf{r}\cdot\mathbf{E}_{\rm ext})\mathbf{r}}{r^2} - \mathbf{E}_{\rm ext}\right]$. According to the ordinary textbooks of electromagnetics, in general, a dipole electric field $\mathbf{E}_{\rm dipole}$ can be rewritten as $\mathbf{E}_{\rm dipole} = \mathbf{E}_{\rm ext} + \frac{3\mathbf{n}(\mathbf{n}\cdot\mathbf{p})-\mathbf{p}}{4\pi\epsilon_0\epsilon_{\rm out}}\frac{1}{r^3}$. By comparing this result with the above formula for $\mathbf{E}_{\rm out}$, one can obtain the induced electric dipole moment $\mathbf{p} = 4\pi\epsilon_0\epsilon_{\rm out}R^3\left(\frac{\epsilon_{\rm in}-\epsilon_{\rm out}}{\epsilon_{\rm in}+2\epsilon_{\rm out}}\right)\mathbf{E}_{\rm ext}$ [55] in the subwavelength plasmonic nanosphere. Here, the subwavelength plasmonic nanosphere can be identified as an electric dipole because it has such a dipole moment $\mathbf{p}$.

Alternatively, by using the outside-nanosphere electric field $\mathbf{E}_{\rm out} = \mathbf{E}_{\rm ext} + \frac{p}{4\pi\epsilon_0\epsilon_{\rm out}r^3}\left[3(\mathbf{n}\cdot\mathbf{e}_z)\mathbf{n} - \mathbf{e}_z\right]$ and by comparing it with $\mathbf{E}_{\rm out} = \mathbf{E}_{\rm ext} + \frac{\epsilon_{\rm in}-\epsilon_{\rm out}}{\epsilon_{\rm in}+2\epsilon_{\rm out}}\frac{R^3}{r^3}E_{\rm ext}\left[3(\mathbf{n}\cdot\mathbf{e}_z)\mathbf{n} - \mathbf{e}_z\right]$ [55], we have the relation $\frac{p}{4\pi\epsilon_0\epsilon_{\rm out}} = \frac{\epsilon_{\rm in}-\epsilon_{\rm out}}{\epsilon_{\rm in}+2\epsilon_{\rm out}}R^3E_{\rm ext}$, and the induced electric dipole moment in the plasmonic nanosphere is $p = 4\pi\epsilon_0\epsilon_{\rm out}R^3\left(\frac{\epsilon_{\rm in}-\epsilon_{\rm out}}{\epsilon_{\rm in}+2\epsilon_{\rm out}}\right)E_{\rm ext}$ [55].

By using the Fröhlich condition $\epsilon_{\rm in} = -2\epsilon_{\rm out}$ [55] (this can be identified as a surface plasmon resonance condition), we have $\mathbf{E}_{\rm in} = \frac{3\epsilon_{\rm out}}{\epsilon_{\rm in}+2\epsilon_{\rm out}}\mathbf{E}_{\rm ext} \to \infty$. Then the external illuminating field $\mathbf{E}_{\rm ext}$ can be very weak, but since $\epsilon_{\rm in} + 2\epsilon_{\rm out} \to 0$, the induced internal electric field $\mathbf{E}_{\rm in}$ (inside the metallic nanoparticle) is still quite strong. Now the magnitude of the induced electric dipole moment in the plasmonic nanosphere is $p = -4\pi\epsilon_0\epsilon_{\rm out}R^3\left(\frac{3\epsilon_{\rm out}}{\epsilon_{\rm in}+2\epsilon_{\rm out}}\right)E_{\rm ext} = -4\pi\epsilon_0\epsilon_{\rm out}R^3E_{\rm in}$. The surface plasmon field outside the subwavelength plasmonic nanosphere is $\mathbf{E}_{\rm out} \to -\frac{R^3}{r^3}\left[\frac{3(\mathbf{E}_{\rm in}\cdot\mathbf{r})\mathbf{r}}{r^2} - \mathbf{E}_{\rm in}\right] = -\frac{R^3}{r^3}E_{\rm in}\left[3(\mathbf{n}\cdot\mathbf{e}_z)\mathbf{n} - \mathbf{e}_z\right]$ [54, 55], and the induced electric dipole moment (vector) in the plasmonic nanosphere is $\mathbf{p} = -4\pi\epsilon_0\epsilon_{\rm out}R^3\mathbf{E}_{\rm in}$.

If the induced electric dipole moment in the plasmonic nanosphere can be written as $\mathbf{p} = \mathbf{a}_z p = p\left(\mathbf{e}_r\cos\theta - \mathbf{e}_\theta\sin\theta\right)$, then the surface plasmon field outside the subwavelength plasmonic nanosphere is $\mathbf{E}_{\rm out} = \mathbf{E}_{\rm ext} + \frac{p}{4\pi\epsilon_0\epsilon_{\rm out}r^3}\left(\mathbf{e}_r 2\cos\theta + \mathbf{e}_\theta\sin\theta\right)$. The above expressions for electric fields or electric scalar potentials inside and outside the subwavelength sphere can be found in some monographes of electrodynamics and surface plasmonics [54, 55].

In the second quantization formulation, the surface plasmon field inside the subwavelength plasmonic nanosphere is given by $\hat{\mathbf{E}}_{\text{in}} = \bar{\mathbf{E}}_{\text{in}}\left(\hat{a}^{\dagger}\mathrm{e}^{\mathrm{i}\omega_{\text{s}}t} + \hat{a}\mathrm{e}^{-\mathrm{i}\omega_{\text{s}}t}\right)$, where $\hat{a}^{\dagger}$ and $\hat{a}$ are the creation and annihilation operators of the surface plasmons, respectively, and hence the operator of the induced electric dipole moment in the nanosphere is given by $\hat{\mathbf{p}} = \mathbf{p}\left(\hat{a}^{\dagger}\mathrm{e}^{\mathrm{i}\omega_{\text{s}}t} + \hat{a}\mathrm{e}^{-\mathrm{i}\omega_{\text{s}}t}\right)$.

5.2.2. Quantization of Surface Plasmons and Its Interaction with Cavity Modes

Since we shall address the issue of plasmon-cavity interaction, in which a metallic nanosphere of a size of one tenth of the wavelength of cavity modes is placed into a quantum cavity, we shall now consider the quantized and quantum-vacuum modes in the cavity (this cavity has a scale of the same order of magnitude as the electromagnetic wavelength, e.g., $L = m\frac{\lambda}{2}$, where the integer m is the cavity mode index).

By using the well-known field quantization formulation [109, 110], in a microcavity, there is a quantized cavity mode with the electric field operator $\hat{\mathbf{E}} = \mathrm{i}E_0\left(\hat{\mathcal{B}}\mathbf{u}^*(\mathbf{r})\mathrm{e}^{-\mathrm{i}\omega t} - \hat{\mathcal{B}}^{\dagger}\mathbf{u}(\mathbf{r})\mathrm{e}^{\mathrm{i}\omega t}\right)$ (without dispersion), where a single-photon field strength is $E_0 = \sqrt{\frac{\hbar\omega}{2\epsilon_0\mathcal{V}}}$ and the effective volume is given by $\mathcal{V} = \int |\mathbf{u}(\mathbf{r})|^2\mathrm{d}V$ (according to the ordinary textbooks of quantum optics) [109, 110]. Here, $\hat{\mathcal{B}}$ and $\hat{\mathcal{B}}^{\dagger}$ are the annihilation and creation operators, respectively, of the cavity-mode photons. It can be easy to explain why $E_0 = \sqrt{\frac{\hbar\omega}{2\epsilon_0\mathcal{V}}}$: The square of the electric field operator, $\hat{\mathbf{E}}^2$, is expressed by $\hat{\mathbf{E}}^2 = -E_0^2\Big(- \hat{\mathcal{B}}\hat{\mathcal{B}}^{\dagger}\mathbf{u}^*(\mathbf{r})\cdot\mathbf{u}(\mathbf{r}) - \hat{\mathcal{B}}^{\dagger}\hat{\mathcal{B}}\mathbf{u}^*(\mathbf{r})\cdot\mathbf{u}(\mathbf{r}) + \hat{\mathcal{B}}^2(\mathbf{u}^*)^2\mathrm{e}^{-\mathrm{i}2\omega t} + (\hat{\mathcal{B}}^{\dagger})^2\mathbf{u}^2\mathrm{e}^{\mathrm{i}2\omega t}\Big)$. If the high-frequency terms are ignored, one can have $\int \epsilon_0\hat{\mathbf{E}}^2\mathrm{d}V = \epsilon_0 E_0^2\int[\hat{\mathcal{B}}\hat{\mathcal{B}}^{\dagger}\mathbf{u}^*(\mathbf{r})\cdot\mathbf{u}(\mathbf{r}) + \hat{\mathcal{B}}^{\dagger}\hat{\mathcal{B}}\mathbf{u}^*(\mathbf{r})\cdot\mathbf{u}(\mathbf{r})]\mathrm{d}V = \epsilon_0 E_0^2(\hat{\mathcal{B}}\hat{\mathcal{B}}^{\dagger} + \hat{\mathcal{B}}^{\dagger}\hat{\mathcal{B}})\mathcal{V} = 2\epsilon_0 E_0^2\mathcal{V}(\hat{\mathcal{B}}^{\dagger}\hat{\mathcal{B}} + \frac{1}{2})$, where $2\epsilon_0 E_0^2\mathcal{V} = \hbar\omega$. Thus, one can have $\int \epsilon_0\hat{\mathbf{E}}^2\mathrm{d}V = (\hat{\mathcal{B}}^{\dagger}\hat{\mathcal{B}} + \frac{1}{2})\hbar\omega$. It should be stressed that the above (conventional) formalism is applicable only to the non-dispersive electromagnetic materials.

The interaction Hamiltonian operator for characterizing the coupling of the metallic-nanosphere surface plasmons to the cavity modes is given by $\hat{V} = -\hat{\mathbf{p}} \cdot \hat{\mathbf{E}}$, the explicit expression of which is $\hat{V} = -\mathrm{i}E_0\big(\mathbf{p}\cdot\mathbf{u}^*\hat{a}^\dagger\hat{\mathcal{B}}\mathrm{e}^{\mathrm{i}(\omega_\mathrm{s}-\omega)t} - \mathbf{p}\cdot\mathbf{u}\hat{a}\hat{\mathcal{B}}^\dagger\mathrm{e}^{-\mathrm{i}(\omega_\mathrm{s}-\omega)t} + \mathbf{p}\cdot\mathbf{u}^*\hat{a}\hat{\mathcal{B}}\mathrm{e}^{-\mathrm{i}(\omega_\mathrm{s}+\omega)t} - \mathbf{p}\cdot\mathbf{u}\hat{a}^\dagger\hat{\mathcal{B}}^\dagger\mathrm{e}^{\mathrm{i}(\omega_\mathrm{s}+\omega)t}\big)$. If the surface plasmon frequency ω_s approaches the cavity mode eigen frequency, i.e., $\omega_\mathrm{s} - \omega \to 0$, we will have

$$\hat{V} = -\mathrm{i}E_0\left(\mathbf{p}\cdot\mathbf{u}^*\hat{a}^\dagger\hat{\mathcal{B}} - \mathbf{p}\cdot\mathbf{u}\hat{a}\hat{\mathcal{B}}^\dagger\right), \tag{5.12}$$

where the rotating wave approximation has been taken into account. In the above, the surface plasmon frequency ω_s is determined by the Fröhlich (surface plasmon resonance) condition $\epsilon_\mathrm{in} = -2\epsilon_\mathrm{out}$ [55] and the cavity mode eigen frequency ω is determined by $L = m\frac{\lambda}{2}$ (the integer m is the cavity mode index).

By using the completeness condition (closure relation) $|n'_\mathrm{sp}\rangle\langle n'_\mathrm{sp}| = 1$ and $|n_\mathrm{sp}\rangle\langle n_\mathrm{sp}| = 1$ (for the Fock states of the quantized nanosphere surface plasmons), the surface-plasmon–cavity-mode interaction potential energy operator $\hat{V}$ can be rewritten as $\hat{V} = \sum_{n'_\mathrm{sp}}\sum_{n_\mathrm{sp}}|n'_\mathrm{sp}\rangle\langle n'_\mathrm{sp}|\hat{V}|n_\mathrm{sp}\rangle\langle n_\mathrm{sp}| = \sum_{n'_\mathrm{sp}}\sum_{n_\mathrm{sp}}\langle n'_\mathrm{sp}|\hat{V}|n_\mathrm{sp}\rangle|n'_\mathrm{sp}\rangle\langle n_\mathrm{sp}|$, where the matrix element $\langle n'_\mathrm{sp}|\hat{V}|n_\mathrm{sp}\rangle$ will be written explicitly as follows: With the help of the relations $[\hat{a}, \hat{a}^\dagger] = 1$, $\hat{a}|n_\mathrm{sp}\rangle = \sqrt{n_\mathrm{sp}}|(n-1)_\mathrm{sp}\rangle$ and $\hat{a}^\dagger|n_\mathrm{sp}\rangle = \sqrt{(n+1)_\mathrm{sp}}|(n+1)_\mathrm{sp}\rangle$ for the metallic-nanosphere sustained localized surface plasmons, we shall calculate the matrix element

$$\begin{aligned}\langle n'_\mathrm{sp}|\hat{V}|n_\mathrm{sp}\rangle &= -\mathrm{i}E_0\left(\mathbf{p}\cdot\mathbf{u}^*\langle n'_\mathrm{sp}|\hat{a}^\dagger|n_\mathrm{sp}\rangle\hat{\mathcal{B}} - \mathbf{p}\cdot\mathbf{u}\langle n'_\mathrm{sp}|\hat{a}|n_\mathrm{sp}\rangle\hat{\mathcal{B}}^\dagger\right)\\ &= -\mathrm{i}E_0\Big(\mathbf{p}\cdot\mathbf{u}^*\sqrt{(n+1)_\mathrm{sp}}\delta_{n'_\mathrm{sp},(n+1)_\mathrm{sp}}\hat{\mathcal{B}} - \mathbf{p}\cdot\mathbf{u}\sqrt{n_\mathrm{sp}}\delta_{n'_\mathrm{sp},(n-1)_\mathrm{sp}}\hat{\mathcal{B}}^\dagger\Big).\end{aligned} \tag{5.13}$$

Thus, the surface-plasmon–cavity-mode interaction potential energy operator is

$$\begin{aligned}\hat{V} &= -\mathrm{i}E_0\Big(\mathbf{p}\cdot\mathbf{u}^*\sum_{n_\mathrm{sp}=0}^{\infty}\sqrt{(n+1)_\mathrm{sp}}\hat{\mathcal{B}}|(n+1)_\mathrm{sp}\rangle\langle n_\mathrm{sp}|\\ &\quad -\mathbf{p}\cdot\mathbf{u}\sum_{n_\mathrm{sp}=1}^{\infty}\sqrt{n_\mathrm{sp}}|(n-1)_\mathrm{sp}\rangle\langle n_\mathrm{sp}|\hat{\mathcal{B}}^\dagger\Big)\\ &= -\mathrm{i}E_0\Big(\mathbf{p}\cdot\mathbf{u}^*\sum_{n_\mathrm{sp}=0}^{\infty}\sqrt{(n+1)_\mathrm{sp}}\hat{\mathcal{B}}|(n+1)_\mathrm{sp}\rangle\langle n_\mathrm{sp}|\\ &\quad -\mathbf{p}\cdot\mathbf{u}\sum_{n_\mathrm{sp}=0}^{\infty}\sqrt{(n+1)_\mathrm{sp}}|n_\mathrm{sp}\rangle\langle (n+1)_\mathrm{sp}|\hat{\mathcal{B}}^\dagger\Big),\end{aligned} \tag{5.14}$$

where $\mathbf{u}$ denotes the polarization and field distribution of the quantum cavity modes. For example, we can choose $\mathbf{u} = \mathbf{u}_0 \sin(\frac{m\pi x}{L})$ with L the cavity length.

Here we consider only the states $|n_{\mathrm{sp}}, (m+1)_{\mathrm{cav}}\rangle$ and $|(n+1)_{\mathrm{sp}}, m_{\mathrm{cav}}\rangle$, which have the same energy eigenvalue. Because of the surface-plasmon–cavity-mode interaction, these two states $|n_{\mathrm{sp}}, (m+1)_{\mathrm{cav}}\rangle$ and $|(n+1)_{\mathrm{sp}}, m_{\mathrm{cav}}\rangle$ will be linearly combined and two orthogonal dressed states will yield. In this case, the surface-plasmon–cavity-mode interaction Hamiltonian operator is

$$\hat{V} = g\hat{\mathcal{B}}|(n+1)_{\mathrm{sp}}\rangle\langle n_{\mathrm{sp}}| + g^*|n_{\mathrm{sp}}\rangle\langle (n+1)_{\mathrm{sp}}|\hat{\mathcal{B}}^\dagger, \tag{5.15}$$

where the nanosphere-mediated surface-plasmon–cavity-mode coupling coefficients are $g = -\mathrm{i}\sqrt{(n+1)_{\mathrm{sp}}}E_0\mathbf{p}\cdot\mathbf{u}^*$ and $g^* = \mathrm{i}\sqrt{(n+1)_{\mathrm{sp}}}E_0\mathbf{p}\cdot\mathbf{u}$. Here, $\sqrt{(n+1)_{\mathrm{sp}}}$ is proportional to the surface plasmon field, $E_0\mathbf{u}$ is the cavity mode, and $\mathbf{p}$ is the induced electric dipole moment of the nanosphere (mediator). The induced electric dipole moment vector $\mathbf{p}$ in the plasmonic nanosphere is $\mathbf{p} = \mathbf{e}_z p$, where $p = 4\pi\epsilon_0\epsilon_{\mathrm{out}}R^3\left(\frac{\epsilon_{\mathrm{in}}-\epsilon_{\mathrm{out}}}{\epsilon_{\mathrm{in}}+2\epsilon_{\mathrm{out}}}\right)E_{\mathrm{ext}}$ [55]. Here, $\mathbf{e}_z$ is a unit vector in the z direction.

It is apparent that $\hat{V}$ describes the surface-plasmon–cavity-mode interaction mediated by the subwavelength plasmonic nanosphere.

In the interaction picture, the full Hamiltonian is given by $H = \frac{1}{2}\hbar\Delta\big(|(n+1)_{\mathrm{sp}}\rangle\langle(n+1)_{\mathrm{sp}}| - |n_{\mathrm{sp}}\rangle\langle n_{\mathrm{sp}}|\big) + \hat{V}$ with the frequency detuning $\Delta = \omega_{\mathrm{s}} - \omega$. We assume g is a real number. The eigenstate of the full Hamiltonian can be taken to be $|\sigma\rangle = c|n_{\mathrm{sp}}, (m+1)_{\mathrm{cav}}\rangle + s|(n+1)_{\mathrm{sp}}, m_{\mathrm{cav}}\rangle$. By using $\hat{\mathcal{B}}|(m+1)_{\mathrm{cav}}\rangle = \sqrt{m_{\mathrm{cav}}}|m_{\mathrm{cav}}\rangle$ and $\hat{\mathcal{B}}^\dagger|m_{\mathrm{cav}}\rangle = \sqrt{(m+1)_{\mathrm{cav}}}|(m+1)_{\mathrm{cav}}\rangle$, we can obtain

$$\begin{aligned} H|\sigma\rangle &= \left(-\frac{\hbar\Delta}{2}c + gs\sqrt{(m+1)_{\mathrm{cav}}}\right)|n_{\mathrm{sp}}, (m+1)_{\mathrm{cav}}\rangle \\ &\quad + \left(\frac{\hbar\Delta}{2}s + gc\sqrt{m_{\mathrm{cav}}}\right)|(n+1)_{\mathrm{sp}}, m_{\mathrm{cav}}\rangle \\ &= \mathcal{E}_\sigma\Big(c|n_{\mathrm{sp}}, (m+1)_{\mathrm{cav}}\rangle + s|(n+1)_{\mathrm{sp}}, m_{\mathrm{cav}}\rangle\Big). \end{aligned} \tag{5.16}$$

From Eq. (5.16), we should have

$$\frac{-\frac{\hbar\Delta}{2}c + gs\sqrt{(m+1)_{\rm cav}}}{\frac{\hbar\Delta}{2}s + gc\sqrt{m_{\rm cav}}} = \frac{c}{s}, \tag{5.17}$$

and

$$\hbar\Delta sc = g\left(s^2\sqrt{(m+1)_{\rm cav}} - c^2\sqrt{m_{\rm cav}}\right). \tag{5.18}$$

We assume $s = \sin\theta$, $c = \cos\theta$, and Eq. (5.18) becomes

$$\frac{\hbar\Delta}{2}\sin(2\theta) = g\Big[\frac{1-\cos(2\theta)}{2}\sqrt{(m+1)_{\rm cav}} - \frac{1+\cos(2\theta)}{2}\sqrt{m_{\rm cav}}\Big]. \tag{5.19}$$

From this equation, we can in principle obtain $\cos\theta$ and $\sin\theta$. Thus, we can have the explicit expression for the eigenstates $|\sigma\rangle = c|n_{\rm sp}, (m+1)_{\rm cav}\rangle + s|(n+1)_{\rm sp}, m_{\rm cav}\rangle$ of the full Hamiltonian operator (for the surface-plasmon–cavity-mode interaction). This is an surface-plasmon–cavity-mode entangled state. In what follows, we shall consider two typical cases, where the cavity mode index $m_{\rm cav}$ is large or small.

i) *Discussion: when $m_{\rm cav}$ is large*:

When the cavity mode index $m_{\rm cav}$ is large, from Eq. (5.19), we have $\frac{\hbar\Delta}{2}\sin(2\theta) = -g\sqrt{(m+1)_{\rm cav}}\cos(2\theta)$ and $\tan(2\theta) = -\frac{2g\sqrt{(m+1)_{\rm cav}}}{\hbar\Delta}$. Then we can obtain $\sin(2\theta) = -\frac{2g\sqrt{(m+1)_{\rm cav}}}{\sqrt{(\hbar\Delta)^2+4g^2(m+1)_{\rm cav}}}$ and $\cos(2\theta) = \frac{\hbar\Delta}{\sqrt{(\hbar\Delta)^2+4g^2(m+1)_{\rm cav}}}$ $(\Delta > 0)$. Note that when the surface-plasmon–cavity-mode coupling coefficient g is large or $\hbar\Delta$ is small, $\theta \to \pi/4$ or $3\pi/4$, i.e., there are dressed states. If, however, g is small or $\hbar\Delta$ is large, no such dressed states will be present.

Now we have $\sin\theta$ and $\cos\theta$ given as follows:

$$\sin\theta = \pm\sqrt{\frac{1-\cos(2\theta)}{2}} = \pm\sqrt{\frac{1-\frac{\hbar\Delta}{\sqrt{(\hbar\Delta)^2+4g^2(m+1)_{\rm cav}}}}{2}},$$

$$\cos\theta = \pm\sqrt{\frac{1+\cos(2\theta)}{2}} = \pm\sqrt{\frac{1+\frac{\hbar\Delta}{\sqrt{(\hbar\Delta)^2+4g^2(m+1)_{\rm cav}}}}{2}}. \tag{5.20}$$

Since $\sin(2\theta) < 0$, $\cos\theta$ and $\sin\theta$ must be opposite in sign, we can choose

$$c_- = \sqrt{\frac{1+\frac{\hbar\Delta}{\sqrt{(\hbar\Delta)^2+4g^2(m+1)_{\rm cav}}}}{2}}, \quad s_- = -\sqrt{\frac{1-\frac{\hbar\Delta}{\sqrt{(\hbar\Delta)^2+4g^2(m+1)_{\rm cav}}}}{2}}. \tag{5.21}$$

The eigenstate of the Hamiltonian operator (for the interacting system of surface-plasmon–cavity-mode coupling) is $|-\rangle = c_-|n_{\rm sp}, (m+1)_{\rm cav}\rangle + s_-|(n+1)_{\rm sp}, m_{\rm cav}\rangle$. Thus, we can obtain the energy eigenvalue

$$\begin{aligned}
\mathcal{E}_- &= \frac{-\frac{\hbar\Delta}{2}c_- + gs_-\sqrt{(m+1)_{\rm cav}}}{c_-} \\
&= -\frac{\hbar\Delta}{2} - g\sqrt{(m+1)_{\rm cav}}\sqrt{\frac{1-\frac{\hbar\Delta}{\sqrt{(\hbar\Delta)^2+4g^2(m+1)_{\rm cav}}}}{1+\frac{\hbar\Delta}{\sqrt{(\hbar\Delta)^2+4g^2(m+1)_{\rm cav}}}}} \\
&= -\frac{\hbar\Delta}{2} - g\sqrt{(m+1)_{\rm cav}}\sqrt{\frac{\sqrt{(\hbar\Delta)^2+4g^2(m+1)_{\rm cav}}-\hbar\Delta}{\sqrt{(\hbar\Delta)^2+4g^2(m+1)_{\rm cav}}+\hbar\Delta}} \\
&= -\frac{\hbar\Delta}{2} - g\sqrt{(m+1)_{\rm cav}}\frac{\sqrt{(\hbar\Delta)^2+4g^2(m+1)_{\rm cav}}-\hbar\Delta}{\sqrt{4g^2(m+1)_{\rm cav}}} \\
&= -\sqrt{\left(\frac{\hbar\Delta}{2}\right)^2 + g^2(m+1)_{\rm cav}}.
\end{aligned} \tag{5.22}$$

It should be noticed that in the above $\sin(2\theta)$ is negative while $\cos(2\theta)$ is positive. Now we shall consider another case, in which $\sin(2\theta)$ is positive while $\cos(2\theta)$ is negative, i.e., $\sin(2\theta) = \frac{2g\sqrt{(m+1)_{\rm cav}}}{\sqrt{(\hbar\Delta)^2+4g^2(m+1)_{\rm cav}}}$ and $\cos(2\theta) = -\frac{\hbar\Delta}{\sqrt{(\hbar\Delta)^2+4g^2(m+1)_{\rm cav}}}$ ($\Delta > 0$). Since $\sin(2\theta) > 0$, $\cos\theta$ and $\sin\theta$ must be the same in sign, we can choose

$$c_+ = \sqrt{\frac{1-\frac{\hbar\Delta}{\sqrt{(\hbar\Delta)^2+4g^2(m+1)_{\rm cav}}}}{2}}, \qquad s_+ = \sqrt{\frac{1+\frac{\hbar\Delta}{\sqrt{(\hbar\Delta)^2+4g^2(m+1)_{\rm cav}}}}{2}}. \tag{5.23}$$

Now the energy eigenstate of the surface-plasmon–cavity-mode interacting system is $|+\rangle = c_+|n_{\rm sp}, (m+1)_{\rm cav}\rangle + s_+|(n+1)_{\rm sp}, m_{\rm cav}\rangle$, and the corresponding energy eigenvalue is given by

$$\begin{aligned}\mathcal{E}_+ &= \frac{-\frac{\hbar\Delta}{2}c_+ + gs_+\sqrt{(m+1)_{\rm cav}}}{c_+} \\ &= -\frac{\hbar\Delta}{2} + g\sqrt{(m+1)_{\rm cav}}\sqrt{\frac{1+\frac{\hbar\Delta}{\sqrt{(\hbar\Delta)^2+4g^2(m+1)_{\rm cav}}}}{1-\frac{\hbar\Delta}{\sqrt{(\hbar\Delta)^2+4g^2(m+1)_{\rm cav}}}}} \\ &= -\frac{\hbar\Delta}{2} + g\sqrt{(m+1)_{\rm cav}}\sqrt{\frac{\sqrt{(\hbar\Delta)^2+4g^2(m+1)_{\rm cav}}+\hbar\Delta}{\sqrt{(\hbar\Delta)^2+4g^2(m+1)_{\rm cav}}-\hbar\Delta}} \\ &= -\frac{\hbar\Delta}{2} + g\sqrt{(m+1)_{\rm cav}}\frac{\sqrt{(\hbar\Delta)^2+4g^2(m+1)_{\rm cav}}+\hbar\Delta}{\sqrt{4g^2(m+1)_{\rm cav}}} \\ &= +\sqrt{\left(\frac{\hbar\Delta}{2}\right)^2 + g^2(m+1)_{\rm cav}}. \end{aligned} \tag{5.24}$$

The result for the present problem is that $|+\rangle$ and $|-\rangle$ can be expressed in a unified form: i.e., we have $c = c_+$, $s = s_+$, and $c_- = s$, $s_- = -c$, and the two normalized orthogonal eigenstates are

$$\begin{aligned}|+\rangle &= c|n_{\rm sp},(m+1)_{\rm cav}\rangle + s|(n+1)_{\rm sp}, m_{\rm cav}\rangle, \\ |-\rangle &= s|n_{\rm sp},(m+1)_{\rm cav}\rangle - c|(n+1)_{\rm sp}, m_{\rm cav}\rangle \\ &= {\rm e}^{{\rm i}\pi}\Big(c|(n+1)_{\rm sp}, m_{\rm cav}\rangle - s|n_{\rm sp},(m+1)_{\rm cav}\rangle\Big). \end{aligned} \tag{5.25}$$

Since the constant phase π in $|-\rangle$ can be omitted, the two eigenstates can be rewritten as

$$\begin{aligned}|+\rangle &= c|n_{\rm sp},(m+1)_{\rm cav}\rangle + s|(n+1)_{\rm sp}, m_{\rm cav}\rangle, \\ |-\rangle &= c|(n+1)_{\rm sp}, m_{\rm cav}\rangle - s|n_{\rm sp},(m+1)_{\rm cav}\rangle. \end{aligned} \tag{5.26}$$

They are orthogonal and normalized states ($\langle+|+\rangle = 1$, $\langle-|-\rangle = 1$ and $\langle+|-\rangle = 0$).

The two eigenstates can be understood alternatively as follows: Note that $\tan(2\theta) = \frac{2\tan\theta}{1-\tan^2\theta}$ can also be $\tan(2\theta) = \frac{2\left(\frac{-1}{\tan\theta}\right)}{1-\left(\frac{-1}{\tan\theta}\right)^2}$. If, therefore, $c|n_{\rm sp},(m+1)_{\rm cav}\rangle + s|(n+1)_{\rm sp}, m_{\rm cav}\rangle$ is one of the eigenstates, then clearly $-s|n_{\rm sp},(m+1)_{\rm cav}\rangle + c|(n+1)_{\rm sp}, m_{\rm cav}\rangle$ is the other eigenstate.

The above results can also be derived in an alternative way: Since $\tan(2\theta) = \frac{2\tan\theta}{1-\tan^2\theta}$, we can obtain $\tan^2\theta + \frac{2}{\tan(2\theta)}\tan\theta - 1 = 0$ and $\tan\theta = -\frac{1}{\tan(2\theta)} \pm \sqrt{\frac{1}{\tan^2(2\theta)} + 1}$. Then the explicit expression for $\tan\theta$ is given by

$$\begin{aligned} \tan\theta &= \frac{\hbar\Delta}{2g\sqrt{(m+1)_{\mathrm{cav}}}} \pm \sqrt{\frac{(\hbar\Delta)^2}{4g^2(m+1)_{\mathrm{cav}}} + 1} \\ &= \frac{\hbar\Delta \pm \sqrt{(\hbar\Delta)^2 + 4g^2(m+1)_{\mathrm{cav}}}}{2g\sqrt{(m+1)_{\mathrm{cav}}}}. \end{aligned} \tag{5.27}$$

The corresponding energy eigenvalues are given by

$$\mathcal{E}_{\pm} = -\frac{\hbar\Delta}{2} + g\sqrt{(m+1)_{\mathrm{cav}}}\tan\theta = \pm\sqrt{\left(\frac{\hbar\Delta}{2}\right)^2 + g^2(m+1)_{\mathrm{cav}}}. \tag{5.28}$$

When $\tan\theta = \frac{\hbar\Delta+\sqrt{(\hbar\Delta)^2+4g^2(m+1)_{\mathrm{cav}}}}{2g\sqrt{(m+1)_{\mathrm{cav}}}}$, we can have

$$\begin{aligned} \cos\theta &= \sqrt{\frac{1 - \frac{\hbar\Delta}{\sqrt{(\hbar\Delta)^2+4g^2(m+1)_{\mathrm{cav}}}}}{2}} \equiv c, \\ \sin\theta &= \sqrt{\frac{1 + \frac{\hbar\Delta}{\sqrt{(\hbar\Delta)^2+4g^2(m+1)_{\mathrm{cav}}}}}{2}} \equiv s. \end{aligned} \tag{5.29}$$

When $\tan\theta = \frac{\hbar\Delta-\sqrt{(\hbar\Delta)^2+4g^2(m+1)_{\mathrm{cav}}}}{2g\sqrt{(m+1)_{\mathrm{cav}}}}$, we can have

$$\begin{aligned} \cos\theta &= \sqrt{\frac{1 + \frac{\hbar\Delta}{\sqrt{(\hbar\Delta)^2+4g^2(m+1)_{\mathrm{cav}}}}}{2}} \equiv s, \\ \sin\theta &= -\sqrt{\frac{1 - \frac{\hbar\Delta}{\sqrt{(\hbar\Delta)^2+4g^2(m+1)_{\mathrm{cav}}}}}{2}} \equiv -c. \end{aligned} \tag{5.30}$$

Thus, the eigenstates are the ones that have been given in Eqs. (5.25) and (5.26) for the interaction between the *metallic-nanosphere quantized surface plasmons* and the *quantum cavity modes*.

ii) *Discussion: when $m_{\rm cav}$ is small:*

When the cavity mode index $m_{\rm cav}$ is small, let us address the above issue again but briefly. In Eq. (5.19), we consider the case of resonance $\Delta = 0$ (the frequency detuning is zero). It can be found that $\cos(2\theta)$ is determined by $\frac{1-\cos(2\theta)}{2}\sqrt{(m+1)_{\rm cav}} = \frac{1+\cos(2\theta)}{2}\sqrt{m_{\rm cav}}$. Then we can obtain $\cos(2\theta) = \frac{\sqrt{(m+1)_{\rm cav}}-\sqrt{m_{\rm cav}}}{\sqrt{(m+1)_{\rm cav}}+\sqrt{m_{\rm cav}}}$. Then by using $\sin\theta = \pm\sqrt{\frac{1-\cos(2\theta)}{2}}$ and $\cos\theta = \pm\sqrt{\frac{1+\cos(2\theta)}{2}}$, where $1-\cos(2\theta) = \frac{2\sqrt{m_{\rm cav}}}{\sqrt{(m+1)_{\rm cav}}+\sqrt{m_{\rm cav}}}$ and $1+\cos(2\theta) = \frac{2\sqrt{(m+1)_{\rm cav}}}{\sqrt{(m+1)_{\rm cav}}+\sqrt{m_{\rm cav}}}$, we can obtain $\tan\theta = \pm\frac{m_{\rm cav}^{\frac{1}{4}}}{(m+1)_{\rm cav}^{\frac{1}{4}}}$. Now the energy eigenvalues corresponding to these two values of $\tan\theta$ are given by

$$\mathcal{E}_{\pm} = \pm g\sqrt{(m+1)_{\rm cav}}\frac{m_{\rm cav}^{\frac{1}{4}}}{(m+1)_{\rm cav}^{\frac{1}{4}}} = \pm g m_{\rm cav}^{\frac{1}{4}}(m+1)_{\rm cav}^{\frac{1}{4}}. \tag{5.31}$$

This is the energy eigenvalues corresponding to the coupling of the metallic-nanosphere surface plasmons to the quantum cavity modes.

5.2.3. The Case of Off-Resonance between the Metallic-Nanosphere Surface Plasmons and the Quantum Cavity Modes

In the above, we have considered the case of Fröhlich condition $\epsilon_{\rm in} = -2\epsilon_{\rm out}$ for the resonance between the metallic-nanosphere surface plasmons and the quantum cavity modes. Now we shall consider the case, in which the Fröhlich condition is not satisfied, i.e., $\epsilon_{\rm in} \neq -2\epsilon_{\rm out}$ [55]. According to the ordinary quantization formulation presented in some textbooks [109, 110] such as advanced quantum mechanics, quantum optics and quantum field theory, it can be found that the cavity-mode quantized field is given by $\hat{\mathbf{E}} = {\rm i}E_0\left(\hat{\mathcal{B}}\mathbf{u}^*(\mathbf{r}){\rm e}^{-{\rm i}\omega t} - \hat{\mathcal{B}}^\dagger\mathbf{u}(\mathbf{r}){\rm e}^{{\rm i}\omega t}\right)$, where $E_0 = \sqrt{\frac{\hbar\omega}{2\epsilon_0\mathcal{V}}}$ and the effective volume $\mathcal{V} = \int|\mathbf{u}(\mathbf{r})|^2{\rm d}V$. The induced electric dipole moment on the plasmonic nanosphere is $\hat{\mathbf{p}} = 4\pi\epsilon_0\epsilon_{\rm out}R^3\left(\frac{\epsilon_{\rm in}-\epsilon_{\rm out}}{\epsilon_{\rm in}+2\epsilon_{\rm out}}\right)\hat{\mathbf{E}}$ [55]. Therefore, in the nanosphere-mediated interaction between the cavity-mode photons and the nanosphere-supported surface plasmons, the interaction Hamiltonian operator is given by $\hat{V} = -\hat{\mathbf{p}}\cdot\hat{\mathbf{E}}$, where $\hat{\mathbf{E}}^2$

is expressed by $\hat{\mathbf{E}}^2 = -E_0^2\Big(- \hat{\mathcal{B}}\hat{\mathcal{B}}^\dagger \mathbf{u}^*(\mathbf{r}) \cdot \mathbf{u}(\mathbf{r}) - \hat{\mathcal{B}}^\dagger\hat{\mathcal{B}} \mathbf{u}^*(\mathbf{r}) \cdot \mathbf{u}(\mathbf{r}) + \hat{\mathcal{B}}^2 (\mathbf{u}^*)^2 \mathrm{e}^{-\mathrm{i}2\omega t} + (\hat{\mathcal{B}}^\dagger)^2 \mathbf{u}^2 \mathrm{e}^{\mathrm{i}2\omega t}\Big)$. Now the interaction Hamiltonian is

$$\begin{aligned}
\hat{V} &= -4\pi\epsilon_0\epsilon_{\mathrm{out}} R^3 \left(\frac{\epsilon_{\mathrm{in}} - \epsilon_{\mathrm{out}}}{\epsilon_{\mathrm{in}} + 2\epsilon_{\mathrm{out}}}\right) \hat{\mathbf{E}}^2 \\
&= -4\pi\epsilon_0\epsilon_{\mathrm{out}} R^3 \left(\frac{\epsilon_{\mathrm{in}} - \epsilon_{\mathrm{out}}}{\epsilon_{\mathrm{in}} + 2\epsilon_{\mathrm{out}}}\right) E_0^2 \left(\hat{\mathcal{B}}\hat{\mathcal{B}}^\dagger + \hat{\mathcal{B}}^\dagger\hat{\mathcal{B}}\right) \mathbf{u}^*(\mathbf{r}) \cdot \mathbf{u}(\mathbf{r}) \\
&= \mathcal{V}_0(\mathbf{r}) \left(\hat{\mathcal{B}}\hat{\mathcal{B}}^\dagger + \hat{\mathcal{B}}^\dagger\hat{\mathcal{B}}\right), \qquad (5.32)
\end{aligned}$$

where $\mathcal{V}_0(\mathbf{r}) = -4\pi\epsilon_0\epsilon_{\mathrm{out}} R^3 \left(\frac{\epsilon_{\mathrm{in}} - \epsilon_{\mathrm{out}}}{\epsilon_{\mathrm{in}} + 2\epsilon_{\mathrm{out}}}\right) E_0^2 \mathbf{u}^*(\mathbf{r}) \cdot \mathbf{u}(\mathbf{r})$. Since the expectation value of $\hat{\mathcal{B}}\hat{\mathcal{B}}^\dagger + \hat{\mathcal{B}}^\dagger\hat{\mathcal{B}}$ is $\langle \hat{\mathcal{B}}\hat{\mathcal{B}}^\dagger + \hat{\mathcal{B}}^\dagger\hat{\mathcal{B}} \rangle = 2m_{\mathrm{cav}} + 1$, we can have

$$V(\mathbf{r}) = (2m_{\mathrm{cav}} + 1)\,\mathcal{V}_0(\mathbf{r}). \qquad (5.33)$$

When the cavity mode index $m_{\mathrm{cav}} = 0$, the quantum-vacuum induced guiding potential is $V(\mathbf{r}) = \mathcal{V}_0(\mathbf{r})$, which occurs in the interaction between the metallic-nanosphere quantum-vacuum surface plasmons and the quantum-vacuum cavity modes.

In the above, we have presented a theoretical framework of nanosphere-mediated interaction between the cavity-mode photons and the subwavelength surface plasmons at quantum level. The essential idea is that the plasmonic nanosphere can be identified as an induced electric dipole $\hat{\mathbf{p}}$, which can be coupled to the quantum cavity mode field $\hat{\mathbf{E}}$, and thus the interaction Hamiltonian operator for characterizing the coupling of the metallic-nanosphere surface plasmons to the cavity modes can be of the form $\hat{V} = -\hat{\mathbf{p}} \cdot \hat{\mathbf{E}}$. Within this theoretical framework, some new issues such as nanosphere quantum levitation (via quantum optically induced force) and plasmonic vacuum-vacuum interaction can be considered. Here, the "vacuum-vacuum interaction" means the coupling of the nanosphere surface plasmon quantum-vacuum modes to the other quantum-vacuum modes in the QED cavity. Clearly, the electromagnetic field is a linear (Abelian) gauge field, a vacuum mode cannot be coupled directly to another vacuum modes. If, however, two quantum-vacuum modes can be mediated by a third one (such as an electric current or a nanoscale metallic sphere), we can say that these two vacuum modes "interact with" each other. As a matter of fact, the "vacuum-vacuum interaction" is actually the "vacuum-nanosphere-vacuum interaction".

5.3. Localized Surface Plasmon Resonance on Subatomic Particles

According to the result (2.45), the induced electric dipole moment of a subwavelength spherical particle with radius a is given by $p = \frac{\epsilon_{\rm in}-\epsilon_{\rm out}}{\epsilon_{\rm in}+2\epsilon_{\rm out}}4\pi\epsilon_{\rm out}E_0a^3$ [55]. Then the interaction energy of the induced electric dipole with the external field E_0 is given by

$$V = -\mathbf{p}\cdot\mathbf{E}_0 = -\frac{\epsilon_{\rm in}-\epsilon_{\rm out}}{\epsilon_{\rm in}+2\epsilon_{\rm out}}4\pi\epsilon_{\rm out}E_0^2a^3. \tag{5.34}$$

This would be ignored when the radius a is as small as possible. But the possible localized surface plasmon excitation may still deserve consideration.

From the equation (2.45), the induced electric dipole moment of the subwavelength sphere can be rewritten as $p = \frac{\epsilon_{\rm in}-\epsilon_{\rm out}}{\epsilon_{\rm in}+2\epsilon_{\rm out}}4\pi\epsilon_{\rm out}E_{\rm ext}a^3 = \frac{\epsilon_{\rm in}-\epsilon_{\rm out}}{\epsilon_{\rm in}+2\epsilon_{\rm out}}3\epsilon_{\rm out}E_{\rm ext}V_{\rm sphere}$ with $V_{\rm sphere} = \frac{4\pi}{3}a^3$. From the electric scalar potential $\Phi_{\rm in}(r,\theta) = -\frac{3\epsilon_{\rm out}}{\epsilon_{\rm in}+2\epsilon_{\rm out}}E_{\rm ext}r\cos\theta$ $(r \le a)$ given in Eq. (2.42), the electric field in the subwavelength sphere is $E_{\rm in} = \frac{3\epsilon_{\rm out}}{\epsilon_{\rm in}+2\epsilon_{\rm out}}E_{\rm ext}$. Now the induced electric dipole moment of the subwavelength sphere is of the form

$$p = (\epsilon_{\rm in}-\epsilon_{\rm out})E_{\rm in}V_{\rm sphere} \to -3\epsilon_{\rm out}E_{\rm in}V_{\rm sphere}, \tag{5.35}$$

where in the last step, we have used the surface plasmon resonance condition $\epsilon_{\rm in} = -2\epsilon_{\rm out}$.

The mode of surface plasmons is a bound state on a subwavelength sphere. When there is an electromagnetic field $E_{\rm ext}$, of which the frequency satisfies the surface plasmon resonance condition $\epsilon_{\rm in} \to -2\epsilon_{\rm out}$, the surface plasmon modes will be drastically excited. According to the field quantization formulation, there must be the localized surface plasmons at quantum vacuum level on the subwavelength sphere. But this conclusion assumes that the external electric field is nonzero. If there is no external electric field, then there are no such surface plasmons at quantum vacuum level, either. In the free vacuum, there is quantum-vacuum zero-point electromagnetic field. However, $E_{\rm vac} = \sqrt{\frac{\hbar\omega}{2\epsilon_0 V}}$ with

$V \rightarrow \infty$ (in the free vacuum), and E_{vac} is negligibly small. Though the denominator $\epsilon_{\mathrm{in}} + 2\epsilon_{\mathrm{out}}$ in the induced electric dipole moment $p = \frac{\epsilon_{\mathrm{in}}-\epsilon_{\mathrm{out}}}{\epsilon_{\mathrm{in}}+2\epsilon_{\mathrm{out}}} 4\pi\epsilon_{\mathrm{out}} E_{\mathrm{ext}} a^3$ of the subwavelength sphere can also be quite small, yet E_{vac} (i.e., E_{ext}) is a smaller quantity of higher order than the denominator $\epsilon_{\mathrm{in}} + 2\epsilon_{\mathrm{out}}$, because $\epsilon_{\mathrm{in}} = \epsilon_{\infty} - \frac{\omega_{\mathrm{p}}^2}{\omega(\omega+\mathrm{i}\gamma)}$, which has an imaginary part, and hence $\epsilon_{\mathrm{in}} + 2\epsilon_{\mathrm{out}}$ cannot be identically zero. In a quantum cavity, however, $E_{\mathrm{vac}} = \sqrt{\frac{\hbar\omega}{2\epsilon_0 V}}$ can be quite large and the surface plasmons at quantum vacuum level can emerge.

As is well known, the electric dipole moment of neutron is related to CP symmetry breaking. We would like to point out that because of the excitation of localized surface plasmons, neutrons may have their induced electric dipole moments in the external electric field (regardless of whether the external electric field is static or dynamical).

For the localized surface plasmons sustained on the proton and neutron, we would have three topics that deserve further consideration: i) *Proton spin problem*: It should be pointed out that the angular quantum number of the localized surface plasmons on the proton and neutron would be $l = 1$. The magnetic vector potential $\mathbf{A}$ can be ignored. Thus, the spin current density $\mathbf{s} = \epsilon_0 \mathbf{E} \times \mathbf{A}$ of the surface plasmons could also be ignored. It is also worth addressing the problem whether the orbital angular momentum of localized surface plasmons makes a contribution to (or has an influence on) the spin of proton and neutron; ii) *Proton and neutron electric dipole problem*: In the nanosphere surface plasmonics, it has been shown that the metallic nanosphere with negative permittivity illuminated by an external optical field can be considered to be an effective electric dipole, because the induced electric field distribution of the metallic nanosphere is exactly the same as an electric dipole, and therefore, we think that the neutron and proton would also have such an (equivalent) electric dipole moment; iii) *Proton radius problem*: Since the electric charge would be rearranged or redistributed inside the proton because of the (localized) surface plasmon resonance (excitation) and an equivalent electric dipole is induced, we believe that the mass radius and charge radius of the proton would change due to the effect of surface plasmon resonance or excitation on the proton.

A nucleus consists of protons and neutrons. According to the Fermi gas model of nucleus, protons can move relatively freely in the nucleus, so the charged particles in the nucleus can be regarded as a system described by an ionic gas model, which has the dielectric coefficient of plasma or ionic gas ($\epsilon_{\mathrm{r}} = 1 - \frac{\omega_{\mathrm{p}}^2}{\omega^2}$). Therefore, we firmly believe that the localized surface plasmon modes can also be excited on a nucleus by the γ photons at appropriate frequencies ($\omega = \frac{\omega_{\mathrm{p}}}{\sqrt{3}}$, where we have used the surface plasmon resonance condition $\epsilon_{\mathrm{r}} + 2 = 0$). Thus, some nuclear properties (e.g., nuclear energy level shift and electric charge redistribution) relevant to the surface plasmon resonance also deserve consideration.

For an atom, there are electron wavefunctions inside it. Though we can define a plasma frequency $\omega_{\mathrm{p}} = \sqrt{\frac{n_{\mathrm{e}} e^2}{\epsilon_0 m_{\mathrm{e}}}}$ with the electron number density $n_{\mathrm{e}} \simeq 10^{31}\ \mathrm{m}^{-3}$, yet it may be not easy to excite the localized surface plasmons: specifically, the atom-internal plasma frequency ω_{p} has the order of magnitude of $10^{17}\ s^{-1}$. By using the surface plasmon resonance condition $\epsilon_{\mathrm{in}} + 2\epsilon_{\mathrm{out}} = 0$ (i.e., $1 - \frac{\omega_{\mathrm{p}}^2}{\omega^2} + 2 = 0$), we obtain the surface plasmon resonance frequency $\omega = \frac{\omega_{\mathrm{p}}}{\sqrt{3}}$. The energy $\hbar\omega$ for creating a plasmon in the atom has the order of 40 eV, which is much higher than the ground state energy $|E_{\mathrm{g}}|$ of the atom. Thus, before the localized surface plasmons are excited, the atom has been ionized. If the outermost electrons of the atom have been stripped and the energy $|E|$ of the inner electrons can be higher than 40 eV, then high-energy photons could possibly excite the localized surface plasmons in this case. We think the localized surface plasmons at quantum vacuum level could also exist in atoms (such quantum-vacuum localized surface plasmons would be created when the inner electrons form the bound energy levels inside the atom).

Can the localized surface plasmons be excited on the electron? From the theory of localized surface plasmons, it can be found that the surface plasmon resonance occurs on the premise that there is a plasma gas or an ion gas inside the spherical particle, so that the spherical particle has a negative dielectric coefficient inside. Without such a negative dielectric coefficient, the strong surface plasmons cannot be excited. Naturally, if a particle has no its own internal structure, that is, it is not a composite particle and has no subparticles inside, then no surface plasmon resonance on it will take place. Then is the electron an elementary particle without its own internal structure? We argue that this problem is related intimately to the (ordinary and anomalous) magnetic dipole moments of leptons [111, 112, 113].

Let us mention the brief situation of anomalous magnetic moments of leptons [114, 115, 116]. According to Dirac's relativistic quantum mechanics, the spinning magnetic dipole moment of a charged point particle without internal structure (e.g., an electron) is given by $\mathbf{m} = g\frac{e}{2m}\mathbf{s}$ with the Landé factor $g = -2$ and e is the elementary charge ($e > 0$). In quantum electrodynamics, where the interaction between electron and quantum vacuum fluctuation field (an effect of single loop Feynman diagram) is taken into account, Schwinger showed that the electron Landé factor $g_{\mathrm{e}} = -2(1 + \frac{\alpha}{2\pi}) = -2.002319$, where α is the electromagnetic fine structure constant. According to some textbooks and theoretical essays of particle physics [117, 118, 119], including some papers on anomalous magnetic dipole moments of leptons [112, 113] (and references therein), one of the early versions for the exact theoretical value of electron Landé factor is $g_{\mathrm{e}} = -2(1 + \frac{\alpha}{2\pi}) = -2.002319304402$, where the electromagnetic fine structure constant is given by $\alpha = \frac{1}{137.03599911}$. The recent particle-physics Standard-Model theoretical datum for the Landé factor of electron is $g_{\mathrm{e}} = -2.0023193043632246$ and the Landé factor of muon lepton is $g_{\mu} = -2.0023318362086$ [111]. The experimental Landé factors of modern version provided by the Committee on Data for Science and Technology (CODATA, http://www.codata.org/ and http://physics.nist.gov/constants) are as follows: $g_{\mathrm{e}} = -2.00231930436256(35)$, $g_{\mu} = -2.0023318418(13)$, $g_{\mathrm{p}} = +5.5856946893(16)$ and $g_{\mathrm{n}} = -3.82608545(90)$ [111].

At present, the agreement between theoretical calculation and experimental measurements of the anomalous magnetic dipole moments of electron and muon [112, 113] are up to about ten and eight significant digits, respectively [117, 118, 119]. This fully demonstrates that *leptons are elementary particles that have no internal structure.* As long as it has an internal subparticle structure, even if the size of the internal structure is only the Planck length, its magnetic dipole moment and anomalous magnetic moment would be far from the theoretical value of Standard Model: specifically, if the electron and muon have internal subparticles, even if there are only two subparticles, then the surface plasmons should be excited on the electron and muon in the high-energy particle scattering processes, so that the electric and magnetic dipole moments of electron and muon unavoidably deviate from the Standard-Model predicted values. As has been shown in the topic of localized surface plasmon resonance, the condition of surface plasmon excitation is $\epsilon_{\mathrm{in}} + 2\epsilon_{\mathrm{out}} = 0$ [55] with $\epsilon_{\mathrm{in}} = 1 - \frac{\omega_{\mathrm{p}}^2}{\omega^2}$ and $\epsilon_{\mathrm{out}} = 1$. It can be found that the resonance frequency is $\omega = \frac{\omega_{\mathrm{p}}}{\sqrt{3}}$.

Here, the plasma frequency is $\omega_{\rm p} = \sqrt{\frac{ne^2}{\epsilon_0 m}}$ and the plasma number density is $n \simeq \frac{1}{a^3}$. If the upper limit of the charged lepton radius was $a \simeq 10^{-17}$ m, the internal plasma frequency $\omega_{\rm p}$ has the order of magnitude of 10^{26} $\rm s^{-1}$. Then $\hbar\omega_{\rm p}$ has the order of 100 GeV. If, therefore, the electron has an internal subparticle structure inside at the size $\sim 10^{-17}$ m, the particle scattering at 100 GeV would excite the surface plasmons on the electron. Although the "localized surface plasmons" is a concept in applied electromagnetism and optics [55], it could also help to solve certain problems in particle physics.

The theoretical values of the QED Landé factor of the electron and muon are highly consistent with the experimental values [112, 113]. If we deduct the electron-vacuum interaction that is a quantum electrodynamic effect, the Landé factor of elementary particle without internal structure must be $|g| = 2$. Then it follows from this viewpoint that the electron and muon are such elementary particles without internal structure. Let us consider a composite particle consisting of two subparticles, of which the spinning magnetic dipole moments are given by $\mathbf{m}_1 = \frac{q_1}{m_1}\mathbf{s}_1$ and $\mathbf{m}_2 = \frac{q_2}{m_2}\mathbf{s}_2$. Then the sum of the two subparticle magnetic dipole moments is $\mathbf{m} = \mathbf{m}_1 + \mathbf{m}_2 = \frac{q_1}{m_1}\mathbf{s}_1 + \frac{q_2}{m_2}\mathbf{s}_2$. When can it become $\mathbf{m} = \frac{q_1+q_2}{m_1+m_2}(\mathbf{s}_1 + \mathbf{s}_2)$? It can be found that only when $\frac{q_1}{m_1} = \frac{q_2}{m_2}$, i.e., $\frac{q_1}{m_1} = \frac{q_2}{m_2} = \frac{q_1+q_2}{m_1+m_2}$, we have $\mathbf{m} = \frac{q_1+q_2}{m_1+m_2}(\mathbf{s}_1 + \mathbf{s}_2)$. If a particle consists of two subparticles, i.e, one has an electric charge and the other has no electric charge, or the two subparticles have different charge-mass ratios, the Landé factor of this composite particle is no longer $|g| = 2$. For example, the proton has the Landé factor $g_{\rm p} = 5.6$, which is not $g = 2$, and this means that the proton is a composite particle, which contains quarks and gluons; The Landé factor for neutron is $g_{\rm n} = -3.8$, and this means that the charge-mass ratio $|\frac{e}{m}|$ of the negative-electric-charge quarks inside the neutron is larger than that of positive-electric-charge quarks (including the sea quarks). (According to some textbooks and theoretical essays of particle physics, the more exact values of the Landé factors of proton and neutron are $g_{\rm p} = 5.585695$ and $g_{\rm n} = -3.826085$, respectively.) Therefore, whether the Landé factor is ± 2 or not can be seen as an most important indicator of whether a spinning particle is elementary or composite.

The ancient Chinese philosopher Chuang Tzu (369 B.C.—286 B.C.) put forward a famous saying, "If you have a one-foot stick, you take half of it each day and you can never finish cutting it to the end for all ages." This philosophical saying reflects the idea that substance is infinitely divisible. But as a matter of fact, from the experimental and theoretical data for magnetic dipole moments of particles (such as electron and muon) it is shown that this philosophical saying is invalid for electron and muon, i.e., electron and muon are elementary particles, and both electron and muon can no longer be separated or broken by any current technology. In other words, maybe any theories of subleptons or subquarks could be abandoned.

6. Surface Electromagnetic Waves Sustained on a Single Interface

In the history, the earliest theories of surface plasmonics were suggested by Sommerfeld, Zenneck and Fano and the theory of localized surface plasmons in metallic nanostructures was suggested by Mie [55]. With the development of optical materials and relevant technology, the surface plasmonics has been extensively employed in many research fields, including chemical and biological measurement techniques.

Here we shall present a theoretical formalism of surface plasmon polaritons sustained on a single, smooth interface. The following mathematical formalism for surface plasmon polaritons sustained on a two-media interface is derived within (based on) the framework of some monographs on surface plasmonics [1, 94]. This detailed derivation has been presented and taught in my classroom lecture for a graduate course "Electromagnetic Wave Theory" since the spring semester of 2012 [120], where two cases, i.e., so-called electrically resonant media [i.e., $(\epsilon_1\epsilon_0, \mu_0)$ and $(\epsilon_2\epsilon_0, \mu_0)$] and magnetically resonant media [i.e., $(\epsilon\epsilon_0, \mu_1\mu_0)$ and $(\epsilon\epsilon_0, \mu_2\mu_0)$], were studied separately in two subsections [120]. But now these two cases are derived in a unified formalism. Now we put this unified derivation for SPPs (surface plasmon polaritons) into this Note for beginners in surface plasmonics and metal optics.

We assume the interface is on the y-z plane and the SPPs propagate in the z direction. The media in the half spaces of $x > 0$ and $x < 0$ have the relative permittivity and permeability (ϵ_1, μ_1) and (ϵ_2, μ_2), respectively. The Maxwell field equations are given by $\nabla \times \mathbf{E} = -\frac{\partial \mathbf{B}}{\partial t}$ (Faraday's electromagnetic induction law) and $\nabla \times \mathbf{H} = \frac{\partial \mathbf{D}}{\partial t}$ (Ampère's circuital law).

The Helmholtz equation is $\nabla^2\mathbf{E} + \mu\epsilon k_0^2\mathbf{E} = 0$, where μ and ϵ denote the relative magnetic permeability and the relative electric permittivity, respectively.

Since the surface waves propagate on the interface but decay exponentially into the depth of $\frac{1}{\alpha}$ perpendicular to the interface, we can directly write the following tentative solution for the surface electromagnetic wave as follows [94]:

$$\begin{aligned}\mathbf{E}_1 &= \mathbf{E}_1^{(0)}\exp(-\alpha_1 x)\exp[\mathrm{i}(\beta z - \omega t)] \qquad (x > 0),\\ \mathbf{E}_2 &= \mathbf{E}_2^{(0)}\exp(+\alpha_2 x)\exp[\mathrm{i}(\beta z - \omega t)] \qquad (x < 0).\end{aligned} \tag{6.1}$$

Here, the attenuation constants in medium 1 (in the region of $x > 0$) and medium 2 (in the region of $x < 0$) are given by $\alpha_1 = \sqrt{\beta^2 - \mu_1\epsilon_1 k_0^2}$ and $\alpha_2 = \sqrt{\beta^2 - \mu_2\epsilon_2 k_0^2}$, where the vacuum wave number is $k_0 = \frac{\omega}{c}$. This can be obtained as follows: According to the Helmholtz equation, we have $k_{1x}^2 + k_{1y}^2 + k_{1z}^2 = \mu_1\epsilon_1 k_0^2$, where $k_{1x} = \mathrm{i}\alpha_1$, $k_{1y} = 0$ and $k_{1z} = \beta$; $k_{2x}^2 + k_{2y}^2 + k_{2z}^2 = \mu_2\epsilon_2 k_0^2$, where $k_{2x} = -\mathrm{i}\alpha_2$, $k_{2y} = 0$ and $k_{2z} = \beta$.

For the single-interface geometry configuration in this problem, the incidence plane is the z-x plane and the field is uniform in the y direction, that is, the first-order derivative of the field with respect to the y coordinate vanishes [1, 94]. Such incident electromagnetic waves can be divided into TM and TE modes. For the TM mode, the non-zero magnetic field component is perpendicular to the direction of propagation (say, in the z direction); The non-zero electric field component of the TE mode is perpendicular to the direction of propagation. In our reference [120], the TM and TE modes for the present surface waves were derived and considered separately, but here we shall treat them together in a unified way.

For the first, we will employ the Gauss law in the electric fields given in Eq. (6.1). By making use of the electric Gauss law $\nabla \cdot \mathbf{E}_1 = 0$, we should calculate the following terms $\partial_i E_{1i}$ (with $i = x, y, z$):

$$\begin{aligned}\partial_x E_{1x} &= -\alpha_1 E_{1x}^{(0)}\exp(-\alpha_1 x)\exp[\mathrm{i}(\beta z - \omega t)],\\ \partial_y E_{1y} &= 0,\\ \partial_z E_{1z} &= \mathrm{i}\beta E_{1z}^{(0)}\exp(-\alpha_1 x)\exp[\mathrm{i}(\beta z - \omega t)].\end{aligned} \tag{6.2}$$

Here, the subscript "1" represents the electric field in medium 1 (in the region of $x > 0$) and x, y, z stand for the spatial coordinates. In the Gauss law $\nabla \cdot \mathbf{E}_1 = 0$, where $\nabla \cdot \mathbf{E}_1 = \partial_x E_{1x} + \partial_y E_{1y} + \partial_z E_{1z}$ and $\partial_x E_{1x} + \partial_y E_{1y} + \partial_z E_{1z} = (-\alpha_1 E_{1x}^{(0)} + \mathrm{i}\beta E_{1z}^{(0)})\exp(-\alpha_1 x)\exp[\mathrm{i}(\beta z - \omega t)]$, we will have

$$-\alpha_1 E_{1x}^{(0)} + \mathrm{i}\beta E_{1z}^{(0)} = 0 \quad \Rightarrow \quad E_{1x}^{(0)} = \frac{\mathrm{i}\beta}{\alpha_1} E_{1z}^{(0)}. \tag{6.3}$$

For the Gauss law in medium 2, we should also calculate the following terms $\partial_i E_{2i}$:

$$\begin{aligned} \partial_x E_{2x} &= +\alpha_2 E_{2x}^{(0)} \exp(+\alpha_2 x)\exp[\mathrm{i}(\beta z - \omega t)], \\ \partial_y E_{2y} &= 0, \\ \partial_z E_{2z} &= \mathrm{i}\beta E_{2z}^{(0)} \exp(+\alpha_2 x)\exp[\mathrm{i}(\beta z - \omega t)]. \end{aligned} \tag{6.4}$$

In the Gauss law $\nabla \cdot \mathbf{E}_2 = 0$, where $\nabla \cdot \mathbf{E}_2 = \partial_x E_{2x} + \partial_y E_{2y} + \partial_z E_{2z}$ and $\partial_x E_{2x} + \partial_y E_{2y} + \partial_z E_{2z} = (+\alpha_2 E_{2x}^{(0)} + \mathrm{i}\beta E_{2z}^{(0)})\exp(+\alpha_2 x)\exp[\mathrm{i}(\beta z - \omega t)]$, we have

$$\alpha_2 E_{2x}^{(0)} + \mathrm{i}\beta E_{2z}^{(0)} = 0 \quad \Rightarrow \quad E_{2x}^{(0)} = -\frac{\mathrm{i}\beta}{\alpha_2} E_{2z}^{(0)}. \tag{6.5}$$

We list the explicit expressions for the field strength of the electric fields in the half spaces of $x > 0$ (material 1) and $x < 0$ (material 2) as follows [94]:

$$\begin{aligned} \mathbf{E}_1 &= \left(+\frac{\mathrm{i}\beta}{\alpha_1} E_{1z}^{(0)}, E_{1y}^{(0)}, E_{1z}^{(0)}\right) \exp(-\alpha_1 x)\exp[\mathrm{i}(\beta z - \omega t)] \qquad (x > 0), \\ \mathbf{E}_2 &= \left(-\frac{\mathrm{i}\beta}{\alpha_2} E_{2z}^{(0)}, E_{2y}^{(0)}, E_{2z}^{(0)}\right) \exp(+\alpha_2 x)\exp[\mathrm{i}(\beta z - \omega t)] \qquad (x < 0). \end{aligned} \tag{6.6}$$

With the help of the curl equation (Ampère circuital law) $\nabla \times \mathbf{E} = \mathrm{i}\omega\mu_0\mu\mathbf{H}$, where μ is the relative permeability, the magnetic field strength can be of the form $\mathbf{H} = -\frac{\mathrm{i}}{\omega\mu_0\mu}\nabla \times \mathbf{E}$, i.e.,

$$\mathbf{H} = -\frac{\mathrm{i}}{\omega\mu_0\mu} \begin{pmatrix} e_x & e_y & e_z \\ \partial_x & \partial_y & \partial_z \\ E_x & E_y & E_z \end{pmatrix}. \tag{6.7}$$

In medium 1 ($x > 0$), the magnetic field can be written as $\mathbf{H}_1 = -\frac{\mathrm{i}}{\omega\mu_0\mu_1}(\partial_y E_{1z} - \partial_z E_{1y}, \partial_z E_{1x} - \partial_x E_{1z}, \partial_x E_{1y} - \partial_y E_{1x}) \rightarrow -\frac{\mathrm{i}}{\omega\mu_0\mu_1}(-$

$\partial_z E_{1y}, \partial_z E_{1x} - \partial_x E_{1z}, \partial_x E_{1y}\big)$. The interface is at $x = 0$. Further calculation yields

$$\begin{aligned}\mathbf{H}_1 &= -\frac{\mathrm{i}}{\omega\mu_0\mu_1}\left(-\mathrm{i}\beta E_{1y}^{(0)}, \mathrm{i}\beta\frac{\mathrm{i}\beta}{\alpha_1}E_{1z}^{(0)} + \alpha_1 E_{1z}^{(0)}, -\alpha_1 E_{1y}^{(0)}\right)\\ &\quad\times\exp(-\alpha_1 x)\exp[\mathrm{i}(\beta z - \omega t)]\\ &= \frac{\mathrm{i}}{\omega\mu_0\mu_1}\left(\mathrm{i}\beta E_{1y}^{(0)}, \frac{k_0^2\mu_1\epsilon_1}{\alpha_1}E_{1z}^{(0)}, \alpha_1 E_{1y}^{(0)}\right)\\ &\quad\times\exp(-\alpha_1 x)\exp[\mathrm{i}(\beta z - \omega t)],\end{aligned} \tag{6.8}$$

where $\frac{-\beta^2}{\alpha_1} + \alpha_1 = \frac{-\beta^2+\alpha_1^2}{\alpha_1} = \frac{-\mu_1\epsilon_1 k_0^2}{\alpha_1}$ has been substituted.

It should be pointed out that the above magnetic field $\mathbf{H}_1$ (in the half space of $x > 0$) fulfills the magnetic Gauss law $\nabla\cdot\mathbf{H}_1 = 0$, i.e., $\partial_x H_{1x} + \partial_z H_{1z} = 0$, of which the explicit expression is given by

$$\frac{\mathrm{i}}{\omega\mu_0\mu_1}\big[-\alpha_1\big(\mathrm{i}\beta E_{1y}^{(0)}\big) + \mathrm{i}\beta\big(\alpha_1 E_{1y}^{(0)}\big)\big]\mathrm{e}^{-\alpha_1 x}\mathrm{e}^{\mathrm{i}(\beta z-\omega t)} = 0. \tag{6.9}$$

In medium 2 ($x < 0$), the magnetic field can be written as $\mathbf{H}_2 = -\frac{\mathrm{i}}{\omega\mu_0\mu_2}\big(\partial_y E_{2z} - \partial_z E_{2y}, \partial_z E_{2x} - \partial_x E_{2z}, \partial_x E_{2y} - \partial_y E_{2x}\big) \rightarrow -\frac{\mathrm{i}}{\omega\mu_0\mu_2}\big(-\partial_z E_{2y}, \partial_z E_{2x} - \partial_x E_{2z}, \partial_x E_{2y}\big)$. The explicit expression for the magnetic field in medium 2 ($x < 0$) is

$$\begin{aligned}\mathbf{H}_2 &= -\frac{\mathrm{i}}{\omega\mu_0\mu_2}\left(-\mathrm{i}\beta E_{2y}^{(0)}, -\mathrm{i}\beta\frac{\mathrm{i}\beta}{\alpha_2}E_{2z}^{(0)} - \alpha_2 E_{2z}^{(0)}, \alpha_2 E_{2y}^{(0)}\right)\\ &\quad\times\exp(+\alpha_2 x)\exp[\mathrm{i}(\beta z - \omega t)]\\ &= \frac{\mathrm{i}}{\omega\mu_0\mu_2}\left(\mathrm{i}\beta E_{2y}^{(0)}, -\frac{k_0^2\mu_2\epsilon_2}{\alpha_2}E_{2z}^{(0)}, -\alpha_2 E_{2y}^{(0)}\right)\\ &\quad\times\exp(+\alpha_2 x)\exp[\mathrm{i}(\beta z - \omega t)],\end{aligned} \tag{6.10}$$

where $\frac{\beta^2}{\alpha_2} - \alpha_2 = \frac{\beta^2-\alpha_2^2}{\alpha_2} = \frac{k_0^2\mu_2\epsilon_2}{\alpha_2}$ has been inserted.

It should be pointed out that the above magnetic field $\mathbf{H}_2$ (in the half space of $x < 0$) fulfills the magnetic Gauss law $\nabla\cdot\mathbf{H}_2 = 0$, i.e., $\partial_x H_{2x} + \partial_z H_{2z} = 0$, of which the explicit expression is given by

$$\frac{\mathrm{i}}{\omega\mu_0\mu_2}\big[\alpha_2\big(\mathrm{i}\beta E_{2y}^{(0)}\big) + \mathrm{i}\beta\big(-\alpha_2 E_{2y}^{(0)}\big)\big]\mathrm{e}^{+\alpha_2 x}\mathrm{e}^{\mathrm{i}(\beta z-\omega t)} = 0. \tag{6.11}$$

Now we will list the result of the magnetic field envelopes, where the shared phase factor $\exp[\mathrm{i}(\beta z - \omega t)]$ has been omitted:

$$\begin{aligned}\tilde{\mathbf{H}}_1 &= \frac{\mathrm{i}}{\omega\mu_0}\left(\mathrm{i}\frac{\beta}{\mu_1}E_{1y}^{(0)}, \frac{k_0^2\epsilon_1}{\alpha_1}E_{1z}^{(0)}, \frac{\alpha_1}{\mu_1}E_{1y}^{(0)}\right)\exp(-\alpha_1 x),\\ \tilde{\mathbf{H}}_2 &= \frac{\mathrm{i}}{\omega\mu_0}\left(\mathrm{i}\frac{\beta}{\mu_2}E_{2y}^{(0)}, -\frac{k_0^2\epsilon_2}{\alpha_2}E_{2z}^{(0)}, -\frac{\alpha_2}{\mu_2}E_{2y}^{(0)}\right)\exp(+\alpha_2 x). \quad (6.12)\end{aligned}$$

When $\mu_1 = \mu_2 = 1$, the solution (6.12) reduces to the ordinary form in the metal-dielectric interface structure [1, 94].

As we know, solving a mathematical problem of a model requires two conditions: field equations and boundary conditions. In the above, we have used the electromagnetic field equations and obtained the general solutions of the electric and magnetic fields. Now we need to use the electromagnetic boundary conditions. The tangential components of the magnetic field $\mathbf{H}$ should be continuous across the interface, i.e., $H_{1y}^{(0)} = H_{2y}^{(0)}$ and $H_{1z}^{(0)} = H_{2z}^{(0)}$. Then we have the following relations:

$$\frac{k_0^2\epsilon_1}{\alpha_1}E_{1z}^{(0)} = -\frac{k_0^2\epsilon_2}{\alpha_2}E_{2z}^{(0)}, \quad \text{or} \quad \frac{\alpha_1}{\mu_1}E_{1y}^{(0)} = -\frac{\alpha_2}{\mu_2}E_{2y}^{(0)}. \quad (6.13)$$

On the other hand, the electric field $\mathbf{E}$ should also be continuous across the interface, i.e., $E_{1y}^{(0)} = E_{2y}^{(0)}$ and $E_{1z}^{(0)} = E_{2z}^{(0)}$. From the equation (6.13), it seems that we should have both $\frac{k_0^2\epsilon_1}{\alpha_1} = -\frac{k_0^2\epsilon_2}{\alpha_2}$ and $\frac{\alpha_1}{\mu_1} = -\frac{\alpha_2}{\mu_2}$. But as a matter of fact, these two conditions cannot simultaneously hold true. Thus, we need to consider two independent cases: i) $E_{1y}^{(0)} = E_{2y}^{(0)} = 0$; ii) $E_{1z}^{(0)} = E_{2z}^{(0)} = 0$.

In the case of $E_{1y}^{(0)} = E_{2y}^{(0)} = 0$, from Eq. (6.12), it can be found that only the y-component of $\mathbf{H}$ is nonzero, i.e., this is a TM mode; In the case of $E_{1z}^{(0)} = E_{2z}^{(0)} = 0$, from Eq. (6.6), it can be found that only the y-component of $\mathbf{E}$ is nonzero, i.e., this is a TE mode.

We will calculate the phase constant β in the z direction for TM and TE modes:

In the TM mode, $E_{1y}^{(0)} = E_{2y}^{(0)} = 0$ and the second relation in Eq. (6.13) is always true, no matter whether $\frac{\alpha_1}{\mu_1} = -\frac{\alpha_2}{\mu_2}$ holds true or not. In the first relation $\frac{\epsilon_1}{\alpha_1} = -\frac{\epsilon_2}{\alpha_2}$ given in Eq. (6.13), by substituting $\alpha_1^2 = \beta^2 - k_0^2\mu_1\epsilon_1$

and $\alpha_2^2 = \beta^2 - k_0^2\mu_2\epsilon_2$ into it, we can obtain

$$\begin{aligned}
&\frac{\epsilon_1^2}{\alpha_1^2} = \frac{\epsilon_2^2}{\alpha_2^2} \quad \Rightarrow \quad \frac{\epsilon_1^2}{\beta^2 - k_0^2\mu_1\epsilon_1} = \frac{\epsilon_2^2}{\beta^2 - k_0^2\mu_2\epsilon_2} \\
\Rightarrow \quad & \epsilon_1^2\left(\beta^2 - k_0^2\mu_2\epsilon_2\right) = \epsilon_2^2\left(\beta^2 - k_0^2\mu_1\epsilon_1\right) \\
\Rightarrow \quad & (\epsilon_2^2 - \epsilon_1^2)\beta^2 = k_0^2(\epsilon_2^2\mu_1\epsilon_1 - \epsilon_1^2\mu_2\epsilon_2) \\
\Rightarrow \quad & \beta = k_0\sqrt{\frac{\epsilon_1\epsilon_2(\mu_1\epsilon_2 - \mu_2\epsilon_1)}{\epsilon_2^2 - \epsilon_1^2}}. \qquad (6.14)
\end{aligned}$$

Now from the envelopes of the fields (6.6) and (6.12), where the shared phase factor $\exp[\mathrm{i}(\beta z - \omega t)]$ is omitted, the result is given by

$$\begin{aligned}
\tilde{\mathbf{E}}_1 &= \left(+\frac{\mathrm{i}\beta}{\alpha_1}E_{1z}^{(0)}, 0, E_{1z}^{(0)}\right)\exp(-\alpha_1 x), \\
\tilde{\mathbf{E}}_2 &= \left(-\frac{\mathrm{i}\beta}{\alpha_2}E_{2z}^{(0)}, 0, E_{2z}^{(0)}\right)\exp(+\alpha_2 x), \qquad (6.15)
\end{aligned}$$

and

$$\begin{aligned}
\tilde{\mathbf{H}}_1 &= \frac{\mathrm{i}}{\omega\mu_0}\left(0, \frac{k_0^2\epsilon_1}{\alpha_1}E_{1z}^{(0)}, 0\right)\exp(-\alpha_1 x), \\
\tilde{\mathbf{H}}_2 &= \frac{\mathrm{i}}{\omega\mu_0}\left(0, -\frac{k_0^2\epsilon_2}{\alpha_2}E_{2z}^{(0)}.0\right)\exp(+\alpha_2 x). \qquad (6.16)
\end{aligned}$$

The equations (6.15) and (6.16) are the spatial distributions of electric and magnetic fields of the TM mode surface wave.

When $\mu_1 = \mu_2 = 1$, from the result (6.14), the phase constant of this TM-mode surface wave is given by

$$\beta = k_0\sqrt{\frac{\epsilon_1\epsilon_2(\mu_1\epsilon_2 - \mu_2\epsilon_1)}{\epsilon_2^2 - \epsilon_1^2}} = k_0\sqrt{\frac{\epsilon_1\epsilon_2}{\epsilon_1 + \epsilon_2}}. \qquad (6.17)$$

This is the well-known ordinary phase constant of the surface plasmon polaritons supported on a single, smooth metal-dielectric interface [1, 94].

In the TE mode, $E_{1z}^{(0)} = E_{2z}^{(0)} = 0$. It follows that when we substitute $\alpha_1^2 = \beta^2 - k_0^2\mu_1\epsilon_1$ and $\alpha_2^2 = \beta^2 - k_0^2\mu_2\epsilon_2$ into $\frac{\alpha_1}{\mu_1} = -\frac{\alpha_2}{\mu_2}$ given in Eq.

(6.13), we can obtain

$$\begin{aligned}
&\frac{\alpha_1^2}{\mu_1^2} = \frac{\alpha_2^2}{\mu_2^2} \quad \Rightarrow \quad \frac{\beta^2 - k_0^2\mu_1\epsilon_1}{\mu_1^2} = \frac{\beta^2 - k_0^2\mu_2\epsilon_2}{\mu_2^2} \\
&\Rightarrow \quad \left(\beta^2 - k_0^2\mu_1\epsilon_1\right)\mu_2^2 = \left(\beta^2 - k_0^2\mu_2\epsilon_2\right)\mu_1^2 \\
&\Rightarrow \quad \beta^2(\mu_2^2 - \mu_1^2) = k_0^2(\mu_1\epsilon_1\mu_2^2 - \mu_2\epsilon_2\mu_1^2) \\
&\Rightarrow \quad \beta = k_0\sqrt{\frac{\mu_1\mu_2(\epsilon_1\mu_2 - \epsilon_2\mu_1)}{\mu_2^2 - \mu_1^2}}. \qquad (6.18)
\end{aligned}$$

Now from the envelopes of the electric and magnetic fields (6.6) and (6.12), where the shared phase factor $\exp[\mathrm{i}(\beta z - \omega t)]$ has been omitted, we have

$$\begin{aligned}
\tilde{\mathbf{E}}_1 &= \left(0, E_{1y}^{(0)}, 0\right)\exp(-\alpha_1 x), \\
\tilde{\mathbf{E}}_2 &= \left(0, E_{2y}^{(0)}, 0\right)\exp(+\alpha_2 x), \qquad (6.19)
\end{aligned}$$

and

$$\begin{aligned}
\tilde{\mathbf{H}}_1 &= \frac{\mathrm{i}}{\omega\mu_0}\left(\mathrm{i}\frac{\beta}{\mu_1}E_{1y}^{(0)}, 0, \frac{\alpha_1}{\mu_1}E_{1y}^{(0)}\right)\exp(-\alpha_1 x), \\
\tilde{\mathbf{H}}_2 &= \frac{\mathrm{i}}{\omega\mu_0}\left(\mathrm{i}\frac{\beta}{\mu_2}E_{2y}^{(0)}, 0, -\frac{\alpha_2}{\mu_2}E_{2y}^{(0)}\right)\exp(+\alpha_2 x). \qquad (6.20)
\end{aligned}$$

The equations (6.19) and (6.20) are the spatial distributions of electric and magnetic fields of the TE mode surface wave.

When $\epsilon_1 = \epsilon_2 = \epsilon_\mathrm{r}$, i.e., the two materials made of the interface are magnetic materials, from the result (6.18), the phase constant of this TE mode reduces to the form

$$\beta = k_0\sqrt{\frac{\mu_1\mu_2(\epsilon_1\mu_2 - \epsilon_2\mu_1)}{\mu_2^2 - \mu_1^2}} = k_0\sqrt{\epsilon_\mathrm{r}}\sqrt{\frac{\mu_1\mu_2}{\mu_1 + \mu_2}}. \qquad (6.21)$$

This is the phase constant of the surface electromagnetic wave modes supported on a single, smooth magnetic interface, where one of the media has a negative magnetic permeability.

For the surface electromagnetic wave confined onto the interface, β should be a real number (i.e., its imaginary part can be negligibly small) and the attenuation constants $\alpha_1 = \sqrt{\beta^2 - \mu_1\epsilon_1 k_0^2}$ and $\alpha_2 = \sqrt{\beta^2 - \mu_2\epsilon_2 k_0^2}$ in the direction perpendicular to the interface should have their real parts. If α_1 and α_2 are purely imaginary numbers, this electromagnetic wave is a radiative mode, i.e., it cannot be bound to the interface.

7. Electromagnetic Spin Current Density Tensor of Surface Plasmon Wave

The spin current in canonical field theory [121, 122] and gravitational gauge theory [123] has been considered in the literature. Here we shall introduce and discuss a concept of "*electromagnetic spin current density tensor*" for beginners in this research field, and then we shall list the nonzero components of *electromagnetic spin current density tensor* of surface plasmon polaritons in a metal-dielectric interface (this has been considered in our previous work [3] and here we only list the main results of the spin current of surface electromagnetic waves).

The angular momentum, including both orbital and spinning angular momenta, is the generator of Lorentz spacetime rotational symmetry, and therefore, the electromagnetic spin current density tensor can also be derived through the Noether theorem of the electromagnetic Lagrangian density under the spacetime rotational symmetry [121].

7.1. The Canonical Field Theory for the Electromagnetic Spin Current Density Tensor (Method I)

The canonical field theory includes the Noether theorem, symmetries and conservation laws [121].

Within the framework of the canonical field theory [121, 122], the canonical energy-momentum tensor of electromagnetic field is given by $T_{\mu\nu} = -F_\mu{}^\lambda \partial_\nu A_\lambda - g_{\mu\nu}\mathcal{L}_{\rm em}$ with the electromagnetic Lagrangian density (in a flat spacetime) $\mathcal{L}_{\rm em} = -\frac{1}{4}F_{\mu\nu}F^{\mu\nu}$. Here, the electromagnetic field tensor is $F_{\mu\nu} = \partial_\mu A_\nu - \partial_\nu A_\mu$. It should be pointed out that the electromagnetic canonical energy-momentum tensor is asymmetric in its indices μ, ν. Furthermore, it is not gauge invariant. As is well known in field theory, the gauge-invariant symmetric energy-momentum tensor should be chosen as $\tau_{\mu\nu} = -F_\mu{}^\lambda F_{\nu\lambda} - g_{\mu\nu}\mathcal{L}_{\rm em}$. The relation between the asymmetric gauge-variant electromagnetic energy-momentum tensor $T_{\mu\nu}$ and the symmetric gauge-invariant electromagnetic energy-momentum tensor $\tau_{\mu\nu}$ is given by

$$T_{\mu\nu} = \tau_{\mu\nu} - \partial_\lambda \left(F_\mu{}^\lambda A_\nu\right). \qquad (7.1)$$

This can be obtained as follows: $T_{\mu\nu} = -F_\mu{}^\lambda (F_{\nu\lambda} + \partial_\lambda A_\nu) - g_{\mu\nu}\mathcal{L}_{\rm em} = \tau_{\mu\nu} - F_\mu{}^\lambda \partial_\lambda A_\nu$. The second term $-F_\mu{}^\lambda \partial_\lambda A_\nu$ can be rewritten as

$-\partial_\lambda\left(F_\mu{}^\lambda A_\nu\right)$ (a *4-D total divergence term*) because of the electromagnetic field equation $\partial_\lambda F_\mu{}^\lambda = 0$. Then we have $T_{\mu\nu} = \tau_{\mu\nu} - \partial_\lambda\left(F_\mu{}^\lambda A_\nu\right)$. Since the volume integral of the total divergence term $-\partial_\lambda\left(F_\mu{}^\lambda A_\nu\right)$ vanishes, these two electromagnetic energy-momentum tensors ($T_{\mu\nu}$ and $\tau_{\mu\nu}$) could actually be equivalent to each other.

The orbital angular momentum can be expressed by using the asymmetric canonical energy-momentum tensor $T^{\lambda\nu} = \tau^{\lambda\nu} - \partial_\sigma(F^{\lambda\sigma}A^\nu)$. The result is given by

$$\begin{aligned} L^{\lambda\mu\nu} &= x^\mu T^{\lambda\nu} - x^\nu T^{\lambda\mu} \\ &= x^\mu\Big[\tau^{\lambda\nu} - \partial_\sigma(F^{\lambda\sigma}A^\nu)\Big] - x^\nu\Big[\tau^{\lambda\mu} - \partial_\sigma(F^{\lambda\sigma}A^\mu)\Big] \\ &= x^\mu\tau^{\lambda\nu} - x^\nu\tau^{\lambda\mu} + \partial_\sigma\big(x^\nu F^{\lambda\sigma}A^\mu - x^\mu F^{\lambda\sigma}A^\nu\big) - S^{\lambda\mu\nu}. \end{aligned} \tag{7.2}$$

This can be rewritten as

$$\begin{aligned} &L^{\lambda\mu\nu} + S^{\lambda\mu\nu} = x^\mu\tau^{\lambda\nu} - x^\nu\tau^{\lambda\mu} + \partial_\sigma\big(x^\nu F^{\lambda\sigma}A^\mu - x^\mu F^{\lambda\sigma}A^\nu\big), \\ &S^{\lambda\mu\nu} = -(F^{\lambda\mu}A^\nu - F^{\lambda\nu}A^\mu), \end{aligned} \tag{7.3}$$

where $S^{\lambda\mu\nu}$ is the electromagnetic spin current density tensor. Here, The 4-D covariant divergence of $\partial_\sigma\Big(x^\nu F^{\lambda\sigma}A^\mu - x^\mu F^{\lambda\sigma}A^\nu\Big)$ automatically vanishes, i.e., $\partial_\lambda\partial_\sigma\Big(x^\nu F^{\lambda\sigma}A^\mu - x^\mu F^{\lambda\sigma}A^\nu\Big) = 0$. Then the orbital angular momentum density tensor is $L^{\lambda\mu\nu} = x^\mu T^{\lambda\nu} - x^\nu T^{\lambda\mu}$ and the total angular momentum density tensor is $x^\mu\tau^{\lambda\nu} - x^\nu\tau^{\lambda\mu}$. However, there is still a 4-D total divergence term $\partial_\sigma\big(x^\nu F^{\lambda\sigma}A^\mu - x^\mu F^{\lambda\sigma}A^\nu\big)$, of which the volume integral $\int\partial_\sigma\big(x^\nu F^{\lambda\sigma}A^\mu - x^\mu F^{\lambda\sigma}A^\nu\big)\mathrm{d}^4x$ may be vanishing.

7.2. The Conservation Law of Angular Momentum for the Electromagnetic Spin Current Density Tensor (Method II)

As is well known, the Lagrangian density of electromagnetic field is of the form $\mathcal{L}_{\mathrm{em}} = -\frac{1}{4}F_{\mu\nu}F^{\mu\nu}$. The canonical energy-momentum tensor in a flat spacetime is given by $T_{\mu\nu} = \frac{\partial\mathcal{L}_{\mathrm{em}}}{\partial\partial^\mu A^\sigma}\partial_\nu A^\sigma - \eta_{\mu\nu}\mathcal{L}_{\mathrm{em}} = -F_{\mu\sigma}\partial_\nu A^\sigma - \eta_{\mu\nu}\mathcal{L}_{\mathrm{em}}$ [121], where we have used the relation $\frac{\partial\mathcal{L}_{\mathrm{em}}}{\partial\partial^\mu A^\sigma} = -F_{\mu\sigma}$. This relation can be derived as follows

$$\begin{aligned}\frac{\partial \mathcal{L}_{\rm em}}{\partial \partial^{\mu} A^{\sigma}} &= -\frac{1}{4}\frac{\partial (F_{\alpha\beta}F^{\alpha\beta})}{\partial \partial^{\mu} A^{\sigma}} = -\frac{1}{2}\frac{\partial F^{\alpha\beta}}{\partial \partial^{\mu} A^{\sigma}}F_{\alpha\beta} \\ &= -\frac{1}{2}\frac{\partial (\partial^{\alpha} A^{\beta} - \partial^{\beta} A^{\alpha})}{\partial \partial^{\mu} A^{\sigma}}F_{\alpha\beta} = -\frac{\partial (\partial^{\alpha} A^{\beta})}{\partial \partial^{\mu} A^{\sigma}}F_{\alpha\beta} \\ &= -\eta^{\alpha}{}_{\mu}\eta^{\beta}{}_{\sigma}F_{\alpha\beta} = -F_{\mu\sigma}. \end{aligned} \tag{7.4}$$

The density tensor of the orbital angular momentum of electromagnetic field is $L^{\lambda\mu\nu} = x^{\mu}T^{\lambda\nu} - x^{\nu}T^{\lambda\mu}$. With the help of the conservation law $\partial_{\lambda}T^{\lambda\mu} = 0$, the 4-D divergence of $L^{\lambda\mu\nu}$ can be obtained: $\partial_{\lambda}L^{\lambda\mu\nu} = \delta_{\lambda}{}^{\mu}T^{\lambda\nu} - \delta_{\lambda}{}^{\nu}T^{\lambda\mu} = T^{\mu\nu} - T^{\nu\mu}$. Since the total angular momentum is conserved, $\partial_{\lambda}\left(L^{\lambda\mu\nu} + S^{\lambda\mu\nu}\right) = 0$, we have $\partial_{\lambda}S^{\lambda\mu\nu} = -\left(T^{\mu\nu} - T^{\nu\mu}\right)$, where $S^{\lambda\mu\nu}$ is the spinning angular momentum density tensor (spin current density tensor). The relation $-\left(T^{\mu\nu} - T^{\nu\mu}\right) = -\left[\left(-F^{\mu\lambda}\partial^{\nu}A_{\lambda}\right) - \left(-F^{\nu\lambda}\partial^{\mu}A_{\lambda}\right)\right]$ can be rearranged as

$$\begin{aligned} -\left(T^{\mu\nu} - T^{\nu\mu}\right) &= -F^{\lambda\mu}(F_{\lambda}{}^{\nu} + \partial^{\nu}A_{\lambda}) + F^{\lambda\nu}(F_{\lambda}{}^{\mu} + \partial^{\mu}A_{\lambda}) \\ = \quad & -F^{\lambda\mu}\partial_{\lambda}A^{\nu} + F^{\lambda\nu}\partial_{\lambda}A^{\mu} = -\partial_{\lambda}\left(F^{\lambda\mu}A^{\nu} - F^{\lambda\nu}A^{\mu}\right), \end{aligned} \tag{7.5}$$

where we have substituted the dynamical equation of the free electromagnetic field $\partial_{\lambda}F^{\lambda\mu} = 0$. Therefore, it follows from Eq. (7.5) and $\partial_{\lambda}S^{\lambda\mu\nu} = -\left(T^{\mu\nu} - T^{\nu\mu}\right)$ that the electromagnetic spin current density tensor should be of the form $S^{\lambda\mu\nu} = -\left(F^{\lambda\mu}A^{\nu} - F^{\lambda\nu}A^{\mu}\right)$.

7.3. The Gravitational Gauge Field Theoretical Method for the Electromagnetic Spin Current Density Tensor (Method III) [123]

We shall choose the SI units instead of the natural units where $\hbar = c = 1$. In SI units, we can choose two formulations of the canonical electromagnetic field theory:

i) The Lagrangian density of the electromagnetic field is $\mathcal{L} = -\frac{1}{4}\varepsilon_0 c F_{\mu\nu}F^{\mu\nu}$ with $F_{0i} = \partial_0 A_i - \partial_i A_0$. Then the electromagnetic field can be defined as $\frac{\mathbf{E}}{c} = -(F^{01}, F^{02}, F^{03})$ and $\mathbf{B} = -(F_{23}, F_{31}, F_{12})$. Then we can have $\mathbf{E} = -\nabla V - \frac{\partial \mathbf{A}}{\partial t}$ and $\mathbf{B} = \nabla \times \mathbf{A}$ (these relations are well known in electromagnetics). In the vector formalism of the

Maxwell electromagnetic field, the spatial vector A^i denotes the conventional 3-D magnetic vector potential, and the scalar electric potential is $A^0 = \frac{V}{c}$. Thus, the Lagrangian density of the electromagnetic field is $\mathcal{L} = \frac{1}{2}\frac{\varepsilon_0}{c}(\mathbf{E}^2 - c^2\mathbf{B}^2)$, in which the Lagrangian density $\mathcal{L}$ has the dimension of density of momentum. On the other hand, the spin-affine connection (gravitational gauge potential) ${\omega_\nu}^{pq}$ has the inverse-length dimension, and then $\frac{\delta(\sqrt{-g}\mathcal{L})}{\delta{\omega_\nu}^{pq}} = \frac{\mathrm{i}}{2}\sqrt{-g}{S^\nu}_{qp}$ [123] would have the dimension of density of a spin. Thus, we have ${S^0}_{qp} = -\varepsilon_0 c\left({F^0}_q A_p - {F^0}_p A_q\right)$, i.e., $\mathbf{S} = \varepsilon_0\mathbf{E}\times\mathbf{A}$. Now within this framework, the component of electromagnetic spin current density tensor ${S^0}_{qp}$ should have the unit of the density of ordinary three-dimensional spin vector. But it should be pointed out that the relation $\frac{\delta(\sqrt{-g}\mathcal{L})}{\delta{\omega_\nu}^{pq}} = \frac{\mathrm{i}}{2}\sqrt{-g}{S^\nu}_{qp}$ has only the meanings of indication of dimension. For the Dirac spinor field, this relation $\frac{\delta(\sqrt{-g}\mathcal{L}_{\mathrm{Dirac}})}{\delta{\omega_\nu}^{pq}} = \frac{\mathrm{i}}{2}\sqrt{-g}{S^\nu}_{qp}$ holds true, i.e., the spin current density of Dirac field is coupled to the gravitational field (Lorentz connection ${\omega_\nu}^{pq}$); However, for the electromagnetic gauge field, we have $\frac{\delta(\sqrt{-g}\mathcal{L}_{\mathrm{em}})}{\delta{\omega_\nu}^{pq}} = 0$, i.e., the spin current density of electromagnetic field cannot be coupled to the gravitational gauge field (Lorentz connection ${\omega_\nu}^{pq}$) [123].

ii) In this alternative unit system, the electromagnetic Lagrangian density can be of the form $\mathcal{L} = -\frac{1}{4}\varepsilon_0 c^2 F_{\mu\nu}F^{\mu\nu}$ or $\mathcal{L} = \frac{1}{2}\varepsilon_0(\mathbf{E}^2 - c^2\mathbf{B}^2)$. Then the dimension of the electromagnetic Lagrangian density $\mathcal{L}$ has the dimension of energy density. Therefore, the three-dimensional volume integral of $\frac{\delta(\sqrt{-g}\mathcal{L})}{\delta{\omega_\nu}^{pq}}$ (which equals $\frac{\mathrm{i}}{2}\sqrt{-g}{S^\nu}_{qp}$) is of the $[\hbar c]$ dimension. Then we have the one of the components in the spin current density of electromagnetic field: ${S^0}_{qp} = -\varepsilon_0 c^2\left({F^0}_q A_p - {F^0}_p A_q\right) \rightarrow \varepsilon_0 c\mathbf{E}\times\mathbf{A}$. In this unit system, the spin current density, $\frac{1}{c}{S^0}_{qp}$, of the electromagnetic field has the unit of the density of ordinary three-dimensional spin vector [3].

7.4. Electromagnetic Spin Current Density Tensor of a Surface Plasmon Polariton Wave Mode [3]

Let us consider a simple geometric configuration for a metal-dielectric single interface, which can sustain the surface plasmon polaritons [94]. Here, the electric permittivity of one of the electromagnetic materials (say, medium 1) is taken to be positive, whereas the electric permittivity of the other optical material (i.e., medium 2) is chosen to be negative (e.g., the metal permittivity $\epsilon_{\mathrm{r}} = 1 - \frac{\omega_{\mathrm{p}}^2}{\omega^2}$ can be negative below the plasma frequency ω_{p}).

By making use of the Maxwell electromagnetic field equations and the boundary conditions, the electric field strength of surface plasmon polaritons in medium 1 (in the region of $x > 0$) and medium 2 (in the region of $x < 0$) can be written as [94]

$$\begin{cases} \mathbf{E}_1 = \mathbf{E}_1^{(0)} \exp\left(-\alpha_1 x\right) \exp\left[\mathrm{i}\left(\beta z - \omega t\right)\right] \quad (x > 0), \\ \mathbf{E}_2 = \mathbf{E}_2^{(0)} \exp\left(+\alpha_2 x\right) \exp\left[\mathrm{i}\left(\beta z - \omega t\right)\right] \quad (x < 0), \end{cases} \tag{7.6}$$

where the x-direction attenuation coefficients are given by $\alpha_1 = \sqrt{\beta^2 - k_0^2\varepsilon_1}$ and $\alpha_2 = \sqrt{\beta^2 - k_0^2\varepsilon_2}$ and the z-direction phase constant is of the form $\beta = k_0\sqrt{\frac{\varepsilon_1\varepsilon_2}{\varepsilon_1+\varepsilon_2}}$. The square of vacuum wave number of the electromagnetic wave is given by $k_0^2 = (\frac{\omega}{c})^2$. It can be found that α_1 and α_2 have positive real parts and the electric field of surface plasmonic wave decays exponentially in the x direction (perpendicular to the optical interface of these two electromagnetic materials). By using the electric Gauss law $\nabla \cdot \mathbf{E} = 0$ and Faraday law of electromagnetic induction $\nabla \times \mathbf{E} = -\frac{\partial \mathbf{B}}{\partial t}$, the electric and magnetic field strengths in medium 1 (in the region of $x > 0$) and medium 2 (in the region of $x < 0$) can be obtained with the help of the electromagnetic boundary conditions [124, 125]. The explicit expressions for the electric and magnetic field strengths are given by [94]

$$\begin{cases} \mathbf{E}_1 = \left(\frac{\mathrm{i}\beta}{\alpha_1} E_{1z}^{(0)}, E_{1y}^{(0)}, E_{1z}^{(0)}\right) \exp\left(-\alpha_1 x\right) \exp\left[\mathrm{i}\left(\beta z - \omega t\right)\right] \quad (x > 0), \\ \mathbf{E}_2 = \left(-\frac{\mathrm{i}\beta}{\alpha_2} E_{2z}^{(0)}, E_{2y}^{(0)}, E_{2z}^{(0)}\right) \exp\left(\alpha_2 x\right) \exp\left[\mathrm{i}\left(\beta z - \omega t\right)\right] \quad (x < 0); \end{cases} \tag{7.7}$$

$$\begin{cases} \mathbf{H}_1 = \frac{\mathrm{i}}{\omega\mu_0}\left(\mathrm{i}\beta E_{1y}^{(0)}, \frac{k_0^2\varepsilon_1}{\alpha_1} E_{1z}^{(0)}, \alpha_1 E_{1y}^{(0)}\right) \\ \qquad \times \exp\left(-\alpha_1 x\right) \exp\left[\mathrm{i}\left(\beta z - \omega t\right)\right] \quad (x > 0), \\ \mathbf{H}_2 = \frac{\mathrm{i}}{\omega\mu_0}\left(\mathrm{i}\beta E_{2y}^{(0)}, -\frac{k_0^2\varepsilon_2}{\alpha_2} E_{2z}^{(0)}, -\alpha_2 E_{2y}^{(0)}\right) \\ \qquad \times \exp\left(\alpha_2 x\right) \exp\left[\mathrm{i}\left(\beta z - \omega t\right)\right] \quad (x < 0). \end{cases} \tag{7.8}$$

In these expressions, $E_y^{(0)}$ and $E_z^{(0)}$ denote the amplitudes of electric field of the surface plasmonic wave in the y and z directions. It can be shown that in the present metal-dielectric structure, the TE modes are absent, i.e, only the TM electromagnetic modes can be present, i.e., $E_{1y}^{(0)} = E_{2y}^{(0)} = 0$ and H_{1y}, H_{2y} are nonzero.

We shall consider the spin current density $s^{\mu\lambda\nu} = F^{\lambda\mu}A^{\nu} - F^{\nu\mu}A^{\lambda}$ of the electromagnetic surface plasmon polaritons [3]. For the component with the superscript index $\mu = 0$ in $s^{\mu\lambda\nu}$, $s^{0\lambda\nu} = F^{\lambda 0}A^{\nu} - F^{\nu 0}A^{\lambda}$, which has clear physical meanings (i.e., the spin density vector of the electromagnetic field) when the indices λ and ν are taken to be x, y or z. It should be pointed out that because of $F^{\lambda 0} = E^{\lambda}$ and $F^{\nu 0} = E^{\nu}$, such an electromagnetic-field spinning angular momentum density vector $s^{0\lambda\nu}$ can be rewritten as $\mathbf{s} = \mathbf{E} \times \mathbf{A}$ with $\mathbf{A}$ denoting the 3-D magnetic vector potential. It is very clear that the spin density formula $\mathbf{s} = \mathbf{E} \times \mathbf{A}$ is an expression in the natural units [122], where ϵ_0, μ_0, c and $\hbar$ are taken to be 1 [126]. This electromagnetic spin density turns in the international system of units (SI) is given by

$$\mathbf{s} = \varepsilon_0 \varepsilon_r \cdot \mathrm{Re}(\mathbf{E}) \times \mathrm{Re}(\mathbf{A}). \tag{7.9}$$

Here, the magnetic vector potential $\mathbf{A}$ and the electric field $\mathbf{E}$ are chosen as the real numbers. Thus, corresponding to the real-number formalism of the electromagnetic field, we need to choose the real parts of the electric and magnetic fields of the metal-dielectric surface plasmonic waves given in the results (7.7) and (7.8).

In what follows, the nonzero components of the electromagnetic spin current density tensor of the present surface plasmonic waves will be given. By using the relations $\mathbf{B} = \nabla \times \mathbf{A}$ or $\mathbf{E} = -\frac{\partial \mathbf{A}}{\partial t}$ in classical electrodynamics, we can obtain the 3-D magnetic vector potential $\mathbf{A}$ of the present metal-dielectric-interface supported surface plasmon polaritons [3]

$$\begin{cases} \mathbf{A}_1 = \frac{1}{\mathrm{i}\omega}\left(\frac{\mathrm{i}\beta}{\alpha_1}E_{1z}^{(0)}, E_{1y}^{(0)}, E_{1z}^{(0)}\right)\mathrm{e}^{-\alpha_1 x}\mathrm{e}^{\mathrm{i}(\beta z - \omega t)} & (x > 0), \\ \mathbf{A}_2 = \frac{1}{\mathrm{i}\omega}\left(-\frac{\mathrm{i}\beta}{\alpha_2}E_{2z}^{(0)}, E_{2y}^{(0)}, E_{2z}^{(0)}\right)\mathrm{e}^{\alpha_2 x}\mathrm{e}^{\mathrm{i}(\beta z - \omega t)} & (x < 0). \end{cases} \tag{7.10}$$

Here, the result (7.7) and (7.8) has been substituted. It should be noted that the spin current density tensor $s^{\mu\lambda\nu}$ has 64 components (i.e., the spacetime coordinate indices $\lambda, \mu\ \nu$ can be taken to be t and x, y, z, and $64 = 4^3$). But in these 64 components, there are only 24 (i.e., 4×6) components are independent because of the antisymmetry in the indices λ and ν (i.e.,

$s^{\mu\lambda\nu} = -s^{\mu\nu\lambda}$ and there are only six pairs of nonzero components $\{\lambda, \nu\}$). But as far as the present metal-dielectric-interface supported surface plasmon polariton wave mode is concerned, there are only seven of these 24 independent components that are nonzero [3]. These nonzero components can be given as follows:

In medium 1 (in the half space of $x > 0$), the explicit expressions for the nonzero spin current density tensor components are of the form: $s^{0zx} = \varepsilon_1\varepsilon_0(E_{1z}A_{1x} - E_{1x}A_{1z}) = \varepsilon_1\varepsilon_0\frac{\beta}{\omega\alpha_1}(E_{1z}^{(0)})^2\exp(-2\alpha_1 x)$, $s^{xxz} = \varepsilon_1\varepsilon_0 B_{1y}A_{1x} = -\varepsilon_1\varepsilon_0\frac{\beta k_0^2\varepsilon_1}{\omega^2\alpha_1^2}(E_{1z}^{(0)})^2\exp(-2\alpha_1 x)\sin(\beta z - \omega t)\cos(\beta z - \omega t)$, $s^{zzx} = -\varepsilon_1\varepsilon_0 B_{1y}A_{1z} = \varepsilon_1\varepsilon_0\frac{k_0^2\varepsilon_1}{\omega^2\alpha_1}(E_{1z}^{(0)})^2\exp(-2\alpha_1 x)\sin^2(\beta z - \omega t)$, $s^{x0x} = -\varepsilon_1\varepsilon_0 E_{1x}A_{1x} = \varepsilon_1\varepsilon_0\frac{\beta^2}{\omega\alpha_1^2}(E_{1z}^{(0)})^2\exp(-2\alpha_1 x)\sin(\beta z - \omega t)\cos(\beta z - \omega t)$, $s^{x0z} = -\varepsilon_1\varepsilon_0 E_{1x}A_{1z} = \varepsilon_1\varepsilon_0\frac{\beta}{\omega\alpha_1}(E_{1z}^{(0)})^2\exp(-2\alpha_1 x)\sin^2(\beta z - \omega t)$, $s^{z0x} = -\varepsilon_1\varepsilon_0 E_{1z}A_{1x} = -\varepsilon_1\varepsilon_0\frac{\beta}{\omega\alpha_1}(E_{1z}^{(0)})^2\exp(-2\alpha_1 x)\cos^2(\beta z - \omega t)$ and $s^{z0z} = -\varepsilon_1\varepsilon_0 E_{1z}A_{1z} = -\varepsilon_1\varepsilon_0\frac{1}{\omega}(E_{1z}^{(0)})^2\exp(-2\alpha_1 x)\sin(\beta z - \omega t)\cos(\beta z - \omega t)$ [3].

In medium 2 (in the half space of $x < 0$), the explicit expressions for the nonzero spin current density tensor components are of the form: $s^{0zx} = \varepsilon_2\varepsilon_0(E_{2z}A_{2x} - E_{2x}A_{2z}) = -\varepsilon_2\varepsilon_0\frac{\beta}{\omega\alpha_2}(E_{1z}^{(0)})^2\exp(2\alpha_2 x)$, $s^{xxz} = \varepsilon_2\varepsilon_0 B_{2y}A_{2x} = -\varepsilon_2\varepsilon_0\frac{\beta k_0^2\varepsilon_2}{\omega^2\alpha_2^2}(E_{1z}^{(0)})^2\exp(2\alpha_2 x)\sin(\beta z - \omega t)\cos(\beta z - \omega t)$, $s^{zzx} = -\varepsilon_2\varepsilon_0 B_{2y}A_{2z} = -\varepsilon_2\varepsilon_0\frac{k_0^2\varepsilon_2}{\omega^2\alpha_2}(E_{1z}^{(0)})^2\exp(2\alpha_2 x)\sin^2(\beta z - \omega t)$, $s^{x0x} = -\varepsilon_2\varepsilon_0 E_{2x}A_{2x} = \varepsilon_2\varepsilon_0\frac{\beta^2}{\omega\alpha_2^2}(E_{1z}^{(0)})^2\exp(2\alpha_2 x)\sin(\beta z - \omega t)\cos(\beta z - \omega t)$, $s^{x0z} = -\varepsilon_2\varepsilon_0 E_{2x}A_{2z} = -\varepsilon_2\varepsilon_0\frac{\beta}{\omega\alpha_2}(E_{1z}^{(0)})^2\exp(2\alpha_2 x)\sin^2(\beta z - \omega t)$, $s^{z0x} = -\varepsilon_2\varepsilon_0 E_{2z}A_{2x} = \varepsilon_2\varepsilon_0\frac{\beta}{\omega\alpha_2}(E_{1z}^{(0)})^2\exp(2\alpha_2 x)\cos^2(\beta z - \omega t)$ and $s^{z0z} = -\varepsilon_2\varepsilon_0 E_{2z}A_{2z} = -\varepsilon_2\varepsilon_0\frac{1}{\omega}(E_{1z}^{(0)})^2\exp(2\alpha_2 x)\sin(\beta z - \omega t)\cos(\beta z - \omega t)$ [3].

In all the above nonzero components, the component that has direct, apparent or obvious physical meanings is s^{0zx}, of which the direction (in the y axis) is perpendicular to the incidence plane. The volume integral $\mathbf{S} = \int \varepsilon_0\varepsilon_r \cdot \mathrm{Re}(\mathbf{E}) \times \mathrm{Re}(\mathbf{A})\mathrm{d}V$ is an electromagnetic spinning angular momentum of the present surface plasmonic wave. Within the framework of second quantization [122], the quantized spin vector operator $\hat{\mathbf{S}}$ of the present surface plasmonic wave can be derived.

We have studied the electromagnetic spin current density tensor of surface plasmon polaritons and the relevant mechanical effects [3]. The electromagnetic spin current density tensor of parity-time symmetric surface electromagnetic modes also deserves consideration.

8. A Parity-Time Symmetric Surface Electromagnetic Wave and Its Generalization

The solution of the Maxwell equations on a metal-dielectric interface can be obtained by using the electromagnetic boundary conditions [94]. Such electromagnetic modes can be referred to as the surface plasmon polaritons [1].

The parity-time symmetry that has been extensively explored in quantum mechanics [23, 24], optics, photonics, classical mechanics [25, 26, 27, 28] and acoustics [29, 30] over the past two decades is a generalization of Hermiticity (Hermitian property) in the ordinary quantum mechanics. In such well-known quantum mechanics, any observable mechanical quantities (such as Hamiltonian, interaction potential, linear and angular momenta) should be Hermitian operators and so their eigenvalues are real numbers. However, in fact, this Hermitian condition can be relaxed, as long as the mechanical quantities of the systems satisfy a certain space-time symmetry, even if they are not Hermitian operators, they also have the spectra of real eigenvalues [23, 24]. Such a symmetry is parity-time symmetry, of which the parity-time transformation (or operation) can be characterised by $x \rightarrow -x$, $t \rightarrow -t$ and $\mathrm{i} \rightarrow -\mathrm{i}$ (i is the imaginary number unit). Let us briefly explain the concept of parity-time symmetry (invariance). In the Schrödinger equation $\hat{H}\Psi(x,t) = \mathrm{i}\hbar\frac{\partial\Psi(x,t)}{\partial t}$, we consider only the stationary state, i.e., $\Psi(x,t) = \psi(x)\mathrm{e}^{-\mathrm{i}\omega t}$. It should ne noted that the energy operator $\mathrm{i}\hbar\frac{\partial}{\partial t}$ and the time-dependent phase factor $\mathrm{e}^{-\mathrm{i}\omega t}$ respect the parity-time symmetry (invariance), namely, under the transformation $t \rightarrow -t$ and $\mathrm{i} \rightarrow -\mathrm{i}$, they preserve invariance of parity-time symmetry. Thus, we can simplify the Schrödinger equation as $\hat{H}(x)\psi(x) = E\psi(x)$, where E is the energy eigenvalue (it is a constant number for a certain eigenstate). Under the space reversal transformation ($x \to -x$), this equation becomes $\hat{H}(-x)\psi(-x) = E\psi(-x)$. On the other hand, we also take the complex conjugate transformation for the Schrödinger equation $\hat{H}(x)\psi(x) = E\psi(x)$ and obtain $\hat{H}^*(x)\psi^*(x) = E^*\psi^*(x)$. By comparing these two equations $\hat{H}(-x)\psi(-x) = E\psi(-x)$ and $\hat{H}^*(x)\psi^*(x) = E^*\psi^*(x)$, it can be found that as long as the Hamiltonian operator can obey $\hat{H}(-x) = \hat{H}^*(x)$ (say, they can be written as $\hat{\mathcal{H}}$), we will have $\psi(-x) = \psi^*(x)$ and $E = E^*$ (real eigenvalue spectrum). This can be understood as follows: Since both $\hat{\mathcal{H}}\psi(-x) = E\psi(-x)$ and $\hat{\mathcal{H}}\psi^*(x) = E^*\psi^*(x)$ are the energy eigenvalue equations, they unavoidably have exactly the same spectrum (i.e., $\psi(-x) = \psi^*(x)$) and their eigenvalues have to be equal to each other ($E = E^*$). The Hamiltonian operator relation $\hat{H}(-x) = \hat{H}^*(x)$ can be referred to as the parity-time symmetry. A system whose Hamiltonian $\hat{H}$ or interaction potential $\hat{V}$ fulfills $\hat{H}(-x) = \hat{H}^*(x)$ or $\hat{V}(-x) = \hat{V}^*(x)$ can be called the parity-time symmetric system. But it should be pointed out that in quantum mechanics, it would be difficult for us to design such a system. In optics, electromagnetics and acoustics, however, the so-called parity-time symmetric systems can be easily designed and fabricated, because the refractive index n in optics, the electric permittivity ϵ in electromagnetics and the dynamical mass density ρ of an elastic medium in acoustics can be considered to be the effective (equivalent) Hamiltonians or interaction potentials. As an illustrative example, we assume the interaction potential of a system in the half space of $x > 0$ is V_1 and the interaction potential in the half space of $x < 0$ is V_2. If V_1 and V_2 are complex conjugate to each other, i.e., $V_2 = V_1^*$, clearly, the interaction potential of such a system obeys the parity-time symmetry ($\hat{V}(-x) = \hat{V}^*(x)$). Although the interaction potential (in the parity-time symmetric system) is no longer a Hermitian operator, the spectrum of the Hamiltonian still has real eigenvalues [23, 24].

The parity-time symmetric potential is defined by $\hat{V}(-x) = \hat{V}^*(x)$. This can be rewritten as $\hat{V}^*(-x) = \hat{V}(x)$. Thus, we can describe the parity-time symmetric potential as follows: Let us have an interaction potential $V(x)$. If we apply the operation of $x \to -x$ and the complex conjugate ($\mathrm{i} \to -\mathrm{i}$) to the interaction potential $V(x)$, it can still go back to the original form $V(x)$, so that we say that such a potential function $V(x)$ is of parity-time symmetry.

As far as the parity-time symmetry of an optical interface is concerned, the relative permittivities of two electromagnetic media are complex conjugate with each other, i.e., $\varepsilon_2 = \varepsilon_1^*$. This interface can enable to sustain parity-time symmetric surface electromagnetic eigen modes. As is well known, the phase constant of a surface electromagnetic wave parallel to the interface is $\beta = \frac{\omega}{c}\sqrt{\frac{\varepsilon_1\varepsilon_2}{\varepsilon_1+\varepsilon_2}}$ [94]. It can be found that when this interface respects the parity-time symmetry (i.e., $\varepsilon_2 = \varepsilon_1^*$), this phase constant β is really a real number [29, 30]. Let the x axis be perpendicular to the interface. The wavefunction of this surface wave confined onto the interface is given by $\psi_1(x) = \mathrm{e}^{-\alpha_1 x} \equiv \mathrm{e}^{-\alpha x}$ (in the half space of $x > 0$) and $\psi_2(x) = \mathrm{e}^{+\alpha_2 x} \equiv \mathrm{e}^{+\alpha^* x}$ (in the half space of $x < 0$), where the attenuation coefficients are $\alpha_1 = \sqrt{\beta^2 - (\frac{\omega}{c})^2\epsilon_1} \equiv \alpha$ and $\alpha_2 = \sqrt{\beta^2 - (\frac{\omega}{c})^2\epsilon_2} \equiv \alpha^*$. It is apparent that the wavefunction fulfills $\psi(-x) = \psi^*(x)$ (parity-time symmetry or invariance). A concept of parity-time symmetric acoustic interface can also be suggested: specifically, the (dynamical) mass densities in the half spaces of $x > 0$ and $x < 0$ are ρ_1 and ρ_2, respectively, and then the phase constant of the *interface-confined acoustic surface wave* is $\frac{\omega}{\sqrt{\kappa}}\sqrt{\frac{\rho_1\rho_2}{\rho_1+\rho_2}}$ (the propagation direction is parallel to the interface), where κ denotes the acoustic-medium bulk modulus. If ρ_1 and ρ_2 obey the parity-time symmetry (i.e., $\rho_2^* = \rho_1$), this phase constant β also has a real eigenvalue spectrum [29, 30].

Here, we will generalize the parity-time symmetry [23, 24] of optical interface structure and the relevant confined surface electromagnetic waves, in which the aforementioned parity-time symmetric condition $\varepsilon_2 = \varepsilon_1^*$ is a special case. For the first, we should suggest some schemes for realizing the complex conjugate optical parameters (in, e.g., multilevel atomic vapors or media). In multilevel atomic vapors, both electric- and magnetic-dipole allowed transitions (including the mixed transition due to dressed states) can occur, and both the complex permittivity and the complex permeability can be achieved in such atomic vapors. Thus, the electric and magnetic parity-time symmetric surface electromagnetic waves can possibly be supported on such parity-time symmetric interfaces [29, 30]. Though the previous paper [29] of ours focuses on the parity-time symmetric surface acoustic wave modes, yet it can be easily extended to the electromagnetic analog of parity-time symmetric surface acoustic waves.

8.1. An Ordinary Parity-Time Symmetric Surface Electromagnetic Wave

8.1.1. A Vee-Configuration Gain Medium

In the parity-time symmetric optical structure, the optical gain is needed. This can be realized in a three-level Vee-configuration atomic vapor, of which the electric permittivity has a negative imaginary part. We shall consider a three-level Vee-configuration atomic system $\{|1\rangle, |2\rangle, |3\rangle\}$, in which level $|1\rangle$ is a ground level, and $|2\rangle, |3\rangle$ are upper (excited) levels. We assume a control optical field drives the $|1\rangle$-$|3\rangle$ transition and a weak probe optical field drives the $|1\rangle$-$|2\rangle$ transition. In such a quantum-coherent atomic system, atomic population trapping can occur in upper (excited) levels $|2\rangle, |3\rangle$ and the ground state level $|1\rangle$ is empty (in this case, levels $|2\rangle$ and $|3\rangle$ are in the dark state, which is not coupled effectively to level $|1\rangle$). Thus, population inversion between levels $|1\rangle$ and $|2\rangle$ could possibly be realized, and this may possibly lead to the permittivity with a negative imaginary part.

The density matrix dynamical equations of three-level and multi-level systems can be found in many references [9, 74]. By using the quantum mechanical Schrödinger equation, the evolutional equation of the density matrix of the present Vee-configuration atomic system is given by

$$\begin{aligned}
\dot{\rho}_{11} &= \gamma_{21}\rho_{22}+\gamma_{31}\rho_{33}-\frac{\mathrm{i}}{2}\Omega_{\mathrm{c}}\rho_{13}-\frac{\mathrm{i}}{2}\Omega_{\mathrm{p}}\rho_{12}+\frac{\mathrm{i}}{2}\Omega_{\mathrm{c}}^{*}\rho_{31}+\frac{\mathrm{i}}{2}\Omega_{\mathrm{p}}^{*}\rho_{21},\\
\dot{\rho}_{22} &= \gamma_{32}\rho_{33}-\gamma_{21}\rho_{22}-\frac{\mathrm{i}}{2}\Omega_{\mathrm{p}}^{*}\rho_{21}+\frac{\mathrm{i}}{2}\Omega_{\mathrm{p}}\rho_{12},\\
\dot{\rho}_{33} &= -\left(\gamma_{32}+\gamma_{31}\right)\rho_{33}-\frac{\mathrm{i}}{2}\Omega_{\mathrm{c}}^{*}\rho_{31}+\frac{\mathrm{i}}{2}\Omega_{\mathrm{c}}\rho_{13},\\
\dot{\rho}_{12} &= -\left(\frac{\gamma_{21}}{2}-\mathrm{i}\Delta_{\mathrm{p}}\right)\rho_{12}+\frac{\mathrm{i}}{2}\Omega_{\mathrm{c}}^{*}\rho_{32}+\frac{\mathrm{i}}{2}\Omega_{\mathrm{p}}^{*}\left(\rho_{22}-\rho_{11}\right),\\
\dot{\rho}_{32} &= -\left[\frac{\gamma_{32}+\gamma_{31}+\gamma_{21}+\gamma_{\mathrm{ph}}}{2}-\mathrm{i}\left(\Delta_{\mathrm{p}}-\Delta_{\mathrm{c}}\right)\right]\rho_{32}-\frac{\mathrm{i}}{2}\Omega_{\mathrm{p}}^{*}\rho_{31}+\frac{\mathrm{i}}{2}\Omega_{\mathrm{c}}\rho_{12},\\
\dot{\rho}_{31} &= -\left(\frac{\gamma_{32}+\gamma_{31}}{2}+\mathrm{i}\Delta_{\mathrm{c}}\right)\rho_{31}-\frac{\mathrm{i}}{2}\Omega_{\mathrm{p}}\rho_{32}-\frac{\mathrm{i}}{2}\Omega_{\mathrm{c}}\left(\rho_{33}-\rho_{11}\right),
\end{aligned} \tag{8.1}$$

where the Rabi frequencies $\Omega_{\mathrm{p}}, \Omega_{\mathrm{c}}$ of the probe light field and the control optical beam are defined by $\Omega_{\mathrm{p}} = \wp_{21}\mathcal{E}_{\mathrm{p}}/\hbar$ and $\Omega_{\mathrm{c}} = \wp_{31}\mathcal{E}_{\mathrm{c}}/\hbar$, respectively. The $\wp_{ij}$'s denote the electric-dipole-transition dipole moments of the atom. The frequency detuning Δ_{p} and Δ_{c} in Eq. (8.1) are defined as $\Delta_{\mathrm{p}} = \omega_{21}-\omega_{\mathrm{p}}$ and $\Delta_{\mathrm{c}} = \omega_{31}-\omega_{\mathrm{c}}$. Here, ω_{21} and ω_{31} are the present atomic level transition frequencies ($\omega_{21} = \omega_2-\omega_1$ and $\omega_{31} = \omega_3-\omega_1$) in relevant level transitions; ω_{p} and ω_{c} denote the angular frequencies of the probe and control optical fields, respectively, which excite the present Vee-configuration atomic system.

In the first three dynamical equations of the diagonal density matrix elements in Eq. (8.1), the terms $-\frac{\mathrm{i}}{2}\Omega_{\mathrm{p}}\rho_{12}$, $\frac{\mathrm{i}}{2}\Omega_{\mathrm{p}}^{*}\rho_{21}$, and $-\frac{\mathrm{i}}{2}\Omega_{\mathrm{p}}^{*}\rho_{21}+\frac{\mathrm{i}}{2}\Omega_{\mathrm{p}}\rho_{12}$, are negligibly small. Now the equations of ρ_{11} and ρ_{22} in Eq. (8.1) are given by

$$\begin{aligned}\dot{\rho}_{11} &= \gamma_{21}\rho_{22}+\gamma_{31}\rho_{33}-\frac{\mathrm{i}}{2}\Omega_{\mathrm{c}}\rho_{13}+\frac{\mathrm{i}}{2}\Omega_{\mathrm{c}}^{*}\rho_{31},\\ \dot{\rho}_{22} &= \gamma_{32}\rho_{33}-\gamma_{21}\rho_{22}.\end{aligned} \tag{8.2}$$

We define $\Gamma_{31}=\gamma_{32}+\gamma_{31}$. The last equation in Eq. (8.1), where the small term $-\frac{\mathrm{i}}{2}\Omega_{\mathrm{p}}\rho_{32}$ can be ignored, can be rewritten as the form

$$\dot{\rho}_{31}=-\left(\frac{\Gamma_{31}}{2}+\mathrm{i}\Delta_{\mathrm{c}}\right)\rho_{31}-\frac{\mathrm{i}}{2}\Omega_{\mathrm{c}}\left(\rho_{33}-\rho_{11}\right). \tag{8.3}$$

We shall consider the steady-state optical properties of the present three-level atomic vapor. The steady solution when $\dot{\rho}_{ij}=0$ can be obtained as follows: The off-diagonal density matrix elements ρ_{31} and ρ_{13} can be expressed as

$$\rho_{31}=-\frac{\mathrm{i}}{2}\frac{\Omega_{\mathrm{c}}\left(\rho_{33}-\rho_{11}\right)}{\frac{\Gamma_{31}}{2}+\mathrm{i}\Delta_{\mathrm{c}}},\qquad \rho_{13}=\frac{\mathrm{i}}{2}\frac{\Omega_{\mathrm{c}}^{*}\left(\rho_{33}-\rho_{11}\right)}{\frac{\Gamma_{31}}{2}-\mathrm{i}\Delta_{\mathrm{c}}}. \tag{8.4}$$

By substituting Eq. (8.4) into the second equation in Eq. (8.2), and by using the rule of conservation of atomic level probability (i.e., $\rho_{11}+\rho_{22}+\rho_{33}=1$), one can obtain

$$\begin{aligned}\dot{\rho}_{11} &= \gamma_{21}\rho_{22}+\gamma_{31}\rho_{33}+\frac{1}{4}\Omega_{\mathrm{c}}^{*}\Omega_{\mathrm{c}}\left(\rho_{33}-\rho_{11}\right)\left(\frac{1}{\frac{\Gamma_{31}}{2}+\mathrm{i}\Delta_{\mathrm{c}}}+\frac{1}{\frac{\Gamma_{31}}{2}-\mathrm{i}\Delta_{\mathrm{c}}}\right)\\ &= \gamma_{21}\rho_{22}+\gamma_{31}\left(1-\rho_{11}-\rho_{22}\right)+\mathcal{A}\left(1-2\rho_{11}-\rho_{22}\right)\\ &= -\left(\gamma_{31}+2\mathcal{A}\right)\rho_{11}+\left(\gamma_{21}-\gamma_{31}-\mathcal{A}\right)\rho_{22}+\left(\gamma_{31}+\mathcal{A}\right),\end{aligned} \tag{8.5}$$

where the parameter $\mathcal{A}$ is given by

$$\mathcal{A}=\frac{\Omega_{\mathrm{c}}^{*}\Omega_{\mathrm{c}}}{4}\left(\frac{1}{\frac{\Gamma_{31}}{2}+\mathrm{i}\Delta_{\mathrm{c}}}+\frac{1}{\frac{\Gamma_{31}}{2}-\mathrm{i}\Delta_{\mathrm{c}}}\right). \tag{8.6}$$

Now it follows from Eqs. (8.2) and (8.5) that the equations of ρ_{11} and ρ_{22} are given by

$$\begin{aligned} \dot{\rho}_{11} &= -\left(\gamma_{31}+2\mathcal{A}\right)\rho_{11}+\left(\gamma_{21}-\gamma_{31}-\mathcal{A}\right)\rho_{22}+\left(\gamma_{31}+\mathcal{A}\right), \\ \dot{\rho}_{22} &= -\gamma_{32}\rho_{11}-(\gamma_{32}+\gamma_{21})\rho_{22}+\gamma_{32}. \end{aligned} \tag{8.7}$$

Let us concentrate our attention on the solution of ρ_{11} and ρ_{22}. The steady equations of ρ_{11} and ρ_{22}, where $\dot{\rho}_{11}=0$ and $\dot{\rho}_{22}=0$, can be arranged as

$$\begin{pmatrix} -\left(\gamma_{31}+2\mathcal{A}\right) & \gamma_{21}-\gamma_{31}-\mathcal{A} \\ -\gamma_{32} & -(\gamma_{32}+\gamma_{21}) \end{pmatrix}\begin{pmatrix} \rho_{11} \\ \rho_{22} \end{pmatrix}=\begin{pmatrix} -(\gamma_{31}+\mathcal{A}) \\ -\gamma_{32} \end{pmatrix}. \tag{8.8}$$

The solution of the matrix equation (8.8) is given by

$$\begin{aligned} \rho_{11} &= \frac{\det\begin{pmatrix} -\left(\gamma_{31}+\mathcal{A}\right) & \gamma_{21}-\gamma_{31}-\mathcal{A} \\ -\gamma_{32} & -(\gamma_{32}+\gamma_{21}) \end{pmatrix}}{\left(\gamma_{31}+2\mathcal{A}\right)\left(\gamma_{32}+\gamma_{21}\right)+\gamma_{32}(\gamma_{21}-\gamma_{31}-\mathcal{A})}, \\ \rho_{22} &= \frac{\det\begin{pmatrix} -\left(\gamma_{31}+2\mathcal{A}\right) & -(\gamma_{31}+\mathcal{A}) \\ -\gamma_{32} & -\gamma_{32} \end{pmatrix}}{\left(\gamma_{31}+2\mathcal{A}\right)\left(\gamma_{32}+\gamma_{21}\right)+\gamma_{32}(\gamma_{21}-\gamma_{31}-\mathcal{A})}. \end{aligned} \tag{8.9}$$

This solution (8.9) can also be rewritten as

$$\begin{aligned} \rho_{11} &= \frac{\left(\gamma_{31}+\mathcal{A}\right)\left(\gamma_{32}+\gamma_{21}\right)+\gamma_{32}\left(\gamma_{21}-\gamma_{31}-\mathcal{A}\right)}{\left(\gamma_{31}+2\mathcal{A}\right)\left(\gamma_{32}+\gamma_{21}\right)+\gamma_{32}(\gamma_{21}-\gamma_{31}-\mathcal{A})}, \\ \rho_{22} &= \frac{\gamma_{32}\mathcal{A}}{\left(\gamma_{31}+2\mathcal{A}\right)\left(\gamma_{32}+\gamma_{21}\right)+\gamma_{32}(\gamma_{21}-\gamma_{31}-\mathcal{A})}. \end{aligned} \tag{8.10}$$

Now we have the population probability at level $|1\rangle$: $\rho_{11} = \frac{(\gamma_{31}+\mathcal{A})\gamma_{21}+\gamma_{32}\gamma_{21}}{(\gamma_{31}+2\mathcal{A})(\gamma_{32}+\gamma_{21})+\gamma_{32}(\gamma_{21}-\gamma_{31}-\mathcal{A})}$, and the probability difference between levels $|2\rangle$ and $|1\rangle$ is given by

$$\begin{aligned} \rho_{22}-\rho_{11} &= \frac{\left(\gamma_{32}-\gamma_{21}\right)\mathcal{A}-\gamma_{21}\left(\gamma_{32}+\gamma_{31}\right)}{\left(\gamma_{31}+2\mathcal{A}\right)\left(\gamma_{32}+\gamma_{21}\right)+\gamma_{32}(\gamma_{21}-\gamma_{31}-\mathcal{A})} \\ &= \frac{\left(\gamma_{32}-\gamma_{21}\right)\mathcal{A}-\gamma_{21}\left(\gamma_{32}+\gamma_{31}\right)}{\left(\gamma_{32}+2\gamma_{21}\right)\mathcal{A}+\gamma_{21}\left(\gamma_{31}+\gamma_{32}\right)}. \end{aligned} \tag{8.11}$$

It can be found that if $\left(\gamma_{32}-\gamma_{21}\right)\mathcal{A}-\gamma_{21}\left(\gamma_{32}+\gamma_{31}\right)>0$, i.e., $\gamma_{32}>\gamma_{21}$ and $\mathcal{A}$ is sufficiently large, we have $\rho_{22}-\rho_{11}>0$ (population inversion).

In an atomic system, if levels $|2\rangle$ and $|3\rangle$ have the same parity, the condition $\gamma_{32} > \gamma_{21}$ cannot be fulfilled, and hence the population inversion between $|1\rangle$ and $|2\rangle$ cannot be realized. Thus, we need to use the quantum-dot systems (instead of the atomic systems). In the quantum-dot systems, the mixed-parity energy levels could be easily obtained and hence the condition $\gamma_{32} > \gamma_{21}$ can be fulfilled.

Note that $-\frac{\mathrm{i}}{2}\Omega_{\mathrm{p}}^{*}\rho_{31}$ is small in Eq. (8.1). We define $\Gamma_{32} = \gamma_{32} + \gamma_{31} + \gamma_{21} + \gamma_{\mathrm{ph}}$, and obtain the equations of ρ_{12} and ρ_{32} as

$$\begin{aligned}\dot{\rho}_{12} &= -\left(\frac{\gamma_{21}}{2} - \mathrm{i}\Delta_{\mathrm{p}}\right)\rho_{12} + \frac{\mathrm{i}}{2}\Omega_{\mathrm{c}}^{*}\rho_{32} + \frac{\mathrm{i}}{2}\Omega_{\mathrm{p}}^{*}\left(\rho_{22} - \rho_{11}\right), \\ \dot{\rho}_{32} &= -\left[\frac{\Gamma_{32}}{2} - \mathrm{i}\left(\Delta_{\mathrm{p}} - \Delta_{\mathrm{c}}\right)\right]\rho_{32} + \frac{\mathrm{i}}{2}\Omega_{\mathrm{c}}\rho_{12}. \end{aligned} \tag{8.12}$$

From the second equation in Eq. (8.12), we have the steady value of ρ_{32}:

$$\rho_{32} = \frac{\mathrm{i}}{2}\frac{\Omega_{\mathrm{c}}\rho_{12}}{\frac{\Gamma_{32}}{2} - \mathrm{i}\left(\Delta_{\mathrm{p}} - \Delta_{\mathrm{c}}\right)}. \tag{8.13}$$

By substituting the result (8.13) into Eq. (8.12), one can obtain

$$-\left(\frac{\gamma_{21}}{2} - \mathrm{i}\Delta_{\mathrm{p}}\right)\rho_{12} - \frac{1}{4}\cdot\frac{\Omega_{\mathrm{c}}^{*}\Omega_{\mathrm{c}}\rho_{12}}{\frac{\Gamma_{32}}{2} - \mathrm{i}\left(\Delta_{\mathrm{p}} - \Delta_{\mathrm{c}}\right)} + \frac{\mathrm{i}}{2}\Omega_{\mathrm{p}}^{*}\left(\rho_{22} - \rho_{11}\right) = 0. \tag{8.14}$$

Now from Eq. (8.14), we obtain the off-diagonal density matrix element ρ_{12}:

$$\rho_{12} = \frac{\frac{\mathrm{i}}{2}\Omega_{\mathrm{p}}^{*}\left(\rho_{22} - \rho_{11}\right)\left[\frac{\Gamma_{32}}{2} - \mathrm{i}\left(\Delta_{\mathrm{p}} - \Delta_{\mathrm{c}}\right)\right]}{\left(\frac{\gamma_{21}}{2} - \mathrm{i}\Delta_{\mathrm{p}}\right)\left[\frac{\Gamma_{32}}{2} - \mathrm{i}\left(\Delta_{\mathrm{p}} - \Delta_{\mathrm{c}}\right)\right] + \frac{1}{4}\Omega_{\mathrm{c}}^{*}\Omega_{\mathrm{c}}}, \tag{8.15}$$

and its complex conjugate ρ_{12} is given by

$$\rho_{21} = \frac{-\frac{\mathrm{i}}{2}\Omega_{\mathrm{p}}\left(\rho_{22} - \rho_{11}\right)\left[\frac{\Gamma_{32}}{2} + \mathrm{i}\left(\Delta_{\mathrm{p}} - \Delta_{\mathrm{c}}\right)\right]}{\left(\frac{\gamma_{21}}{2} + \mathrm{i}\Delta_{\mathrm{p}}\right)\left[\frac{\Gamma_{32}}{2} + \mathrm{i}\left(\Delta_{\mathrm{p}} - \Delta_{\mathrm{c}}\right)\right] + \frac{1}{4}\Omega_{\mathrm{c}}^{*}\Omega_{\mathrm{c}}}. \tag{8.16}$$

By using the *microscopic (electric) polarizability* of an atom $\alpha_{\mathrm{e}} = \frac{2\wp_{12}\rho_{21}}{\varepsilon_0\mathcal{E}_{\mathrm{p}}} = \frac{2|\wp_{12}|^2\rho_{21}}{\varepsilon_0\hbar\Omega_{\mathrm{p}}}$ due to the $|1\rangle$-$|2\rangle$ electric-dipole allowed transition driven by the weak probe field, we can obtain

$$\mathcal{N}\alpha_{\mathrm{e}} = -\mathrm{i}\frac{\mathcal{N}|\wp_{12}|^2}{\varepsilon_0\hbar}\frac{\left(\rho_{22} - \rho_{11}\right)\left[\frac{\Gamma_{32}}{2} + \mathrm{i}\left(\Delta_{\mathrm{p}} - \Delta_{\mathrm{c}}\right)\right]}{\left(\frac{\gamma_{21}}{2} + \mathrm{i}\Delta_{\mathrm{p}}\right)\left[\frac{\Gamma_{32}}{2} + \mathrm{i}\left(\Delta_{\mathrm{p}} - \Delta_{\mathrm{c}}\right)\right] + \frac{1}{4}\Omega_{\mathrm{c}}^{*}\Omega_{\mathrm{c}}}, \tag{8.17}$$

where $\mathcal{N}$ denotes the atomic concentration (total number of atoms per unit volume) in the atomic vapor. The electric permittivity of the present atomic vapor is given by $\varepsilon_{\mathrm{r}} = 1 + \frac{\mathcal{N}\alpha_{\mathrm{e}}}{1-\frac{\mathcal{N}\alpha_{\mathrm{e}}}{3}}$, where the local field effect (dipole-dipole mutual polarization characterized by the Clausius-Mossotti-Lorenz-Lorentz relation [41]) has been taken into account.

In the dimensionless quantity $\mathcal{N}\alpha_{\mathrm{e}}$, if $\rho_{22} - \rho_{11} < 0$, the imaginary part is positive; When $\rho_{22} - \rho_{11} > 0$, the imaginary part of $\mathcal{N}\alpha_{\mathrm{e}}$ is negative. Therefore, the electric permittivity $\varepsilon_{\mathrm{r}} = 1 + \frac{\mathcal{N}\alpha_{\mathrm{e}}}{1-\frac{\mathcal{N}\alpha_{\mathrm{e}}}{3}}$ of the present atomic vapor can be tunable and its imaginary part can be either positive or negative. This property is required in realizing the parity-time symmetric dielectric structure.

8.1.2. An Ordinary Parity-Time Symmetric Electromagnetic Surface Wave Mode

A simple configuration of medium structure for sustaining or supporting the eigenmodes (confined states) of surface electromagnetic waves can be a single interface consisting of two media with permittivity in opposite signs [1, 94]. For example, the permittivity of one medium is positive, and the permittivity of the other medium is negative. This is the ordinary (conventional) surface plasmon polaritons, where a metal can serve as the medium of negative permittivity [1, 94].

Within the theoretical framework of surface plasmon polaritons on metal-dielectric single interface [94], the electric field of the sustained electromagnetic surface wave modes on the $\mathcal{P}$-$\mathcal{T}$ symmetric optical interface (see Fig. 28 for its schematic diagram) is of the form

$$\begin{aligned} \mathbf{E}_{+} &= \left(\frac{\mathrm{i}\beta}{\alpha_{+}}E^{(0)}, 0, E^{(0)}\right)\exp\left(-\alpha_{+}x\right)\exp\left[\mathrm{i}(\beta z-\omega t)\right] \ (x>0), \\ \mathbf{E}_{-} &= \left(-\frac{\mathrm{i}\beta}{\alpha_{-}}E^{(0)}, 0, E^{(0)}\right)\exp\left(\alpha_{-}x\right)\exp\left[\mathrm{i}(\beta z-\omega t)\right] \ (x<0). \end{aligned} \tag{8.18}$$

This surface electromagnetic field eigen mode propagating in the z direction and its amplitude decay exponentially in the x positive and negative directions. The magnetic field of the sustained surface electromagnetic

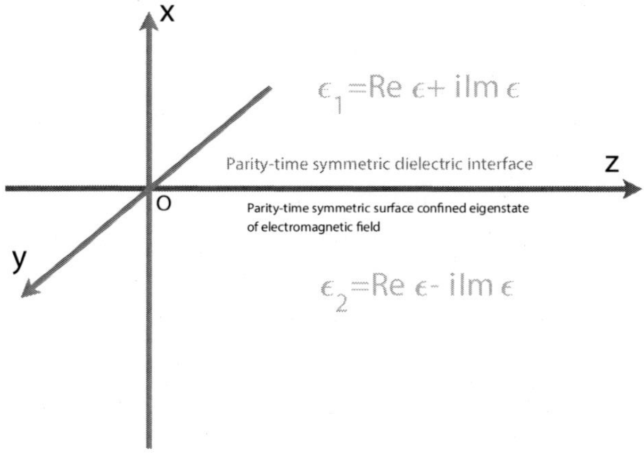

Figure 28. The schematic diagram of electromagnetic parity-time symmetric interface. In the half spaces of $x > 0$ and $x < 0$, the electric permittivities ϵ_1 and ϵ_2 are complex conjugate, and a confined (bound) surface eigenstate with a real spectrum for the phase constant can be obtained.

field eigen modes on this $\mathcal{P}$-$\mathcal{T}$ symmetric interface is given by

$$\begin{aligned} \mathbf{H}_{+} &= \frac{\mathrm{i}}{\omega\mu_0}\left(0, \frac{k_0^2\varepsilon_+}{\alpha_+}E^{(0)}, 0\right)\exp\left(-\alpha_+ x\right)\exp\left[\mathrm{i}(\beta z - \omega t)\right] \ (x > 0), \\ \mathbf{H}_{-} &= \frac{\mathrm{i}}{\omega\mu_0}\left(0, -\frac{k_0^2\varepsilon_-}{\alpha_-}E^{(0)}, 0\right)\exp\left(\alpha_- x\right)\exp\left[\mathrm{i}(\beta z - \omega t)\right] \ (x < 0). \end{aligned} \quad (8.19)$$

Then the envelope of the electric field is of the form

$$\begin{aligned} \tilde{\mathbf{E}}_{+}(x) &= \left(\frac{\mathrm{i}\beta}{\alpha_+}E^{(0)}, 0, E^{(0)}\right)\exp\left(-\alpha_+ x\right) \quad (x > 0), \\ \tilde{\mathbf{E}}_{-}(x) &= \left(-\frac{\mathrm{i}\beta}{\alpha_-}E^{(0)}, 0, E^{(0)}\right)\exp\left(\alpha_- x\right) \quad (x < 0), \end{aligned} \quad (8.20)$$

and the envelope of the magnetic field of the sustained surface electromagnetic field eigen modes is

$$\begin{aligned}\tilde{\mathbf{H}}_{+}(x) &= \frac{\mathrm{i}}{\omega\mu_0}\left(0, \frac{k_0^2\varepsilon_+}{\alpha_+}E^{(0)}, 0\right)\exp\left(-\alpha_+ x\right) \quad (x>0),\\ \tilde{\mathbf{H}}_{-}(x) &= \frac{\mathrm{i}}{\omega\mu_0}\left(0, -\frac{k_0^2\varepsilon_-}{\alpha_-}E^{(0)}, 0\right)\exp\left(\alpha_- x\right) \quad (x<0).\end{aligned} \tag{8.21}$$

Since the two dielectrics comprise a $\mathcal{P}$-$\mathcal{T}$ symmetric interface, their electric permittivity $\varepsilon_\pm$ and the attenuation coefficient $\alpha_\pm$ satisfy the relations

$$\varepsilon_-^* = \varepsilon_+, \qquad \alpha_-^* = \alpha_+, \tag{8.22}$$

and the envelopes of the electric and magnetic fields of the sustained surface electromagnetic field eigen modes fulfill

$$\tilde{\mathbf{E}}_-^*(-x) = \tilde{\mathbf{E}}_+(x), \qquad \tilde{\mathbf{H}}_-^*(-x) = \tilde{\mathbf{H}}_+(x). \tag{8.23}$$

In strong dispersion, the instantaneous values of electric and magnetic field energy densities are $\frac{1}{2}\varepsilon_0\frac{\partial(\omega\varepsilon_{\mathrm{r}})}{\partial\omega}\mathbf{E}^2$ and $\frac{1}{2}\mu_0\frac{\partial(\omega\mu_{\mathrm{r}})}{\partial\omega}\mathbf{H}^2$, respectively [108]. In what follows, we shall consider the case of weak dispersion, where the instantaneous values of the electric and magnetic field energy densities are $\frac{1}{2}\varepsilon_0\varepsilon_{\mathrm{r}}\mathbf{E}^2$ and $\frac{1}{2}\mu_0\mu_{\mathrm{r}}\mathbf{H}^2$, respectively. In the complex formalism of electromagnetic waves, the average values of the electric and magnetic field energy densities are $\frac{1}{4}\mathrm{Re}\left(\varepsilon_0\varepsilon_{\mathrm{r}}\mathbf{E}^*\mathbf{E}\right)$ and $\frac{1}{4}\mathrm{Re}\left(\mu_0\mu_{\mathrm{r}}\mathbf{H}^*\mathbf{H}\right)$, respectively.

The electric field energy density w_{e}^+ and the magnetic field energy density w_{m}^+ in the region of $x>0$ are given by

$$\begin{aligned}w_{\mathrm{e}}^+ &= \frac{1}{2}\mathrm{Re}\left(\frac{1}{2}\varepsilon_+\varepsilon_0\mathbf{E}_+^*\mathbf{E}_+\right)\\ &= \frac{1}{4}\mathrm{Re}[\varepsilon_+\varepsilon_0(\frac{\beta^2}{\alpha_+^*\alpha_+}+1)](E^{(0)})^2\exp\left[-\left(\alpha_+ + \alpha_+^*\right)x\right];\end{aligned} \tag{8.24}$$

$$\begin{aligned}w_{\mathrm{m}}^+ &= \frac{1}{2}\mathrm{Re}\left(\frac{1}{2}\mu_0\mathbf{H}_+^*\mathbf{H}_+\right)\\ &= \frac{1}{2}\mathrm{Re}[\frac{1}{2}\mu_0(\frac{1}{\omega\mu_0})^2\frac{k_0^2\varepsilon_+}{\alpha_+}\frac{k_0^2\varepsilon_+^*}{\alpha_+^*}](E^{(0)})^2\exp\left[-\left(\alpha_+ + \alpha_+^*\right)x\right]\\ &= \frac{1}{4}\mathrm{Re}\left[\varepsilon_+\varepsilon_0\left(\frac{\varepsilon_+^* k_0^2}{\alpha_+^*\alpha_+}\right)\right](E^{(0)})^2\exp\left[-\left(\alpha_+ + \alpha_+^*\right)x\right].\end{aligned} \tag{8.25}$$

The sum of the electric and magnetic energy densities is $w^{\pm} = w_{\mathrm{e}}^{\pm} + w_{\mathrm{m}}^{\pm}$, where the total energy density w^{+} in the region of $x > 0$ and the total energy density w^{-} in the region of $x < 0$ take the form

$$\begin{aligned} w^{+} &= \frac{1}{4}\mathrm{Re}\left[\varepsilon_{+}\varepsilon_{0}\left(\frac{\beta^2 + \varepsilon_{+}^{*}k_0^2}{\alpha_{+}^{*}\alpha_{+}} + 1\right)\right](E^{(0)})^2 \\ &\quad \times \exp\left[-\left(\alpha_{+} + \alpha_{+}^{*}\right)x\right] \qquad (x > 0), \\ w^{-} &= \frac{1}{4}\mathrm{Re}\left[\varepsilon_{-}\varepsilon_{0}\left(\frac{\beta^2 + \varepsilon_{-}^{*}k_0^2}{\alpha_{-}^{*}\alpha_{-}} + 1\right)\right](E^{(0)})^2 \\ &\quad \times \exp\left[+\left(\alpha_{-} + \alpha_{-}^{*}\right)x\right] \qquad (x < 0). \end{aligned} \tag{8.26}$$

Since $\alpha_{-}^{*} = \alpha_{+}$, $\varepsilon_{-}^{*} = \varepsilon_{+}$ and $\int_{-\infty}^{0} w^{-}\mathrm{d}x = \int_{0}^{+\infty} w^{+}\mathrm{d}x$, where

$$\begin{aligned} \int_{0}^{+\infty} w^{+}\mathrm{d}x &= \frac{1}{4}\mathrm{Re}\left[\varepsilon_{+}\varepsilon_{0}\left(\frac{\beta^2 + \varepsilon_{+}^{*}k_0^2}{\alpha_{+}^{*}\alpha_{+}} + 1\right)\right]\frac{(E^{(0)})^2}{\alpha_{+} + \alpha_{+}^{*}}, \\ \int_{-\infty}^{0} w^{-}\mathrm{d}x &= \frac{1}{4}\mathrm{Re}\left[\varepsilon_{-}\varepsilon_{0}\left(\frac{\beta^2 + \varepsilon_{-}^{*}k_0^2}{\alpha_{-}^{*}\alpha_{-}} + 1\right)\right]\frac{(E^{(0)})^2}{\alpha_{-} + \alpha_{-}^{*}}, \end{aligned} \tag{8.27}$$

the total electromagnetic energy of surface electromagnetic wave modes, which is $\mathcal{E} = \left(\int_{0}^{+\infty} w^{+}\mathrm{d}x + \int_{-\infty}^{0} w^{-}\mathrm{d}x\right) l_y l_z$, is of the form

$$\mathcal{E} = \frac{1}{2}\mathrm{Re}\left[\varepsilon_{+}\varepsilon_{0}\left(\frac{\beta^2 + \varepsilon_{+}^{*}k_0^2}{\alpha_{+}^{*}\alpha_{+}} + 1\right)\right]\frac{(E^{(0)})^2 l_y l_z}{\alpha_{+} + \alpha_{+}^{*}}, \tag{8.28}$$

where l_y and l_z are the length scales in the $\hat{y}$- and $\hat{z}$ directions, respectively.

As far as the electromagnetic spin current density is concerned, from the eigen mode solution (8.18), the 3-D magnetic vector potential $\mathbf{A}$ of the surface electromagnetic waves can be obtained. Then by using $\bar{\mathbf{S}} = \frac{1}{2}\mathrm{Re}(\mathbf{E}^{*} \times \mathbf{A})$, the average spin current density of the parity-time symmetric surface electromagnetic waves can be derived. But we shall not continue to calculate this quantity in this subsection.

8.2. A Generalized Parity-Time Symmetry in Surface Plasmonics

In the electromagnetics, TM-mode surface plasmon polaritons propagating in the z direction satisfies the Helmholtz equation $(\frac{\partial^2}{\partial x^2} + \frac{\partial^2}{\partial z^2} + \varepsilon\frac{\omega^2}{c^2})H_y =$

0, or $\frac{\partial^2}{\partial x^2}H_y + \varepsilon\frac{\omega^2}{c^2}H_y = \beta^2 H_y$, which can be rewritten as $\frac{1}{\varepsilon H_y}\frac{\partial^2}{\partial x^2}H_y + \frac{\omega^2}{c^2} = \frac{\beta^2}{\varepsilon}$. From this equation, we define $\Phi = \ln H_y$, and by using the relation $\frac{1}{H_y}\frac{\partial^2}{\partial x^2}H_y = \frac{\partial^2\Phi}{\partial x^2} + (\frac{\partial\Phi}{\partial x})^2$, in the region of $x > 0$, where the medium permittivity is ε_+, the field equation of the auxiliary quantity Φ and its $\mathcal{P}$-$\mathcal{T}$ transformation ($x \to -x$, $t \to -t$ and $\mathrm{i} \to -\mathrm{i}$) are given by

$$\frac{1}{\varepsilon_+}\left[\frac{\partial^2\Phi_+(x)}{\partial x^2} + \left(\frac{\partial\Phi_+(x)}{\partial x}\right)^2\right] + \frac{\omega^2}{c^2} = \frac{\beta^2}{\varepsilon_+}, \quad (8.29a)$$

$$\frac{1}{\varepsilon_+^*}\left[\frac{\partial^2\Phi_+^*(-x)}{\partial x^2} + \left(\frac{\partial\Phi_+^*(-x)}{\partial x}\right)^2\right] + \frac{\omega^2}{c^2} = \frac{(\beta^2)^*}{\varepsilon_+^*}, \quad (8.29b)$$

and in the region of $x < 0$, the field equation of Φ and its $\mathcal{P}$-$\mathcal{T}$ transformation are given by

$$\frac{1}{\varepsilon_-}\left[\frac{\partial^2\Phi_-(x)}{\partial x^2} + \left(\frac{\partial\Phi_-(x)}{\partial x}\right)^2\right] + \frac{\omega^2}{c^2} = \frac{\beta^2}{\varepsilon_-}, \quad (8.30a)$$

$$\frac{1}{\varepsilon_-^*}\left[\frac{\partial^2\Phi_-^*(-x)}{\partial x^2} + \left(\frac{\partial\Phi_-^*(-x)}{\partial x}\right)^2\right] + \frac{\omega^2}{c^2} = \frac{(\beta^2)^*}{\varepsilon_-^*}. \quad (8.30b)$$

Note that $\mathrm{i}\frac{\partial}{\partial t}$ and $\mathrm{e}^{-\mathrm{i}\omega t}$ are invariant under the parity-time symmetry transformation.

In the ordinary $\mathcal{P}$-$\mathcal{T}$ symmetry, i.e., $\varepsilon_+^* = \varepsilon_-$, $\varepsilon_-^* = \varepsilon_+$, $\Phi_-^*(-x) = \Phi_+(x)$, $\Phi_+^*(-x) = \Phi_-(x)$, and $(\beta^2)^* = \beta^2$. ($\mathrm{Re}\varepsilon_+ = \mathrm{Re}\varepsilon_-$, $\mathrm{Im}\varepsilon_+ = -\mathrm{Im}\varepsilon_-$), Eq. (8.29a) and Eq. (8.30b) are the same one, and Eq. (8.29b) and (8.30a) are also the same one. Then we have the relations: $\frac{1}{\varepsilon_+}\left[\frac{\partial^2\Phi_+(x)}{\partial x^2} + \left(\frac{\partial\Phi_+(x)}{\partial x}\right)^2\right] = \frac{1}{\varepsilon_-^*}\left[\frac{\partial^2\Phi_-^*(-x)}{\partial x^2} + \left(\frac{\partial\Phi_-^*(-x)}{\partial x}\right)^2\right]$ and $\frac{1}{\varepsilon_-}\left[\frac{\partial^2\Phi_-(x)}{\partial x^2} + \left(\frac{\partial\Phi_-(x)}{\partial x}\right)^2\right] = \frac{1}{\varepsilon_+^*}\left[\frac{\partial^2\Phi_+^*(-x)}{\partial x^2} + \left(\frac{\partial\Phi_+^*(-x)}{\partial x}\right)^2\right]$.

Now we shall generalize the ordinary $\mathcal{P}$-$\mathcal{T}$ symmetry. In this generalized $\mathcal{P}$-$\mathcal{T}$ symmetry, $\frac{1}{\varepsilon_+}\left[\frac{\partial^2\Phi_+(x)}{\partial x^2} + \left(\frac{\partial\Phi_+(x)}{\partial x}\right)^2\right] \neq \frac{1}{\varepsilon_-^*}\left[\frac{\partial^2\Phi_-^*(-x)}{\partial x^2} + \left(\frac{\partial\Phi_-^*(-x)}{\partial x}\right)^2\right]$ and $\frac{1}{\varepsilon_-}\left[\frac{\partial^2\Phi_-(x)}{\partial x^2} + \left(\frac{\partial\Phi_-(x)}{\partial x}\right)^2\right] \neq \frac{1}{\varepsilon_+^*}\left[\frac{\partial^2\Phi_+^*(-x)}{\partial x^2} + \left(\frac{\partial\Phi_+^*(-x)}{\partial x}\right)^2\right]$. We consider a generalized case, in which (8.29a)–(8.30b) and (8.29b)–(8.30a) are nonzero, but (8.29a)–(8.30b) and (8.29b)–(8.30a) are equal. By using the relation

(8.29a)−(8.29b)+(8.30a)−(8.30b)= 0, we can obtain

$$\begin{aligned} & \frac{1}{\varepsilon_+}\left[\frac{\partial^2\Phi_+(x)}{\partial x^2}+\left(\frac{\partial\Phi_+(x)}{\partial x}\right)^2\right]+\frac{1}{\varepsilon_-}\left[\frac{\partial^2\Phi_-(x)}{\partial x^2}+\left(\frac{\partial\Phi_-(x)}{\partial x}\right)^2\right] \\ = \; & \frac{1}{\varepsilon_+^*}\left[\frac{\partial^2\Phi_+^*(-x)}{\partial x^2}+\left(\frac{\partial\Phi_+^*(-x)}{\partial x}\right)^2\right]+\frac{1}{\varepsilon_-^*}\left[\frac{\partial^2\Phi_-^*(-x)}{\partial x^2}+\left(\frac{\partial\Phi_-^*(-x)}{\partial x}\right)^2\right]. \end{aligned} \tag{8.31}$$

Now it follows from Eqs. (8.29b), (8.30b) and (8.31) that the following relation

$$\frac{\beta^2}{\varepsilon_+}+\frac{\beta^2}{\varepsilon_-}-\left(\frac{(\beta^2)^*}{\varepsilon_+^*}+\frac{(\beta^2)^*}{\varepsilon_-^*}\right)=0 \tag{8.32}$$

can be derived. In order that the phase constant β of the surface electromagnetic modes is a real number, i.e., $(\beta^2)^* = \beta^2$, we should place a constraint condition

$$\frac{1}{\varepsilon_+}+\frac{1}{\varepsilon_-}-\left(\frac{1}{\varepsilon_+^*}+\frac{1}{\varepsilon_-^*}\right)=0. \tag{8.33}$$

It follows that $\frac{1}{\varepsilon_+}-\frac{1}{\varepsilon_+^*}=\frac{1}{\varepsilon_-^*}-\frac{1}{\varepsilon_-}$ and $\frac{\varepsilon_+^*-\varepsilon_+}{\varepsilon_+^*\varepsilon_+}=\frac{\varepsilon_--\varepsilon_-^*}{\varepsilon_-^*\varepsilon_-}$, and then this relation yields

$$\varepsilon_+^*\varepsilon_+\mathrm{Im}\varepsilon_-+\varepsilon_-^*\varepsilon_-\mathrm{Im}\varepsilon_+=0, \tag{8.34}$$

i.e., it can be rearranged as

$$\left[(\mathrm{Re}\varepsilon_+)^2+(\mathrm{Im}\varepsilon_+)^2\right]\mathrm{Im}\varepsilon_-+\left[(\mathrm{Re}\varepsilon_-)^2+(\mathrm{Im}\varepsilon_-)^2\right]\mathrm{Im}\varepsilon_+=0. \tag{8.35}$$

Obviously, the ordinary $\mathcal{P}$-$\mathcal{T}$ symmetry $\varepsilon_+^*=\varepsilon_-$ is a special case of Eq. (8.35). It follows from the relation (8.33) that $\frac{1}{\varepsilon_+}+\frac{1}{\varepsilon_-}$ is a real number, i.e., $\frac{\varepsilon_++\varepsilon_-}{\varepsilon_+\varepsilon_-}$ is a real number. Now the phase constant of the present surface electromagnetic modes is given by

$$\beta=\frac{\omega}{c}\sqrt{\frac{\varepsilon_+\varepsilon_-}{\varepsilon_++\varepsilon_-}}, \tag{8.36}$$

which has a real eigenvalue spectrum. In such ordinary parity-time surface electromagnetic modes (8.33), $\frac{1}{\varepsilon_+}$ and $\frac{1}{\varepsilon_-}$ should be complex conjugate in order to respect this optical parity-time symmetry. But now this condition can be relaxed and we will adopt a more loose condition: $\frac{1}{\varepsilon_+}+\frac{1}{\varepsilon_-}=\frac{1}{\varepsilon_+^*}+\frac{1}{\varepsilon_-^*}$.

From the equation (8.35), we can assume ε_+ and ε_- satisfy

$$\varepsilon_+ = \bar{\varepsilon}\frac{e^{i\varphi_+}}{c_+}, \qquad \varepsilon_- = \bar{\varepsilon}\frac{e^{-i\varphi_-}}{c_-}, \tag{8.37}$$

where c_+, c_- and $\bar{\varepsilon}$ are real numbers. Now such permittivities ε_+ and ε_- fulfill the constraint condition (8.35):

$$\varepsilon_+^*\varepsilon_+ = \frac{\bar{\varepsilon}^2}{c_+^2}, \qquad \varepsilon_-^*\varepsilon_- = \frac{\bar{\varepsilon}^2}{c_-^2}, \tag{8.38}$$

and

$$-\frac{\bar{\varepsilon}^2}{c_+^2}\frac{\sin\varphi_-}{c_-} + \frac{\bar{\varepsilon}^2}{c_-^2}\frac{\sin\varphi_+}{c_+} = 0. \tag{8.39}$$

Thus, we have $\frac{\sin\varphi_-}{c_+} = \frac{\sin\varphi_+}{c_-}$, and we obtain $c_+ = \sin\varphi_-$ and $c_- = \sin\varphi_+$. Now the permittivity of the two dielectrics comprising the generalized $\mathcal{P}$-$\mathcal{T}$ symmetric interface for supporting surface electromagnetic wave modes is given by

$$\varepsilon_+ = \bar{\varepsilon}\frac{e^{i\varphi_+}}{\sin\varphi_-}, \qquad \varepsilon_- = \bar{\varepsilon}\frac{e^{-i\varphi_-}}{\sin\varphi_+}. \tag{8.40}$$

The expression for $\frac{1}{\varepsilon_+} + \frac{1}{\varepsilon_-}$ is of the form

$$\frac{1}{\varepsilon_+} + \frac{1}{\varepsilon_-} = \frac{1}{\bar{\varepsilon}}\left(e^{-i\varphi_+}\sin\varphi_- + e^{i\varphi_-}\sin\varphi_+\right), \tag{8.41}$$

which is a real number. The phase constant of the surface electromagnetic wave modes sustained on the present generalized $\mathcal{P}$-$\mathcal{T}$ symmetric interface is given by

$$\beta = \frac{\omega}{c}\sqrt{\bar{\varepsilon}}\sqrt{\frac{1}{e^{-i\varphi_+}\sin\varphi_- + e^{i\varphi_-}\sin\varphi_+}} = \frac{\omega}{c}\sqrt{\bar{\varepsilon}}\sqrt{\frac{1}{\sin(\varphi_+ + \varphi_-)}}. \tag{8.42}$$

This phase constant is a real number because of the present generalized parity-time symmetry in the optical material interface structure.

9. Supersymmetric Parity-Time Invariance in Optical Materials

9.1. Review of Supersymmetric Quantum Mechanics

The method of supersymmetry can be employed in the ordinary quantum physics, including quantum optics (i.e., multiphoton-atom interaction) [127, 128, 129, 130]. One of such examples is the multi-photon Jaynes-Cummings model. The framework of supersymmetric quantum mechanics, which was suggested by Witten, can be used to treat some quantum systems [131]. In this issue, we first discuss the main theoretical features of supersymmetric quantum mechanics and then extend it to a case of parity-time invariance. For the first, we will refer to Liang and Wei's monograph [131] for an overview of supersymmetric quantum mechanics.

We consider a one-dimensional harmonic oscillator of the Hamiltonian operator $\hat{H} = \frac{\hat{p}^2}{2\mu} + \frac{1}{2}\mu\omega^2 x^2$. In the occupation number representation, according to the well-known ordinary quantum mechanics, the creation and annihilation operators are given by

$$\hat{a} = \sqrt{\frac{\mu\omega}{2\hbar}}\left(x + \frac{\mathrm{i}}{\mu\omega}\hat{p}\right) = \sqrt{\frac{\mu\omega}{2\hbar}}\left(x + \frac{\hbar}{\mu\omega}\frac{\partial}{\partial x}\right),$$
$$\hat{a}^{\dagger} = \sqrt{\frac{\mu\omega}{2\hbar}}\left(x - \frac{\mathrm{i}}{\mu\omega}\hat{p}\right) = \sqrt{\frac{\mu\omega}{2\hbar}}\left(x - \frac{\hbar}{\mu\omega}\frac{\partial}{\partial x}\right). \tag{9.1}$$

The creation and annihilation operators obey the bosonic commutator $[a, a^{\dagger}] = 1$. If we define $\xi = \sqrt{\frac{\mu\omega}{\hbar}}x$, the creation and annihilation operators are

$$\hat{a} = \frac{1}{\sqrt{2}}\left(\xi + \frac{\partial}{\partial\xi}\right), \qquad \hat{a}^{\dagger} = \frac{1}{\sqrt{2}}\left(\xi - \frac{\partial}{\partial\xi}\right). \tag{9.2}$$

They can also be rewritten as

$$\hat{A} = -\mathrm{i}\hat{a} = \frac{1}{\sqrt{2}}\left(-\mathrm{i}\frac{\partial}{\partial\xi} - \mathrm{i}\xi\right),$$
$$\hat{A}^{\dagger} = \mathrm{i}\hat{a}^{\dagger} = \frac{1}{\sqrt{2}}\left(-\mathrm{i}\frac{\partial}{\partial\xi} + \mathrm{i}\xi\right). \tag{9.3}$$

They obey the bosonic commuting relation $[\hat{A}, \hat{A}^{\dagger}] = 1$.

For simplicity, we assume $\hbar = 1$, $\mu = 1$ and $\omega = 1$. A Hamiltonian operator $\hat{H} = \frac{1}{2}\left[\hat{p}^2 + W^2(x)\right] - \frac{1}{2}\sigma_3 W'(x)$ with $W'(x) = \frac{\partial W(x)}{\partial x}$ is constructed and the supercharge operators are $\hat{Q} = \hat{A}\hat{\lambda}$ and $\hat{Q}^\dagger = \hat{A}^\dagger\hat{\lambda}^\dagger$, where $\hat{A} = \frac{1}{\sqrt{2}}\left[\hat{p} - \mathrm{i}W(x)\right]$ and $\hat{A}^\dagger = \frac{1}{\sqrt{2}}\left[\hat{p} + \mathrm{i}W(x)\right]$, can be defined [131]. Then the commuting relation of $\hat{A}$ and $\hat{A}^\dagger$ is $[\hat{A}, \hat{A}^\dagger] = W'(x)$ [131]. It can be found that such $\hat{A}$, $\hat{A}^\dagger$ and the commuting relation is a generalization of the relevant operators in the harmonic oscillator. If $[\hat{A}, \hat{A}^\dagger] = 1$, this means that $W(x) = x + C$ with C an integral constant. We choose $W(x) = x$ and $\hat{\lambda}$, $\hat{\lambda}^\dagger$ are given by [131]

$$\hat{\lambda} = \sigma_- = \begin{pmatrix} 0 & 0 \\ 1 & 0 \end{pmatrix}, \qquad \hat{\lambda}^\dagger = \sigma_+ = \begin{pmatrix} 0 & 1 \\ 0 & 0 \end{pmatrix}. \tag{9.4}$$

Here, $\{\hat{\lambda}, \hat{\lambda}^\dagger\} = 1$, i.e., $\hat{\lambda}^\dagger$ and $\hat{\lambda}$ are the creation and annihilation operators, respectively, of fermions. It can be shown that $[\hat{Q}, \hat{H}] = [\hat{Q}^\dagger, \hat{H}] = 0$, $\{\hat{Q}, \hat{Q}\} = \{\hat{Q}^\dagger, \hat{Q}^\dagger\} = 0$ [131]. Let us calculate the commuting relation $[\hat{Q}^\dagger, \hat{Q}]$:

$$\begin{aligned}
[\hat{Q}^\dagger, \hat{Q}] &= \hat{A}^\dagger\hat{\lambda}^\dagger\hat{A}\hat{\lambda} - \hat{A}\hat{\lambda}\hat{A}^\dagger\hat{\lambda}^\dagger = \hat{A}^\dagger\hat{A}\frac{\hat{\mathcal{I}} + \sigma_3}{2} - \hat{A}\hat{A}^\dagger\frac{\hat{\mathcal{I}} - \sigma_3}{2} \\
&= \frac{1}{2}[\hat{A}^\dagger, \hat{A}] + \frac{1}{2}\{\hat{A}, \hat{A}^\dagger\}\sigma_3 \\
&= \left(\frac{1}{2}\{\hat{A}, \hat{A}^\dagger\} - \frac{1}{2}\sigma_3[\hat{A}, \hat{A}^\dagger]\right)\sigma_3 = \hat{H}\sigma_3.
\end{aligned} \tag{9.5}$$

The Hamiltonian can also be rewritten as $\hat{H} = \{\hat{Q}, \hat{Q}^\dagger\}$ [131]. This can be shown as follows: $\{\hat{Q}^\dagger, \hat{Q}\} = \hat{A}^\dagger\hat{\lambda}^\dagger\hat{A}\hat{\lambda} + \hat{A}\hat{\lambda}\hat{A}^\dagger\hat{\lambda}^\dagger = \hat{A}^\dagger\hat{A}\frac{\hat{\mathcal{I}}+\sigma_3}{2} + \hat{A}\hat{A}^\dagger\frac{\hat{\mathcal{I}}-\sigma_3}{2} = \hat{A}^\dagger\hat{A} + \hat{A}\hat{A}^\dagger - [\hat{A}, \hat{A}^\dagger]\frac{\sigma_3}{2} = \frac{1}{2}\left[\hat{p}^2 + W^2(x)\right] - \frac{1}{2}\sigma_3 W'(x) = \hat{H}$. All the above commutators form an sl(1,1) superalgebra [131].

The Hamiltonian operator $\hat{H}$ can be expanded as

$$\begin{aligned}
\hat{H} &= \frac{1}{2}\{\hat{A}, \hat{A}^\dagger\} - \frac{1}{2}\sigma_3[\hat{A}, \hat{A}^\dagger] \\
&= \begin{pmatrix} \hat{A}^\dagger\hat{A} & 0 \\ 0 & \hat{A}\hat{A}^\dagger \end{pmatrix} = \begin{pmatrix} \hat{H}_+ & 0 \\ 0 & \hat{H}_- \end{pmatrix},
\end{aligned} \tag{9.6}$$

where the two diagonal matrix elements are given by [131]

$$\begin{aligned}\hat{H}_+ &= \frac{1}{2}\left(-\frac{\mathrm{d}^2}{\mathrm{d}x^2} + V_+(x)\right),\\ \hat{H}_- &= \frac{1}{2}\left(-\frac{\mathrm{d}^2}{\mathrm{d}x^2} + V_-(x)\right).\end{aligned} \tag{9.7}$$

Here, $V_\pm(x) = W^2(x) \mp W'(x)$.

We shall now consider the energy eigenvalue equation $\hat{H}(x)\Psi(x) = E\Psi(x)$, where

$$\Psi(x) = \begin{pmatrix} \psi_\mathrm{u} \\ \psi_\mathrm{d} \end{pmatrix}, \quad \Psi_\mathrm{u}(x) = \begin{pmatrix} \psi_\mathrm{u} \\ 0 \end{pmatrix}, \quad \Psi_\mathrm{d}(x) = \begin{pmatrix} 0 \\ \psi_\mathrm{d} \end{pmatrix} \tag{9.8}$$

and the relations between ψ_u and ψ_d are $\psi_\mathrm{u} = \frac{1}{\sqrt{E}}\hat{A}^\dagger\psi_\mathrm{d}$ and $\psi_\mathrm{d} = \frac{1}{\sqrt{E}}\hat{A}\psi_\mathrm{u}$ [131]. One can establish the relation between ψ_d and ψ_u through these relations. If one defines $\psi_\mathrm{u} = \frac{1}{\sqrt{E}}\hat{A}^\dagger\psi_\mathrm{d}$, then it can be obtained:

$$\begin{aligned}\hat{A}\psi_\mathrm{u} &= \frac{1}{\sqrt{E}}\hat{A}\hat{A}^\dagger\psi_\mathrm{d} = \frac{1}{\sqrt{E}}\left([\hat{A},\hat{A}^\dagger] + \hat{A}^\dagger\hat{A}\right)\psi_\mathrm{d}\\ &= \frac{1}{\sqrt{E}}\left(\hat{A}^\dagger\hat{A} + W'(x)\right)\psi_\mathrm{d} = \frac{1}{\sqrt{E}}H_-\psi_\mathrm{d} = \sqrt{E}\psi_\mathrm{d}.\end{aligned} \tag{9.9}$$

Thus, one can have $\psi_\mathrm{d} = \frac{1}{\sqrt{E}}\hat{A}\psi_\mathrm{u}$. On the other hand, $\hat{A}^\dagger\psi_\mathrm{d}$ can be expanded as

$$\begin{aligned}\hat{A}^\dagger\psi_\mathrm{d} &= \frac{1}{\sqrt{E}}\hat{A}^\dagger\hat{A}\psi_\mathrm{u} = \frac{1}{\sqrt{E}}\frac{\{\hat{A}^\dagger,\hat{A}\} - [\hat{A},\hat{A}^\dagger]}{2}\psi_\mathrm{u}\\ &= \frac{1}{\sqrt{E}}\hat{H}_+\psi_\mathrm{u} = \sqrt{E}\psi_\mathrm{u}.\end{aligned} \tag{9.10}$$

Then from this result, we can have $\psi_\mathrm{u} = \frac{1}{\sqrt{E}}\hat{A}^\dagger\psi_\mathrm{d}$.

In the algebra of supersymmetry, there are some other essential relations: $\hat{Q}\Psi_\mathrm{d} = 0$, $\hat{Q}^\dagger\Psi_\mathrm{u} = 0$ and $\hat{Q}^\dagger\Psi_\mathrm{d} = \sqrt{E}\Psi_\mathrm{u}$, $\hat{Q}\Psi_\mathrm{u} = \sqrt{E}\Psi_\mathrm{d}$, where the supercharge operators $\hat{Q} = \hat{A}\hat{\lambda}$ and $\hat{Q}^\dagger = \hat{A}^\dagger\hat{\lambda}^\dagger$ [131]. These can be shown as follows:

$$\hat{Q}\Psi_{\rm d} = \hat{A}\begin{pmatrix} 0 & 0 \\ 1 & 0 \end{pmatrix}\begin{pmatrix} 0 \\ \psi_{\rm d} \end{pmatrix} = 0, \quad \hat{Q}^{\dagger}\Psi_{\rm u} = \hat{A}^{\dagger}\begin{pmatrix} 0 & 1 \\ 0 & 0 \end{pmatrix}\begin{pmatrix} \psi_{\rm u} \\ 0 \end{pmatrix} = 0,$$
$$\hat{Q}^{\dagger}\Psi_{\rm d} = \hat{A}^{\dagger}\begin{pmatrix} 0 & 1 \\ 0 & 0 \end{pmatrix}\begin{pmatrix} 0 \\ \psi_{\rm d} \end{pmatrix} = \hat{A}^{\dagger}\begin{pmatrix} \psi_{\rm d} \\ 0 \end{pmatrix} = \sqrt{E}\begin{pmatrix} \psi_{\rm u} \\ 0 \end{pmatrix} = \sqrt{E}\Psi_{\rm u},$$
$$\hat{Q}\Psi_{\rm u} = \hat{A}\begin{pmatrix} 0 & 0 \\ 1 & 0 \end{pmatrix}\begin{pmatrix} \psi_{\rm u} \\ 0 \end{pmatrix} = \hat{A}\begin{pmatrix} 0 \\ \psi_{\rm u} \end{pmatrix} = \sqrt{E}\begin{pmatrix} 0 \\ \psi_{\rm d} \end{pmatrix} = \sqrt{E}\Psi_{\rm d}. \quad (9.11)$$

Now we shall calculate $\hat{Q}^{\dagger}\Psi$ and $\hat{Q}\Psi$:

$$\hat{Q}^{\dagger}\Psi = \hat{A}^{\dagger}\begin{pmatrix} 0 & 1 \\ 0 & 0 \end{pmatrix}\begin{pmatrix} \psi_{\rm u} \\ \psi_{\rm d} \end{pmatrix} = \hat{A}^{\dagger}\begin{pmatrix} \psi_{\rm d} \\ 0 \end{pmatrix} = \sqrt{E}\begin{pmatrix} \psi_{\rm u} \\ 0 \end{pmatrix} = \sqrt{E}\Psi_{\rm u},$$
$$\hat{Q}\Psi = \hat{A}\begin{pmatrix} 0 & 0 \\ 1 & 0 \end{pmatrix}\begin{pmatrix} \psi_{\rm u} \\ \psi_{\rm d} \end{pmatrix} = \hat{A}\begin{pmatrix} 0 \\ \psi_{\rm u} \end{pmatrix} = \sqrt{E}\begin{pmatrix} 0 \\ \psi_{\rm d} \end{pmatrix} = \sqrt{E}\Psi_{\rm d}. \quad (9.12)$$

It can then be found that $\hat{Q}^{\dagger}+\hat{Q}$ and $\hat{H}$ have the same eigenstates and that their commutating relation vanishes:

$$\left(\hat{Q}^{\dagger}+\hat{Q}\right)\Psi = \sqrt{E}\Psi, \qquad \hat{H}\Psi = E\Psi, \qquad [\hat{Q}^{\dagger}+\hat{Q}, \hat{H}] = 0. \quad (9.13)$$

In the above supersymmetric quantum mechanics, the supersymmetric Hamiltonian operator is given by $\hat{H} = \frac{1}{2}\left[\hat{p}^2 + W^2(x)\right] - \frac{1}{2}\sigma_3 W'(x)$ with $W'(x) = \frac{\partial W(x)}{\partial x}$, which is the description of a non-relativistic particle. The algebra of supersymmetry can also be realized in a k-photon Jaynes-Cummings model, where k photons are absorbed or emitted in one transition [127, 128, 129, 130].

9.2. Supersymmetric Parity-Time Invariance

Now we shall generalize the ordinary *parity-time invariance* to a *supersymmetric parity-time invariance*.

As has been given in the above, in the units $\hbar = 1$, $\mu = 1$ and $\omega = 1$, the supersymmetric Hamiltonian operator is given by $\hat{H} = \frac{1}{2}\left[\hat{p}^2 + W^2(x)\right] - \frac{1}{2}\sigma_3 W'(x)$ with $W'(x) = \frac{\partial W(x)}{\partial x}$ and the supercharge operators are $\hat{Q} = \hat{A}\hat{\lambda}$ and $\hat{Q}^{\dagger} = \hat{A}^{\dagger}\hat{\lambda}^{\dagger}$, where $\hat{A}$ and $\hat{A}^{\dagger}$ are defined as [131]

$$\hat{A} = \frac{1}{\sqrt{2}}\left[\hat{p} - {\rm i}W(x)\right], \quad \hat{A}^{\dagger} = \frac{1}{\sqrt{2}}\left[\hat{p} + {\rm i}W^*(x)\right]. \quad (9.14)$$

In our supersymmetric parity-time invariance presented here, $W(x)$ is an *anti-parity-time symmetric quantity*, i.e., $W^*(-x) = -W(x)$, namely, the real part of $W(x)$ is an odd function of spatial position, and the imaginary part of $W(x)$ is an even function of spatial position ($\mathrm{Re}W(x) = -\mathrm{Re}W(-x)$, $\mathrm{Im}W(x) = \mathrm{Im}W(-x)$). It can be found that under $\mathcal{P}$-$\mathcal{T}$ transformation, $\hat{A}$ and $\hat{A}^\dagger$ preserves the $\mathcal{P}$-$\mathcal{T}$ invariance, i.e.,

$$\hat{A} \to \frac{1}{\sqrt{2}}\left[\hat{p} + \mathrm{i}W^*(-x)\right] = \frac{1}{\sqrt{2}}\left[\hat{p} - \mathrm{i}W(x)\right] = \hat{A},$$
$$\hat{A}^\dagger \to \frac{1}{\sqrt{2}}\left[\hat{p} - \mathrm{i}(W^*(-x))^*\right] = \frac{1}{\sqrt{2}}\left[\hat{p} + \mathrm{i}W^*(x)\right] = \hat{A}^\dagger. \quad (9.15)$$

By using the relation (9.15), we have $[\hat{A}, \hat{A}^\dagger] = \frac{1}{2}\partial_x[W(x) + W^*(x)] = \frac{1}{2}\partial_x[W(x) - W(-x)]$. We can define another pair of annihilation and creation operators $\hat{B}$ and $\hat{B}^\dagger$:

$$\hat{B} = \frac{1}{\sqrt{2}}\left[\hat{p} - \mathrm{i}W^*(x)\right], \qquad \hat{B}^\dagger = \frac{1}{\sqrt{2}}\left[\hat{p} + \mathrm{i}W(x)\right]. \quad (9.16)$$

From the relations (9.15) and (9.16), we can obtain $[\hat{B}, \hat{B}^\dagger] = \frac{1}{2}\partial_x[W^*(x) + W(x)] = \frac{1}{2}\partial_x[W^*(x) - W^*(-x)]$ as well as $[\hat{A}, \hat{B}^\dagger] = \partial_x W(x)$, $[\hat{B}, \hat{A}^\dagger] = \partial_x W^*(x)$.

Under the parity transformation, $\hat{A} \to -\hat{B}$, $\hat{A}^\dagger \to -\hat{B}^\dagger$. Then $\hat{A}^\dagger\hat{B}\psi = \frac{1}{\sqrt{2}}\left[\hat{p} + \mathrm{i}W^*(x)\right]\frac{1}{\sqrt{2}}\left[\hat{p} - \mathrm{i}W^*(x)\right]\psi$, which is $\frac{1}{2}\left(\hat{p}^2 + (W^*)^2 - \mathrm{i}[\hat{p}, W^*]\right)\psi$. An alternative method for calculating $\hat{A}^\dagger\hat{B}$ is given as follows: With the help of the relation

$$\hat{A}^* = \frac{1}{\sqrt{2}}\left[-\hat{p} + \mathrm{i}W^*(x)\right] = -\hat{B},$$
$$(\hat{B}^\dagger)^* = \frac{1}{\sqrt{2}}\left[-\hat{p} - \mathrm{i}W^*(x)\right] = -\hat{A}^\dagger, \quad (9.17)$$

we have

$$\hat{A}^\dagger\hat{B} = \frac{1}{2}\left(\hat{p}^2 + (W^*)^2\right) - \frac{1}{2}\frac{\partial}{\partial x}W^*(x). \quad (9.18)$$

Other operator products $\hat{B}\hat{A}^\dagger$, $\hat{B}^\dagger\hat{A}$ and $\hat{A}\hat{B}^\dagger$ can be obtained. We list all these results:

$$\begin{aligned}
\hat{A}^{\dagger}\hat{B} &= \frac{1}{2}\left(\hat{p}^2+(W^*)^2\right)-\frac{1}{2}\frac{\partial}{\partial x}W^*(x),\\
\hat{B}\hat{A}^{\dagger} &= \frac{1}{2}\left(\hat{p}^2+(W^*)^2\right)+\frac{1}{2}\frac{\partial}{\partial x}W^*(x),\\
\hat{B}^{\dagger}\hat{A} &= \frac{1}{2}\left(\hat{p}^2+W^2\right)-\frac{1}{2}\frac{\partial}{\partial x}W(x),\\
\hat{A}\hat{B}^{\dagger} &= \frac{1}{2}\left(\hat{p}^2+W^2\right)+\frac{1}{2}\frac{\partial}{\partial x}W(x).
\end{aligned} \tag{9.19}$$

The operators $\hat{B}^{\dagger}\hat{A}$ and $\hat{A}\hat{B}^{\dagger}$ form a Hamiltonian

$$\hat{H}=\begin{pmatrix} \hat{B}^{\dagger}\hat{A} & 0 \\ 0 & \hat{A}\hat{B}^{\dagger} \end{pmatrix}=\frac{1}{2}\left(\hat{p}^2+W^2\right)-\frac{1}{2}\sigma_3\frac{\partial}{\partial x}W(x). \tag{9.20}$$

The other Hamiltonian can be of the form

$$\hat{H}^*=\begin{pmatrix} \hat{A}^{\dagger}\hat{B} & 0 \\ 0 & \hat{B}\hat{A}^{\dagger} \end{pmatrix}=\frac{1}{2}\left(\hat{p}^2+(W^*)^2\right)-\frac{1}{2}\sigma_3\frac{\partial}{\partial x}W^*(x). \tag{9.21}$$

Let us determine whether the supersymmetric Hamiltonian $\hat{H}$ is of parity-time symmetry or not. In the above, it has been shown that if the operation of $x\to -x$ and the complex conjugate (i $\to$ −i) to the Hamiltonian $\hat{H}(x)$ and it can still go back to the original form $\hat{H}(x)$, we can say that this Hamiltonian $\hat{H}(x)$, which agrees with the relation $\hat{H}^*(-x)=\hat{H}(x)$, is of parity-time symmetry. We consider the two terms W^2 and $\sigma_3\frac{\partial}{\partial x}W(x)$ given in Eq. (9.20):

$$\begin{aligned}
&W^2(x)\to[W^*(-x)]^2\to[-W(x)]^2=W^2(x);\\
&\sigma_3\frac{\partial}{\partial x}W(x)\to\sigma_3\frac{\partial}{\partial(-x)}W^*(-x)=-\sigma_3\frac{\partial}{\partial x}[-W(x)]=\sigma_3\frac{\partial}{\partial x}W(x).
\end{aligned} \tag{9.22}$$

It can be found that the two terms $W^2(x)$ and $\sigma_3\frac{\partial}{\partial x}W(x)$ in the Hamiltonian (9.20) preserve the parity-time symmetry. Thus, the Hamiltonian (9.20) is of parity-time symmetry.

We define the supercharge operators

$$\hat{Q}_a=\hat{A}\hat{\lambda},\qquad \hat{Q}_b^{\dagger}=\hat{B}^{\dagger}\hat{\lambda}^{\dagger},\qquad \hat{Q}_b=\hat{B}\hat{\lambda},\qquad \hat{Q}_a^{\dagger}=\hat{A}^{\dagger}\hat{\lambda}^{\dagger}. \tag{9.23}$$

Based on these relations, we can obtain some other relations:

$$\hat{Q}_a^* = \hat{A}^*\hat{\lambda} = -\hat{Q}_b, \qquad (\hat{Q}_b^\dagger)^* = (\hat{B}^\dagger)^*\hat{\lambda}^\dagger = -\hat{A}^\dagger\hat{\lambda}^\dagger = -\hat{Q}_a^\dagger,$$
$$\hat{Q}_b^* = \hat{B}^*\hat{\lambda} = -\hat{A}\hat{\lambda} = -\hat{Q}_a, \ (\hat{Q}_a^\dagger)^* = (\hat{A}^\dagger)^*\hat{\lambda}^\dagger = -\hat{B}^\dagger\hat{\lambda}^\dagger = -\hat{Q}_b^\dagger. \tag{9.24}$$

Then we can use the supercharge operators to construct the Hamiltonian operators

$$\hat{Q}_b^\dagger\hat{Q}_a + \hat{Q}_a\hat{Q}_b^\dagger = \hat{B}^\dagger\hat{\lambda}^\dagger\hat{A}\hat{\lambda} + \hat{A}\hat{\lambda}\hat{B}^\dagger\hat{\lambda}^\dagger = \begin{pmatrix} \hat{B}^\dagger\hat{A} & 0 \\ 0 & \hat{A}\hat{B}^\dagger \end{pmatrix} = \hat{H},$$
$$\hat{Q}_a^\dagger\hat{Q}_b + \hat{Q}_b\hat{Q}_a^\dagger = \hat{A}^\dagger\hat{\lambda}^\dagger\hat{B}\hat{\lambda} + \hat{B}\hat{\lambda}\hat{A}^\dagger\hat{\lambda}^\dagger = \begin{pmatrix} \hat{A}^\dagger\hat{B} & 0 \\ 0 & \hat{B}\hat{A}^\dagger \end{pmatrix} = \hat{H}^*. \tag{9.25}$$

These two Hamiltonian operators can also be rewritten as

$$\hat{H} = \frac{1}{2}\{\hat{A}, \hat{B}^\dagger\} - \frac{1}{2}\sigma_3[\hat{A}, \hat{B}^\dagger], \qquad \hat{H}^* = \frac{1}{2}\{\hat{B}, \hat{A}^\dagger\} - \frac{1}{2}\sigma_3[\hat{B}, \hat{A}^\dagger]. \tag{9.26}$$

Since $\hat{H} = \hat{Q}_b^\dagger\hat{Q}_a + \hat{Q}_a\hat{Q}_b^\dagger$, we have

$$[\hat{Q}_b^\dagger, \hat{H}] = \hat{Q}_b^\dagger\left(\hat{Q}_b^\dagger\hat{Q}_a + \hat{Q}_a\hat{Q}_b^\dagger\right) - \left(\hat{Q}_b^\dagger\hat{Q}_a + \hat{Q}_a\hat{Q}_b^\dagger\right)\hat{Q}_b^\dagger = 0,$$
$$[\hat{Q}_a, \hat{H}] = \hat{Q}_a\left(\hat{Q}_b^\dagger\hat{Q}_a + \hat{Q}_a\hat{Q}_b^\dagger\right) - \left(\hat{Q}_b^\dagger\hat{Q}_a + \hat{Q}_a\hat{Q}_b^\dagger\right)\hat{Q}_a = 0. \tag{9.27}$$

9.3. Discussions on the Parity-Time Symmetry

We will consider a *parity-time symmetric infinitely deep potential well*, where the potential $V = +\infty$ when $x > a$ and $x < -a$. The Schrödinger equation is given by $-\frac{\hbar^2}{2m}\frac{\partial^2}{\partial x^2}\psi + V\psi = E\psi$, or $\frac{\partial^2}{\partial x^2}\psi = -\frac{2m}{\hbar^2}(E - V)\psi$. Thus, $\frac{\partial^2}{\partial x^2}\psi_+ = -k_+^2\psi_+$ with $V_+ = V_1 + \mathrm{i}V_2$ when x is in the range $[0, +a]$ and $\frac{\partial^2}{\partial x^2}\psi_- = -k_-^2\psi_-$ with $V_- = V_1 - \mathrm{i}V_2$ when x is in the range $[-a, 0]$. Here, $k_+^2 = \frac{2m}{\hbar^2}(E - V_+)$ and $k_-^2 = \frac{2m}{\hbar^2}(E - V_-)$.

For the electron matter waves, it is difficult to realize the potential energy V with imaginary parts. But for a light wave, such a potential energy term is very easy to be achieved, because the Helmholtz equation

$\frac{\partial^2}{\partial x^2}E + \epsilon_{\rm r}\frac{\omega^2}{c^2}E = 0$, where the permittivity $\epsilon_{\rm r}$ can have a positive imaginary part (optical loss) or a negative imaginary part (optical gain).

The solution ψ_+ of the wave in the region $[0, +a]$ and the solution ψ_- of the wave in the region $[-a, 0]$ are given by

$$\psi_+ = A_+ \cos(k_+ x + \varphi_+), \quad \psi_- = A_- \cos(k_- x + \varphi_-). \tag{9.28}$$

It seems that there are five parameters to be determined: $\varphi_\pm$, $A_\pm$ and E ($k_\pm$ contains only the unknown parameter E). But since there is a normalization condition $\int_{-a}^{+a}\psi^*\psi {\rm d}x = 1$, which can also enable to determine one of the amplitudes $A_\pm$, this means that there are four parameters left (to be determined). Thus, we need four boundary conditions: i) $\psi_+(x = +a) = 0$; ii) $\psi_-(x = -a) = 0$; iii) $\psi_+(x = 0) = \psi_-(x = 0)$; iv) $\psi'_+(x = 0) = \psi'_-(x = 0)$ (the first derivative of the wavefunction at $x = 0$ should be continuous). The fourth condition can be derived as follows: The integral of $\frac{\partial^2}{\partial x^2}\psi = -k^2\psi$ at $[-\epsilon, +\epsilon]$ (with ϵ an infinitesimal parameter) is $\frac{\partial}{\partial x}\psi|_{-\epsilon}^{+\epsilon} = -\int_{-\epsilon}^{0} k_-^2\psi_- {\rm d}x - \int_0^{+\epsilon} k_+^2\psi_+ {\rm d}x$. Since $\psi_- = \psi_+$ at $x = 0$, the term on the right-handed side is $-(k_-^2 + k_+^2)\epsilon\psi(x = 0) \to 0$ when ϵ is an infinitesimal parameter. This means that $\frac{\partial}{\partial x}\psi|_{-\epsilon}^{+\epsilon} \to 0$, i.e., the first order derivative of the wave function ψ is continuous at $x = 0$ ($\frac{\partial}{\partial x}\psi_+(x = 0) = \frac{\partial}{\partial x}\psi_-(x = 0)$). In a word, when the potential function changes suddenly at $x = 0$, the wavefunction ψ and its first-order derivative $\frac{\partial\psi}{\partial x}$ have to be continuous, respectively, at $x = 0$.

According to the above four boundary conditions, we can obtain

$$k_+ a + \varphi_+ = \left(n + \frac{1}{2}\right)\pi, \qquad -k_- a + \varphi_- = \left(n' + \frac{1}{2}\right)\pi,$$
$$A_+ \cos\varphi_+ = A_- \cos\varphi_-, \qquad A_+ k_+ \sin\varphi_+ = A_- k_- \sin\varphi_-. \tag{9.29}$$

If the potential $V = 0$ inside the potential well, the wave function is $\psi = A\cos(kx + \varphi)$. Here, $ka + \varphi = (n + \frac{1}{2})\pi$ at $x = +a$, whereas $-ka + \varphi = (n' + \frac{1}{2})\pi$ at $x = -a$. From this equation, we obtain $k = \frac{n-n'}{2a}\pi$ and $\varphi = \frac{n+n'+1}{2}\pi$. Now the wave function is given by $\psi = A\cos(\frac{n-n'}{2a}\pi x + \frac{n+n'+1}{2}\pi) = A\cos(\frac{n-n'}{2a}\pi x + \frac{n-n'}{2}\pi + \frac{2n'+1}{2}\pi) = A\cos(\frac{n-n'}{2a}\pi x + \frac{n-n'}{2}\pi)\cos(\frac{2n'+1}{2}\pi) - A\sin(\frac{n-n'}{2a}\pi x + \frac{n-n'}{2}\pi)\sin(\frac{2n'+1}{2}\pi) = -A(-1)^{n'}\sin(\frac{n-n'}{2a}\pi x + \frac{n-n'}{2}\pi)$. On the other hand, as is well known to any students who learn quantum mechanics, the wave function of an infinitely deep potential well (with the potential $V = +\infty$ when the range is $x > 2a$ and $x < 0$) is $\widehat{\psi} = B\sin(\frac{l\pi}{2a}X)$. By using the replacement $X \to x+a$, where $x \in [-a, +a]$ and $X \in [0, +2a]$, we have $\widehat{\psi} = B\sin[\frac{l\pi}{2a}(x+a)] = B\sin(\frac{l}{2a}\pi x+\frac{l\pi}{2})$. It can be found that the aforementioned wavefunction $\psi = -A(-1)^{n'}\sin(\frac{n-n'}{2a}\pi x + \frac{n-n'}{2}\pi)$ has exactly the same form as the present wavefunction $\widehat{\psi} = B\sin(\frac{l}{2a}\pi x + \frac{l\pi}{2})$, where $l = n - n'$ and $B = -A(-1)^{n'}$.

The other typical system in quantum mechanics is the parity-time symmetric harmonic oscillator. The one-dimensional parity-time symmetric harmonic oscillator has the Hamiltonian operator $\hat{H} = \frac{\hat{p}^2}{2m} + \frac{1}{2}m\omega^2 \mathrm{e}^{+\mathrm{i}\theta}x^2$ when $x > 0$ and $\hat{H} = \frac{\hat{p}^2}{2m} + \frac{1}{2}m\omega^2\mathrm{e}^{-\mathrm{i}\theta}x^2$ when $x < 0$, or $\hat{H} = \frac{\hat{p}^2}{2m} + \frac{1}{2}m\omega^2(x + \mathrm{i}\kappa)^2$ (where the $\frac{1}{2}m\omega^2(x + \mathrm{i}\kappa)^2$ has an even-function real part and an odd-function imaginary part). Though the solution can in principle be obtained, the solving process will be extremely complicated and we will not proceed to solve the parity-time symmetric harmonic oscillator equation.

10. Comparison between Acoustic and Electromagnetic Field Equations

In an elastic (acoustic) material, a sonic wave equation, i.e., the Navier-Poisson-Lamé equation, is written as $\rho\frac{\partial^2 \mathbf{d}}{\partial t^2} = (\lambda + \mu)\nabla\nabla\cdot\mathbf{d} + \mu\nabla^2\mathbf{d} + \mathbf{f}$ [132], where $\mathbf{d}$ denotes the elastic displacement vector and λ, μ are the elastic solid parameters (i.e., the Lamé parameters). These two elastic solid parameters can be expressed as $\lambda = \frac{\sigma Y}{(1+\sigma)(1-2\sigma)}$ and $\mu = \frac{Y}{2(1+\sigma)}$. Here, Y denotes the Young modulus and σ is the Poisson ratio of the elastic (acoustic) material [133].

The Navier-Poisson-Lamé equation [132, 133] can be rearranged as the equation

$$\rho\frac{\partial^2 \mathbf{d}}{\partial t^2} + \mu\nabla\times(\nabla\times\mathbf{d}) - (\lambda + 2\mu)\nabla\nabla\cdot\mathbf{d} = \mathbf{f}. \tag{10.1}$$

We will show that this elastic field equation [132, 133] has exactly the same structure as the Maxwell electromagnetic field equation. The Ampère circuital equation is $\frac{1}{\mu_0}\nabla\times\mathbf{B} = \epsilon_0\frac{\partial \mathbf{E}}{\partial t} + \mathbf{J}$, where $\mathbf{E} = -\frac{\partial \mathbf{A}}{\partial t} - \nabla V$ and

$\mathbf{B} = \nabla \times \mathbf{A}$. This equation can be rearranged as

$$\epsilon_0 \frac{\partial^2 \mathbf{A}}{\partial t^2} + \frac{1}{\mu_0} \nabla \times (\nabla \times \mathbf{A}) + \epsilon_0 \nabla \frac{\partial}{\partial t} V = \mathbf{J}. \tag{10.2}$$

It can be seen that there is some similarity between the Navier-Poisson-Lamé equation (10.1) and the Ampère-Maxwell equation (10.2).

In what follows, we shall discuss some aspects of the similarity between these two equations, including the results of my efforts in this area over the past few years.

10.1. Lorenz Gauge Condition and Maxwell Vector Field

In the Navier-Poisson-Lamé equation [132, 133], we define $\nabla \cdot \mathbf{d} = -\frac{1}{c_\mathrm{s}} \frac{\partial d^0}{\partial t}$ with the sonic speed $c_\mathrm{s} = \sqrt{\frac{\mu}{\rho}}$, and this Navier-Poisson-Lamé equation can be rewritten as

$$\rho \frac{\partial^2 \mathbf{d}}{\partial t^2} + \mu \nabla \times (\nabla \times \mathbf{d}) + \left(\frac{\lambda + 2\mu}{c_\mathrm{s}^2}\right) \nabla \frac{\partial (d^0 c_\mathrm{s})}{\partial t} = \mathbf{f}. \tag{10.3}$$

As is well known in electrodynamics, the Lorenz gauge condition is $\partial_\mu A^\mu = 0$, i.e., $\nabla \cdot \mathbf{A} + \frac{1}{c} \frac{\partial A^0}{\partial t} = 0$, where the zeroth component of the 4-D electromagnetic potential is $A^0 = \frac{V}{c}$ (V is the electric scalar potential). Thus, for the sonic wave, we can define $d^0 = \frac{V_\mathrm{s}}{c_\mathrm{s}}$, where d^0 is the zeroth component of a 4-D elastic displacement d^ν and V_s is a sonic (acoustic) scalar potential. Now the equation (10.3) becomes the form

$$\rho \frac{\partial^2 \mathbf{d}}{\partial t^2} + \mu \nabla \times (\nabla \times \mathbf{d}) + \left(\frac{\lambda + 2\mu}{\mu}\right) \rho \nabla \frac{\partial}{\partial t} V_\mathrm{s} = \mathbf{f}. \tag{10.4}$$

It can be found that when the sonic material parameters fulfill $\lambda + \mu = 0$, i.e., $\frac{\lambda + 2\mu}{\mu} = 1$, the Navier-Poisson-Lamé equation (10.4) has exactly the same structure as the Ampère-Maxwell equation (10.2). Here, the acoustic-electromagnetic correspondence relations are given by

$$\begin{array}{llll} \mathbf{d} \leftrightarrow \mathbf{A}, & d^0 \leftrightarrow A^0, & V_\mathrm{s} \leftrightarrow V, & V_\mathrm{s} = d^0 c_\mathrm{s} \Leftrightarrow V = A^0 c, \\ \rho \leftrightarrow \epsilon_0, & \dfrac{1}{\mu_0} \leftrightarrow \mu, & \mathbf{f} \leftrightarrow \mathbf{J}, & c_\mathrm{s} = \sqrt{\dfrac{\mu}{\rho}} \Leftrightarrow c = \sqrt{\dfrac{1}{\mu_0 \epsilon_0}}. \end{array} \tag{10.5}$$

Now we shall consider the acoustic Gauss law by comparing it with the electric Gauss law:

$$\epsilon_0 \nabla \cdot \mathbf{E} = \rho_{\mathrm{e}} \quad \Rightarrow \quad \epsilon_0 \nabla \cdot \left(-\frac{\partial \mathbf{A}}{\partial t} - \nabla V \right) = \rho_{\mathrm{e}};$$

$$\rho \nabla \cdot \mathbf{E}_{\mathrm{s}} = \rho_{\mathrm{s}} \quad \Rightarrow \quad \rho \nabla \cdot \left(-\frac{\partial \mathbf{d}}{\partial t} - \nabla V_{\mathrm{s}} \right) = \rho_{\mathrm{s}}, \tag{10.6}$$

where the equation in the first line is the electric Gauss law and the equation in the second line is the acoustic Gauss law. By using the introduced condition $\nabla \cdot \mathbf{d} = -\frac{1}{c_{\mathrm{s}}} \frac{\partial d^0}{\partial t} = -\frac{1}{c_{\mathrm{s}}^2} \frac{\partial V_{\mathrm{s}}}{\partial t}$, the acoustic Gauss law given in Eq. (10.6) can be rearranged as a sonic (acoustic) d'Alembert wave equation

$$\frac{1}{c_{\mathrm{s}}^2} \frac{\partial^2}{\partial t^2} V_{\mathrm{s}} - \nabla^2 V_{\mathrm{s}} = \frac{\rho_{\mathrm{s}}}{\rho}, \tag{10.7}$$

where the sonic (acoustic) source density ρ_{s} fulfills $\frac{\partial \rho_{\mathrm{s}}}{\partial t} + \nabla \cdot \mathbf{f} = 0$, which is an acoustic analog of the electric charge conservation law $\frac{\partial \rho_{\mathrm{e}}}{\partial t} + \nabla \cdot \mathbf{J} = 0$. The 3-D divergence of the equation (10.4) can also yield the sonic (acoustic) d'Alembert wave equation (10.7). Since the sonic scalar potential V_{s} is determined by the longitudinal acoustic displacement vector $\mathbf{d}_{\mathrm{l}}$ (note that $\nabla \cdot \mathbf{d}_{\mathrm{t}} = 0$ with $\mathbf{d}_{\mathrm{t}}$ the transverse acoustic displacement vector), it seems that the sonic source density ρ_{s} is a source of the longitudinal acoustic wave only, i.e., for the transverse acoustic wave, there is no such a source ρ_{s}. It is the force density $\mathbf{f}$ that is the source of the transverse acoustic wave. The introduced condition $\nabla \cdot \mathbf{d} = -\frac{1}{c_{\mathrm{s}}} \frac{\partial d^0}{\partial t} = -\frac{1}{c_{\mathrm{s}}^2} \frac{\partial V_{\mathrm{s}}}{\partial t}$ can be identified as the acoustic analog of the electromagnetic Lorenz gauge condition ($\partial_\mu A^\mu = 0$).

The sonic "electric field" is $\mathbf{E}_{\mathrm{s}} = -\frac{\partial \mathbf{d}}{\partial t} - \nabla V_{\mathrm{s}}$ and the sonic "magnetic induction field" is $\mathbf{B}_{\mathrm{s}} = \nabla \times \mathbf{d}$. Then the sonic Faraday law of "electromagnetic induction" is $\nabla \times \mathbf{E}_{\mathrm{s}} = -\frac{\partial \mathbf{B}_{\mathrm{s}}}{\partial t}$ and the sonic "magnetic Gauss law" $\nabla \cdot \mathbf{B}_{\mathrm{s}} = 0$. It should be pointed out that the sonic Faraday law of "electromagnetic induction" and the sonic "magnetic Gauss law" are the mathematical identities (i.e., the sonic "Bianchi identities").

Since the acoustic or elastic wave equation has the same mathematical structure as the Maxwell electromagnetic wave equation, in acoustic metamaterials with negative dynamical mass density [134, 135, 136], an acoustic analog of surface plasmon polaritons can also be confined to an interface

[137, 138]. As is well known in surface plasmonics or metal nanophotonics, on a single, smooth interface between a dielectric with positive relative permittivity $\epsilon_1^{(\mathrm{r})}$ and a metal with negative relative permittivity $\epsilon_2^{(\mathrm{r})}$, the phase constant (the wave vector parallel to the dielectric-metal interface) of surface plasmon polaritons is $\beta = \omega\sqrt{\mu_0\epsilon_0}\sqrt{\frac{\epsilon_1^{(\mathrm{r})}\epsilon_2^{(\mathrm{r})}}{\epsilon_1^{(\mathrm{r})}+\epsilon_2^{(\mathrm{r})}}} = \omega\sqrt{\mu_0}\sqrt{\frac{\epsilon_1\epsilon_2}{\epsilon_1+\epsilon_2}}$, where $\epsilon_i = \epsilon_i^{(\mathrm{r})}\epsilon_0$ with $i = 1, 2$ [94]. According to the electromagnetic-acoustic correspondence relation $\epsilon \leftrightarrow \rho$ and $\frac{1}{\mu_0} \leftrightarrow \kappa$ (κ has the same status as $\lambda + 2\mu$ or μ when $\lambda + \mu = 0$), such an acoustic surface wave can also be sustained on an interface under proper conditions. The phase constant (in the propagation direction parallel to the acoustic interface) of the present confined acoustic surface wave is given by $\beta = \frac{\omega}{\sqrt{\kappa}}\sqrt{\frac{\rho_1\rho_2}{\rho_1+\rho_2}}$. If the acoustic interface is of parity-time symmetry, i.e., $\rho_2 = \rho_1^*$, the factor $\sqrt{\frac{\rho_1\rho_2}{\rho_1+\rho_2}}$ is still a real number, and such a mode can be referred to as "parity-time symmetric acoustic surface wave" [29, 30].

Now we shall discuss the sonic electric field $\mathbf{E}_\mathrm{s} = -\frac{\partial \mathbf{d}}{\partial t} - \nabla V_\mathrm{s}$ given in the sonic Gauss law given in Eq. (10.6). Let the sonic material parameters fulfill $\lambda + \mu = 0$, i.e., $\frac{\lambda+2\mu}{\mu} = 1$. We ignore the transverse wave term $\mu\nabla \times (\nabla \times \mathbf{d})$, and the Navier-Poisson-Lamé equation (10.4) reduces to the form

$$\frac{\partial^2 \mathbf{d}_{(\mathrm{l})}}{\partial t^2} + \nabla\frac{\partial}{\partial t}V_\mathrm{s} = \frac{\mathbf{f}}{\rho}. \tag{10.8}$$

Since the force density is $\mathbf{f} = \frac{\partial \mathbf{p}_{\mathrm{moment}}}{\partial t} = \rho\frac{\partial \mathbf{v}}{\partial t}$, where $\mathbf{p}_{\mathrm{moment}}$ is the momentum density of the elastic medium mass element, the equation (10.8) can be rewritten as

$$\frac{\partial^2 \mathbf{d}_{(\mathrm{l})}}{\partial t^2} + \nabla\frac{\partial}{\partial t}V_\mathrm{s} = \frac{\partial \mathbf{v}}{\partial t} \quad \Rightarrow \quad \frac{\partial \mathbf{d}_{(\mathrm{l})}}{\partial t} + \nabla V_\mathrm{s} = \mathbf{v}, \tag{10.9}$$

i.e., $\mathbf{u}_{(\mathrm{l})} + \nabla V_\mathrm{s} = \mathbf{v}$. We have obtained the Gauss law of elastic field: $\rho\nabla\cdot\left(-\frac{\partial \mathbf{d}}{\partial t} - \nabla V_\mathrm{s}\right) = \rho_\mathrm{s}$ given in Eq. (10.6), where the sonic electric field $\mathbf{E}_\mathrm{s} = -\frac{\partial \mathbf{d}}{\partial t} - \nabla V_\mathrm{s}$ can be given by

$$-\frac{\partial \mathbf{d}}{\partial t} - \nabla V_\mathrm{s} = -\frac{\partial \mathbf{d}_{(\mathrm{t})}}{\partial t} - \frac{\partial \mathbf{d}_{(\mathrm{l})}}{\partial t} - \nabla V_\mathrm{s} = -\frac{\partial \mathbf{d}_{(\mathrm{t})}}{\partial t} - \mathbf{v}. \tag{10.10}$$

Since $\nabla\cdot\frac{\partial \mathbf{d}_{(\mathrm{t})}}{\partial t}=0$, the Gauss law of elastic field $\rho\nabla\cdot\left(-\frac{\partial \mathbf{d}}{\partial t}-\nabla V_{\mathrm{s}}\right)=\rho_{\mathrm{s}}$ becomes $-\rho\nabla\cdot\mathbf{v}=\rho_{\mathrm{s}}$. The time derivative of the two sides yields the relation $-\rho\nabla\cdot\frac{\partial \mathbf{v}}{\partial t}=\frac{\partial \rho_{\mathrm{s}}}{\partial t}$, which can be rewritten as $-\nabla\cdot\mathbf{f}=\frac{\partial \rho_{\mathrm{s}}}{\partial t}$, because $\mathbf{p}_{\mathrm{moment}}=\rho\mathbf{v}$ is a momentum density, and its rate of change is the force density $\mathbf{f}$.

The relation $\frac{\partial \rho_{\mathrm{s}}}{\partial t}+\nabla\cdot\mathbf{f}=0$ can be identified as an acoustic analog of the electric charge conservation law $\frac{\partial \rho_{\mathrm{e}}}{\partial t}+\nabla\cdot\mathbf{J}=0$.

This "conservation" law $\frac{\partial \rho_{\mathrm{s}}}{\partial t}+\nabla\cdot\mathbf{f}=0$ can be verified (by the Navier-Poisson-Lamé equation) as follows: With the help of the Navier-Poisson-Lamé equation $\rho\frac{\partial^2 \mathbf{d}}{\partial t^2}+\mu\nabla\times(\nabla\times\mathbf{d})+\left(\frac{\lambda+2\mu}{\mu}\right)\rho\nabla\frac{\partial}{\partial t}V_{\mathrm{s}}=\mathbf{f}$ given in Eq. (10.4) and the Gauss law of elastic field $\rho\nabla\cdot\left(-\frac{\partial \mathbf{d}}{\partial t}-\nabla V_{\mathrm{s}}\right)=\rho_{\mathrm{s}}$ given in Eq. (10.6), where $\lambda+\mu=0$ will be adopted, we can obtain

$$\frac{\partial \rho_{\mathrm{s}}}{\partial t}+\nabla\cdot\mathbf{f}=\rho\nabla\cdot\left(-\frac{\partial^2 \mathbf{d}}{\partial t^2}-\nabla\frac{\partial V_{\mathrm{s}}}{\partial t}\right)+\rho\nabla\cdot\frac{\partial^2 \mathbf{d}}{\partial t^2}+\rho\nabla^2\frac{\partial}{\partial t}V_{\mathrm{s}}=0.\quad(10.11)$$

In Eq. (10.10), the sonic electric field can be given by

$$\mathbf{E}_{\mathrm{s}}=-\frac{\partial \mathbf{d}}{\partial t}-\nabla V_{\mathrm{s}}=-\frac{\partial \mathbf{d}_{(\mathrm{t})}}{\partial t}-\mathbf{v}.\quad(10.12)$$

If the force density $\mathbf{f}=0$, the "sonic charge density" $\rho_{\mathrm{s}}=0$ and the fluid mass element velocity could be $\mathbf{v}=0$, and the sonic electric field is $\mathbf{E}_{\mathrm{s}}=-\frac{\partial \mathbf{d}_{(\mathrm{t})}}{\partial t}$, i.e., there is no $-\frac{\partial \mathbf{d}_{(\mathrm{l})}}{\partial t}$ in the sonic electric field.

10.2. Non-Lorenz Gauge Condition and a Generalized Vector Field

In the above, we have used the Lorenz condition $\partial_\mu A^\mu=0$. But we can also discard this condition. In this case, in order that we can find certain correspondence relations between the elastic wave and a generalized 4-D vector field A^ν, we shall introduce a Lagrangian density: $\mathcal{L}_A=-\frac{1}{4}\alpha F_{\mu\nu}F^{\mu\nu}-\frac{1}{2}\beta(\partial_\mu A^\mu)^2-J^\nu A_\nu$. This Lagrangian density has been used in quantum electrodynamics. The Euler-Lagrange equation for this generalized vector field A^ν is given by

$$\alpha\partial_\mu F^{\mu\nu}+\beta\partial^\nu\partial_\mu A^\mu=J^\nu\quad(10.13)$$

with the (electromagnetic field tensor) $F^{\mu\nu} = \partial^\mu A^\nu - \partial^\nu A^\mu$. Here, Einstein's rule of "*summation over repeated coordinate indices*" has been implied in the field equation. We choose the spacetime metric convention as $\eta = \mathrm{diag}[+1, -1, -1, -1]$ ("diag" represents the diagonal matrix elements). Let the parameter $\alpha = 1$. The equation (10.13) can be rearranged as

$$\epsilon_0 \frac{\partial}{\partial t}(-\mathbf{E}) + \frac{1}{\mu_0}\nabla \times (\nabla \times \mathbf{A}) - \beta \nabla \partial_0 A^0 - \beta \nabla \nabla \cdot \mathbf{A} = \mathbf{J}, \quad (10.14)$$

where the electric field $-\mathbf{E} = \frac{\partial \mathbf{A}}{\partial t} + \nabla(A^0 c)$ ($A^0 c = V$ with V the electric scalar potential) has been used. Then the equation (10.14) can be expanded as the form

$$\epsilon_0 \frac{\partial^2 \mathbf{A}}{\partial t^2} + \frac{1}{\mu_0}\nabla \times (\nabla \times \mathbf{A}) + \left(\epsilon_0 - \frac{\beta}{c^2}\right) \nabla \frac{\partial}{\partial t}(A^0 c) - \beta \nabla \nabla \cdot \mathbf{A} = \mathbf{J}. \quad (10.15)$$

If $\epsilon_0 - \frac{\beta}{c^2} = 0$, i.e., $\beta = \frac{1}{\mu_0}$, the equation (10.15) reduces to the form

$$\epsilon_0 \frac{\partial^2 \mathbf{A}}{\partial t^2} + \frac{1}{\mu_0}\nabla \times (\nabla \times \mathbf{A}) - \frac{1}{\mu_0}\nabla \nabla \cdot \mathbf{A} = \mathbf{J}. \quad (10.16)$$

Note that in this case, the Lorenz condition $\partial_\mu A^\mu = 0$ no longer holds. But compared with the Navier-Poisson-Lamé equation $\rho \frac{\partial^2 \mathbf{d}}{\partial t^2} + \mu \nabla \times (\nabla \times \mathbf{d}) - (\lambda + 2\mu) \nabla \nabla \cdot \mathbf{d} = \mathbf{f}$ given in Eq. (10.1), when $\lambda + \mu = 0$, the equation (10.16) and the special case of the Navier-Poisson-Lamé equation have the same structure.

In the equation (10.13), when the coordinate index $\nu = 0$, the zeroth-component equation $\alpha \partial_\mu F^{\mu 0} + \beta \partial^0 \partial_\mu A^\mu = J^0$ can be written as the form

$$\begin{aligned} &\epsilon_0 \nabla \cdot \left(-\frac{\partial \mathbf{A}}{\partial t} - \nabla V\right) + \frac{1}{\mu_0}\frac{1}{c^2}\frac{\partial}{\partial t}(\partial_\mu A^\mu) = \rho_{\mathrm{e}} \\ \Rightarrow \quad &-\epsilon_0 \left[\frac{\partial}{\partial t}\left(\nabla \cdot \mathbf{A} + \frac{1}{c}\frac{\partial A^0}{\partial t}\right) - \frac{1}{c}\frac{\partial^2 A^0}{\partial t^2}\right] \\ &+ \frac{1}{\mu_0}\frac{1}{c^2}\frac{\partial}{\partial t}(\partial_\mu A^\mu) - \epsilon_0 \nabla^2 V = \rho_{\mathrm{e}}. \end{aligned} \quad (10.17)$$

Since $\frac{1}{\mu_0}\frac{1}{c^2} = \epsilon_0$, $\partial_\mu A^\mu = \nabla \cdot \mathbf{A} + \frac{1}{c}\frac{\partial A^0}{\partial t}$ and $A^0 = \frac{V}{c}$, we have the d'Alembert wave equation

$$\frac{1}{c^2}\frac{\partial^2 V}{\partial t^2} - \nabla^2 V = \frac{\rho_{\mathrm{e}}}{\epsilon_0}. \quad (10.18)$$

In the above, we have not used the Lorenz gauge condition $\partial_\mu A^\mu = 0$, but we have adopted a particular relation $\epsilon_0 - \frac{\beta}{c^2} = 0$. In the following analysis, we shall consider a case: $\partial_\mu A^\mu = G$, i.e., $\partial_0 A^0 = G - \nabla \cdot \mathbf{A}$. Now the equation (10.15) can be rewritten as

$$\epsilon_0 \frac{\partial^2 \mathbf{A}}{\partial t^2} + \frac{1}{\mu_0} \nabla \times (\nabla \times \mathbf{A}) + \left(\epsilon_0 - \frac{\beta}{c^2}\right) c^2 \nabla \frac{\partial A^0}{c \partial t} - \beta \nabla \nabla \cdot \mathbf{A} = \mathbf{J}. \quad (10.19)$$

By substituting the relation $\partial_0 A^0 = G - \nabla \cdot \mathbf{A}$ into Eq. (10.19), we have

$$\begin{aligned} &\epsilon_0 \frac{\partial^2 \mathbf{A}}{\partial t^2} + \frac{1}{\mu_0} \nabla \times (\nabla \times \mathbf{A}) \\ &+ \left(\epsilon_0 c^2 - \beta\right) \nabla (G - \nabla \cdot \mathbf{A}) - \beta \nabla \nabla \cdot \mathbf{A} = \mathbf{J}. \end{aligned} \quad (10.20)$$

It can be rearranged as

$$\epsilon_0 \frac{\partial^2 \mathbf{A}}{\partial t^2} + \frac{1}{\mu_0} \nabla \times (\nabla \times \mathbf{A}) - \frac{1}{\mu_0} \nabla \nabla \cdot \mathbf{A} + \left(\frac{1}{\mu_0} - \beta\right) \nabla G = \mathbf{J}. \quad (10.21)$$

By comparing it with the Navier-Poisson-Lamé equation $\rho \frac{\partial^2 \mathbf{d}}{\partial t^2} + \mu \nabla \times (\nabla \times \mathbf{d}) - (\lambda + 2\mu) \nabla \nabla \cdot \mathbf{d} = \mathbf{f}$ given in Eq. (10.1), when $\lambda + \mu = 0$, we can find that the equation (10.21) and the special case of the Navier-Poisson-Lamé equation have the same structure. An additional condition is $\mathbf{f} = \mathbf{f}_1 + \xi \nabla G$.

From the equation (10.21), it can be seen that no matter whether $\beta = \frac{1}{\mu_0}$ or $G = 0$, this term $\left(\frac{1}{\mu_0} - \beta\right) \nabla G = 0$, then this reduces to the former case.

Since 2016 to 2021, I was studying the similarities between elasticity (or solid state acoustics) and electromagnetic field theory: specifically, I tried to find the similarities between the Navier-Poisson-Lamé equation and the Maxwell electromagnetic field equation, but because I was trying to eliminate an extra constraint $\lambda + \mu = 0$, that is, in order to give up $\lambda + \mu = 0$, it took me a lot of time to make the Navier-Poisson-Lamé equation mathematically identical to the Maxwell electromagnetic field equation.

10.3. Force Density in a Fluid

In the above discussions, we have imposed a constraint on the solid material parameters $\lambda + \mu = 0$, which is a special case that is rarely seen. Now we will consider the case of $\lambda+\mu\neq 0$. Here, we assume the force density is given by $\mathbf{f}=\bar{\mathbf{f}}-\nabla p_{\rm th}-\nabla p_{\rm s}$, where $p_{\rm th}$ and $p_{\rm s}$ denote the thermal pressure and the sonic (acoustic) pressure, respectively. The sonic (acoustic) pressure is given by $p_{\rm s}=-\kappa\nabla\cdot\mathbf{d}$ with κ the bulk modulus of fluid (gas or liquid). Thus, the Navier-Poisson-Lamé equation in a liquid-crystal like material is

$$\rho\frac{\partial^2\mathbf{d}}{\partial t^2}+\mu\nabla\times(\nabla\times\mathbf{d})-(\lambda+2\mu+\kappa)\nabla\nabla\cdot\mathbf{d}+\nabla p_{\rm th}=\bar{\mathbf{f}}. \quad (10.22)$$

Now under the condition of the following correspondence relations

$$\lambda+2\mu+\kappa\leftrightarrow\frac{1}{\mu_0},\quad p_{\rm th}\leftrightarrow\left(\frac{1}{\mu_0}-\beta\right)G,\quad \bar{\mathbf{f}}\leftrightarrow\mathbf{J}, \quad (10.23)$$

the equation (10.21) and the equation (10.22) have the same mathematical structure. Since $G=\partial_\mu A^\mu$ in Eq. (10.21), the thermal pressure in Eq. (10.22) is $p_{\rm th}\propto\partial_\mu d^\mu$.

10.4. The Source Term of the Acoustic Transverse Wave

. We consider a case of $\bar{\mathbf{f}}=0$ in Eq. (10.22). By taking the curl and divergence of the equation (10.22), one can obtain

$$\rho\frac{\partial^2(\nabla\times\mathbf{d})}{\partial t^2}+\mu\nabla\times\nabla\times(\nabla\times\mathbf{d})=0,$$
$$\rho\frac{\partial^2\nabla\cdot\mathbf{d}}{\partial t^2}-(\lambda+2\mu+\kappa)\nabla^2\nabla\cdot\mathbf{d}+\nabla^2 p_{\rm th}=0. \quad (10.24)$$

Clearly, $\nabla\times\mathbf{d}$ and $\nabla\cdot\mathbf{d}$ are the transverse and the longitudinal acoustic fields, respectively. It seems that $p_{\rm th}$ is not the source of the transverse acoustic field, but rather of the longitudinal acoustic field. However, this is just an illusion. As a matter of fact, the vanishing-divergence component in $\nabla p_{\rm th}$ would also make contribution to the transverse acoustic field: specifically, $\nabla p_{\rm th}=\nabla p_{\rm th}^{(\rm z)}+\nabla p_{\rm th}^{(\rm nz)}$ with $\nabla^2 p_{\rm th}^{(\rm z)}=0$ (zero divergence of $\nabla p_{\rm th}^{(\rm z)}$) and $\nabla^2 p_{\rm th}^{(\rm nz)}\neq 0$ (nonzero divergence of $\nabla p_{\rm th}^{(\rm nz)}$). For example, a simple form for $p_{\rm th}^{(\rm z)}$ can be given by $p_{\rm th}^{(\rm z)}=ax+by+cz$, of which the gradient is $\nabla p_{\rm th}^{(\rm z)}=(a,b,c)$. Then it fulfills the zero-divergence condition $\nabla^2 p_{\rm th}^{(\rm z)}=0$.

Now the equation (10.22) can be decomposed into two relations:

$$\rho\frac{\partial^2\mathbf{d}}{\partial t^2}+\mu\nabla\times(\nabla\times\mathbf{d})+\nabla p_{\rm th}^{(\rm z)}=0,$$
$$\rho\frac{\partial^2\mathbf{d}}{\partial t^2}-(\lambda+2\mu+\kappa)\nabla\nabla\cdot\mathbf{d}+\nabla p_{\rm th}^{(\rm nz)}=0, \qquad (10.25)$$

where the first equation is the transverse acoustic field equation and the second equation is the longitudinal acoustic field equation. In the first equation (transverse acoustic field equation), $\nabla\cdot\mathbf{d}=0$, i.e., $\mathbf{d}$ is the transverse elastic displacement vector field; In the second equation (longitudinal acoustic field equation), $\nabla\times\mathbf{d}=0$, i.e., $\mathbf{d}$ is the longitudinal elastic displacement vector field.

The curl of the transverse acoustic field equation in Eq. (10.25) is $\rho\frac{\partial^2(\nabla\times\mathbf{d})}{\partial t^2}+\mu\nabla\times\nabla\times(\nabla\times\mathbf{d})=0$. In a case of static field (i.e., $\frac{\partial^2\mathbf{d}}{\partial t^2}=0$), we have $\mu\nabla\times\nabla\times(\nabla\times\mathbf{d})=0$, where there seems to be no contribution of $p_{\rm th}^{(\rm z)}$ in $\nabla\times\nabla\times(\nabla\times\mathbf{d})$. But in $\nabla\times(\nabla\times\mathbf{d})$, $\nabla\times\mathbf{d}$ and the acoustic displacement vector $\mathbf{d}$, there is still the contribution of $p_{\rm th}^{(\rm z)}$. In the case of time dependence, $p_{\rm th}^{(\rm z)}$ can still make contribution to $\nabla\times(\nabla\times\mathbf{d})$, $\nabla\times\mathbf{d}$ and the acoustic displacement vector $\mathbf{d}$; There may be the contribution of $p_{\rm th}^{(\rm z)}$ in both $\rho\frac{\partial^2(\nabla\times\mathbf{d})}{\partial t^2}$ and $\mu\nabla\times\nabla\times(\nabla\times\mathbf{d})$, but the contribution in them cancel out exactly, and hence no effective or net contribution appears in the equation $\rho\frac{\partial^2(\nabla\times\mathbf{d})}{\partial t^2}+\mu\nabla\times\nabla\times(\nabla\times\mathbf{d})=0$.

10.5. The Lagrangian Density of Elastic Displacement Vector Field

According to the theory of elasticity [133], we can write the Lagrangian density of free elastic field as

$$\mathcal{L}_{\rm elas}=\frac{1}{2}\rho\left(\frac{\rm d\mathbf{d}}{\rm dt}\right)^2+\frac{\mu}{4}f_{ij}f^{ij}-\mu\partial_i d_j\partial^i d^j-\frac{\lambda}{2}(\partial_i d^i)^2, \qquad (10.26)$$

where d^j with $j=1,2,3$ is the elastic displacement vector field and $\mathbf{d}=(d^1,d^2,d^3)$. The elastic field tensors are given by $f_{ij}=\partial_i d_j-\partial_j d_i$ and $f^{ij}=\partial^i d^j-\partial^j d^i$, where the spatial coordinate indices i and j run over from 1 to 3. The spatial metric is $\eta_{ij}={\rm diag}[-1,-1,-1]$ and thus $d_i=\eta_{ij}d^j=-d^i$ and $\partial_i=\eta_{ij}\partial^j=-\partial^i$. The kinetic term $\frac{1}{2}\rho(\frac{\rm d\mathbf{d}}{\rm dt})^2$ can be rewritten as $\frac{1}{2}\rho(\frac{\mathrm{d}\mathbf{d}}{\mathrm{d}t})^2 = -\frac{1}{2}\rho\frac{\mathrm{d}d_i}{\mathrm{d}t}\frac{\mathrm{d}d^i}{\mathrm{d}t}$ with the indices $i, j = 1, 2, 3$. By using the Euler-Lagrange equation

$$\partial_i \frac{\partial \mathcal{L}_{\mathrm{elas}}}{\partial \partial_i d_j} + \partial_t \frac{\partial \mathcal{L}_{\mathrm{elas}}}{\partial \partial_t d_j} = 0, \qquad (10.27)$$

where

$$\begin{aligned} \partial_i \frac{\partial \mathcal{L}_{\mathrm{elas}}}{\partial \partial_i d_j} &= \mu\partial_i f^{ij} - 2\mu\partial_i\partial^i d^j - \lambda\partial^j(\partial_i d^i) \\ &= \mu\partial_i f^{ij} - 2\mu\partial_i(\partial^i d^j - \partial^j d^i + \partial^j d^i) - \lambda\partial^j(\partial_i d^i) \\ &= -\mu\partial_i f^{ij} - (2\mu + \lambda)\partial^j(\partial_i d^i) \end{aligned} \qquad (10.28)$$

and

$$\partial_t \frac{\partial \mathcal{L}_{\mathrm{elas}}}{\partial \partial_t d_j} = -\rho\frac{\mathrm{d}^2 d^j}{\mathrm{d}t^2}, \qquad (10.29)$$

we can obtain the free elastic field equation

$$\rho\frac{\mathrm{d}^2 d^j}{\mathrm{d}t^2} + \mu\partial_i f^{ij} + (2\mu + \lambda)\partial^j(\partial_i d^i) = 0. \qquad (10.30)$$

We should point out that $\mu\partial_i f^{ij}$ is the transverse elastic wave term and $(2\mu + \lambda)\partial^j(\partial_i d^i)$ is the longitudinal elastic wave term. The first (kinetic) term on the left-handed side in Eq. (10.30) is $\rho\frac{\mathrm{d}^2 d^j}{\mathrm{d}t^2} = \rho\frac{\mathrm{d}^2 d^j_{(\mathrm{t})}}{\mathrm{d}t^2} + \rho\frac{\mathrm{d}^2 d^j_{(\mathrm{l})}}{\mathrm{d}t^2}$. Here, the transverse elastic field $d^j_{(\mathrm{t})}$ obeys $\partial_j d^j_{(\mathrm{t})} = 0$ (i.e., $\nabla \cdot \mathbf{d}_{(\mathrm{t})} = 0$) and the longitudinal elastic field $d_{(\mathrm{l})j}$ obeys $\partial_i d_{(\mathrm{l})j} - \partial_j d_{(\mathrm{l})i} = 0$ (i.e., $\nabla \times \mathbf{d}_{(\mathrm{l})} = 0$). When we calculate the 3-D divergence of Eq. (10.30), we can obtain

$$\rho\frac{\mathrm{d}^2 \partial_j d^j}{\mathrm{d}t^2} + (2\mu + \lambda)\partial_j\partial^j(\partial_i d^i) = 0, \qquad (10.31)$$

where $\mu\partial_j\partial_i f^{ij} \equiv 0$ has been inserted. This is the longitudinal elastic wave equation. From this equation (10.31), it can be seen that the sonic speed of the longitudinal elastic wave is $c^{\mathrm{s}}_{(\mathrm{l})} = \sqrt{\frac{2\mu+\lambda}{\rho}}$.

From the free elastic field equation (10.30), we can obtain

$$\rho\frac{\mathrm{d}^2(\partial^i d^j - \partial^j d^i)}{\mathrm{d}t^2} + \mu\partial_k(\partial^i f^{kj} - \partial^j f^{ki}) = 0. \qquad (10.32)$$

Here, $\partial^i f^{kj} - \partial^j f^{ki} = \partial^i(\partial^k d^j - \partial^j d^k) - \partial^j(\partial^k d^i - \partial^i d^k) = \partial^k(\partial^i d^j - \partial^j d^i)$. Then the equation (10.32) can be rewritten as

$$\rho \frac{\mathrm{d}^2 f^{ij}}{\mathrm{d}t^2} + \mu \partial_k \partial^k f^{ij} = 0. \tag{10.33}$$

This is the transverse elastic wave equation. From this equation, it can be seen that the sonic speed of the transverse elastic wave is $c^{\mathrm{s}}_{(\mathrm{t})} = \sqrt{\frac{\mu}{\rho}}$.

According to the elastic mechanics [132], the source-free Navier-Poisson-Lamé equation can be given by

$$\rho \frac{\partial^2 \mathbf{d}}{\partial t^2} + \mu \nabla \times (\nabla \times \mathbf{d}) - (\lambda + 2\mu) \nabla \nabla \cdot \mathbf{d} = 0, \tag{10.34}$$

where the force density $\mathbf{f}$ on the right-handed side is absent. It can be found that the free elastic field equation (10.30) is consistent with the source-free Navier-Poisson-Lamé equation (10.34).

It should be noted that in the Lagrangian density $\mathcal{L}_{\text{elas}} = -\frac{1}{2}\rho \frac{\mathrm{d}d_i}{\mathrm{d}t}\frac{\mathrm{d}d^i}{\mathrm{d}t} + \frac{\mu}{4} f_{ij} f^{ij} - \mu \partial_i d_j \partial^i d^j - \frac{\lambda}{2}(\partial_i d^i)^2$ given in Eq. (10.26), the term $-\mu \partial_i d_j \partial^i d^j$ contains the contribution of both transverse and longitudinal elastic fields. We can decompose this term into two parts: transverse and longitudinal components. This Lagrangian density can be rearranged as a compact form

$$\begin{aligned} \mathcal{L}_{\text{elas}} &= -\frac{1}{2}\rho \frac{\mathrm{d}d_i}{\mathrm{d}t}\frac{\mathrm{d}d^i}{\mathrm{d}t} + \frac{\mu}{4} f_{ij} f^{ij} - \mu(\partial_i d_j - \partial_j d_i + \partial_j d_i)\partial^i d^j - \frac{\lambda}{2}(\partial_i d^i)^2 \\ &= -\frac{1}{2}\rho \frac{\mathrm{d}d_i}{\mathrm{d}t}\frac{\mathrm{d}d^i}{\mathrm{d}t} - \frac{\mu}{4} f_{ij} f^{ij} - \frac{2\mu + \lambda}{2}(\partial_i d^i)^2 + \text{D.T.}, \end{aligned} \tag{10.35}$$

where $-\mu(\partial_i d_j - \partial_j d_i + \partial_j d_i)\partial^i d^j = -\mu(\partial_i d_j - \partial_j d_i)\partial^i d^j - \mu \partial_j d_i \partial^i d^j = -\frac{\mu}{2} f_{ij} f^{ij} - \mu \partial_j d_i \partial^i d^j$ and $\partial_j d_i \partial^i d^j = \partial^i[(\partial_j d_i) d^j] - (\partial_j \partial^i d_i) d^j = \text{D.T.} - \partial_j[(\partial^i d_i) d^j] + \partial^i d_i \partial^j d_j$ have been used. In this Lagrangian density (10.35), $-\frac{\mu}{4} f_{ij} f^{ij}$ and $-\frac{2\mu+\lambda}{2}(\partial_i d^i)^2$ are the transverse and the longitudinal wave terms, respectively.

10.6. The 3-D Elastic Lagrangian Density Written in a 4-D Form

According to the elastic field equation [132] or the theory of elasticity and its vector-form Lagrangian density [133], the 3-D elastic Lagrangian density in a solid can be given by

$$\begin{aligned} \mathcal{L}_{\text{elas}} &= \frac{1}{2}\rho\left(\frac{\partial \mathbf{d}}{\partial t}\right)^2 - \frac{1}{2}\mu\left(\nabla\times\mathbf{d}\right)^2 - \frac{1}{2}(\lambda+2\mu)\left(\nabla\cdot\mathbf{d}\right)^2 \\ &= -\frac{1}{2}\rho\frac{\partial d_j}{\partial t}\frac{\partial d^j}{\partial t} - \frac{1}{4}\mu(\partial_i d_j - \partial_j d_i)(\partial^i d^j - \partial^j d^i) - \frac{1}{2}(\lambda+2\mu)(\partial_j d^j)^2, \end{aligned} \tag{10.36}$$

where the spatial metric is chosen to be $\eta_{ij} = \mathrm{diag}[-1,-1,-1]$ with $i,j = 1,2,3$ and the elastic displacement vector is $\mathbf{d} = \{d^1, d^2, d^3\} = \{-d_1, -d_2, -d_3\}$. In $-\frac{1}{2}\rho\frac{\partial d_j}{\partial t}\frac{\partial d^j}{\partial t}$, the elastic displacement is a sum of transverse and longitudinal components: $d^j = d^j_{(\mathrm{t})} + d^j_{(\mathrm{l})}$. Although in $-\frac{1}{4}\mu(\partial_i d_j - \partial_j d_i)(\partial^i d^j - \partial^j d^i)$ and $-\frac{1}{2}(\lambda+2\mu)(\partial_j d^j)^2$, the sum $d^j = d^j_{(\mathrm{t})} + d^j_{(\mathrm{l})}$ is still valid, yet in $-\frac{1}{4}\mu(\partial_i d_j - \partial_j d_i)(\partial^i d^j - \partial^j d^i)$ there is only the contribution of the transverse field $d^j_{(\mathrm{t})}$, and in $-\frac{1}{2}(\lambda+2\mu)(\partial_j d^j)^2$, there is only the contribution of the longitudinal field $d^j_{(\mathrm{l})}$.

We simplify the above elastic field Lagrangian density, i.e., by using $\lambda+\mu=0$, the equation (10.36) can be written as

$$\mathcal{L}_{\text{elas}} = -\frac{1}{2}\rho\frac{\partial d_j}{\partial t}\frac{\partial d^j}{\partial t} - \frac{1}{4}\mu(\partial_i d_j - \partial_j d_i)(\partial^i d^j - \partial^j d^i) - \frac{1}{2}\mu(\partial_j d^j)^2. \tag{10.37}$$

Let us introduce an elastic scalar potential d^0 through $\partial_0 d^0 = -\partial_j d^j$, i.e., $\partial_\mu d^\mu = 0$ (Lorenz type gauge condition), where the Latin index j runs over 1 to 3, and the Greek index μ runs over 0 and 1 to 3. The time derivative operator is $\partial_0 = \frac{1}{c_{\mathrm{s}}}\frac{\partial}{\partial t}$ with $c_{\mathrm{s}} = \sqrt{\frac{\mu}{\rho}}$. Now the equation (10.37) is given by

$$\mathcal{L}_{\text{elas}} = \mu\left[-\frac{1}{2}\partial_0 d_j\partial^0 d^j - \frac{1}{4}(\partial_i d_j - \partial_j d_i)(\partial^i d^j - \partial^j d^i) - \frac{1}{2}(\partial_j d^j)^2\right]. \tag{10.38}$$

Now we shall discuss the sum of $-\frac{1}{2}\partial_0 d_j\partial^0 d^j$ and $-\frac{1}{2}(\partial_j d^j)^2$:

$$\begin{aligned} &-\frac{1}{2}\partial_0 d_j\partial^0 d^j - \frac{1}{2}(\partial_j d^j)^2 = -\frac{1}{2}\left(\partial_0 d_j\partial^0 d^j + \partial_0 d^0\partial_0 d^0\right) \\ &= -\frac{1}{2}\partial_0 d_\mu\partial^0 d^\mu + \left\{\left[\partial_\mu(d_0\partial^0 d^\mu) - d_0\partial^0\partial_\mu d^\mu\right] + \mathrm{D.T.}\right\}, \end{aligned} \tag{10.39}$$

where $\left\{\left[\partial_\mu(d_0\partial^0 d^\mu) - d_0\partial^0\partial_\mu d^\mu\right] + \mathrm{D.T.}\right\}$, which is identically zero, is an added term, because $\mathrm{D.T.} = -\partial_\mu(d_0\partial^0 d^\mu)$ is a 4-D total divergence term and $-d_0\partial^0\partial_\mu d^\mu \equiv 0$. Since $[\partial_\mu(d_0\partial^0 d^\mu) - d_0\partial^0\partial_\mu d^\mu] = \partial_\mu d_0\partial^0 d^\mu$, the equation (10.39) can be rearranged as

$$\begin{aligned} -\frac{1}{2}\partial_0 d_j\partial^0 d^j - \frac{1}{2}(\partial_j d^j)^2 &= -\frac{1}{2}\partial_0 d_\mu\partial^0 d^\mu + \partial_\mu d_0\partial^0 d^\mu + \text{D.T.} \\ = \quad & -\frac{1}{2}\partial_0 d_\mu\partial^0 d^\mu + \partial_\mu d_0\partial^0 d^\mu - \frac{1}{2}\partial_\mu d_0\partial^\mu d^0 + \frac{1}{2}\partial_\mu d_0\partial^\mu d^0 + \text{D.T.} \\ = \quad & -\frac{1}{2}(\partial_0 d_\mu - \partial_\mu d_0)(\partial^0 d^\mu - \partial^\mu d^0) + \frac{1}{2}\partial_\mu d_0\partial^\mu d^0 + \text{D.T.}, \end{aligned} \tag{10.40}$$

which can also be rewritten as

$$-\frac{1}{2}\partial_0 d_j\partial^0 d^j - \frac{1}{2}(\partial_j d^j)^2 = -\frac{1}{4}f_{0\mu}f^{0\mu} - \frac{1}{4}f_{\mu 0}f^{\mu 0} + \frac{1}{2}\partial_\mu d_0\partial^\mu d^0 + \text{D.T.}. \tag{10.41}$$

Then the elastic Lagrangian density (10.38) can be rewritten as

$$\begin{aligned} \mathcal{L}_{\text{elas}} &= \mu\Big(-\frac{1}{4}f_{0\mu}f^{0\mu} - \frac{1}{4}f_{\mu 0}f^{\mu 0} - \frac{1}{4}f_{ij}f^{ij} + \frac{1}{2}\partial_\mu d_0\partial^\mu d^0 + \text{D.T.}\Big) \\ &= -\frac{\mu}{4}f_{\mu\nu}f^{\mu\nu} + \frac{1}{2}\partial_\mu d_0\partial^\mu d^0 + \text{D.T.}, \end{aligned} \tag{10.42}$$

where $-\frac{\mu}{4}f_{\mu\nu}f^{\mu\nu}$, which has exactly the same structure as the electromagnetic Lagrangian density, is the Lagrangian density of the elastic transverse field, and $\frac{1}{2}\partial_\mu d_0\partial^\mu d^0$ is the Lagrangian density of the elastic longitudinal field. Though the contribution of acoustic longitudinal field can be included as d^0 into $-\frac{\mu}{4}f_{\mu\nu}f^{\mu\nu}$, yet there is still its own Lagrangian density $\frac{1}{2}\partial_\mu d_0\partial^\mu d^0$.

10.7. The Navier-Poisson-Lamé Equation of Spatiotemporal Elastic Displacement in a Spacetime Crystal

In the literature, a concept of "*spacetime crystal*", sometimes also referred to as "*Planck crystal*" or "*world crystal*" [139, 140, 141, 142], has been suggested because of various physical reasons such as noncommutative geometry, loop quantum gravity, defect (dislocation and disclination) gauge field theory as well as lattice gauge field theory. We can generalize the 3-D Navier-Poisson-Lamé equation [132, 133] in the ordinary elastic materials to a *4-D spatiotemporal crystal* in the long-wavelength limit. The equation of the spatiotemporal elastic displacement vector field λ^ν is given by

$$\rho\frac{\mathrm{d}^2\lambda^\nu}{\mathrm{d}\tau^2} + b_1\partial^\nu\left(\partial_\mu\lambda^\mu\right) + b\left(\partial_\mu\lambda^{\mu\nu} + m_\lambda^2\lambda^\nu\right) - \sigma\partial^\nu p = f^\nu, \tag{10.43}$$

where the elastic field tensor is $\lambda^{\mu\nu} = \partial^{\mu}\lambda^{\nu} - \partial^{\nu}\lambda^{\mu}$ and the pressure gradient is $\partial^{\nu}p = s\partial^{\nu}T + n\partial^{\nu}\mu$. Here, p is the thermal pressure, s is the entropy density, T is the temperature, n is the thermodynamical particle number density and μ is the chemical potential. $\mathrm{d}\tau$ is the proper time element (in a 4-D spacetime). The thermodynamical pressure can be decomposed into two parts: $p = p_{\mathrm{l}} + p_{\mathrm{t}}$, where p_{l} and p_{t} satisfy $\partial_{\nu}\partial^{\nu}p_{\mathrm{l}} \neq 0$ and $\partial_{\nu}\partial^{\nu}p_{\mathrm{t}} = 0$, i.e., p_{l} and p_{t} are the elastic displacement sources of the longitudinal and the transverse spatiotemporal elastic displacements, respectively. If we focus only on the transverse spatiotemporal elastic displacement vector field, which agrees with $\partial_{\nu}\lambda_{\mathrm{t}}^{\nu} = 0$, then we have

$$\rho\frac{\mathrm{d}^2\lambda_{\mathrm{t}}^{\nu}}{\mathrm{d}\tau^2} + b\left(\partial_{\mu}\lambda_{\mathrm{t}}^{\mu\nu} + m_{\lambda}^2\lambda_{\mathrm{t}}^{\nu}\right) - \sigma\partial^{\nu}p_{\mathrm{t}} = f_{\mathrm{t}}^{\nu}. \tag{10.44}$$

For simplicity, we can ignore the external force density source f_{t}^{ν}, and we consider the case of "static" spatiotemporal elastic displacement, i.e., $\frac{\mathrm{d}^2\lambda_{\mathrm{t}}^{\nu}}{\mathrm{d}\tau^2} = 0$. Thus, the equation (10.44) reduces to the form

$$b\left(\partial_{\mu}\lambda_{\mathrm{t}}^{\mu\nu} + m_{\lambda}^2\lambda_{\mathrm{t}}^{\nu}\right) = \sigma\left(s_{\mathrm{t}}\partial^{\nu}T_{\mathrm{t}} + n_{\mathrm{t}}\partial^{\nu}\mu_{\mathrm{t}}\right). \tag{10.45}$$

In a gravitational gauge field theory suggested by us [52, 53], we can show that the spin current density tensor of the spatiotemporal elastic displacement serves as one of the gravitating sources, which might play a role of an equivalent dark matter [50]. But this topic is beyond the scope of this chapter and we will not continue to discuss this problem.

11. Conclusions on the Surface Plasmonics in Electromagnetics and Quantum Optics

The present chapter focused on five topics in classical and quantum surface plasmonics. These topics include: i) Free-electron Drude model in metal optics; ii) Tunable quantum-coherence assisted negative permittivity and surface plasmon excitation in EIT-prism couplers; iii) Quantized surface plasmonic modes at metal surfaces; iv) Parity-time symmetric electromagnetic surface wave modes; v) Comparison of the equations between the Maxwell electromagnetic field and the elastic field.

As is well known, the electromagnetic properties of optical materials can be described by the Drude-Lorentz model. In some cases, the local field effect associated with the neighboring electric dipole mutual polarization has been taken into account in the Drude-Lorentz model. However, for the metals, the electric permittivity obtained by the Drude model without the local field effect is in good agreement with the optical observations [37, 38]. This is a mystery in the Drude model. In this chapter, this problem is solved, i.e., if the plasma oscillation effect is taken into consideration, the effect of local field is precisely counterbalanced by the effect of plasma oscillation, and as a result, the effect of local field cannot be exhibited in the Drude model. It is for this reason that the Drude model without the local field correction agrees very well with the experimental observation results of optical materials.

In surface plasmonics, the control and manipulation of optical characteristics of electromagnetic materials is essentially important. In this chapter, we give an overview of our previous work on multilevel quantum-coherent atomic media, where the atomic density matrix element evolution, tunable electric permittivity or magnetic permeability, and high-dispersion properties have been indicated [15, 16, 21, 72, 73]. These multilevel atomic media could be utilized in surface plasmon resonance, optical parity-time symmetry design, and high-dispersion field quantization. Of course, these schemes also have their own disadvantages, such as large optical loss and substance instability.

The classical optical/electromagnetic properties in localized surface plasmons (on metallic nanospheres) [55] and surface plasmon polaritons (on metal-dielectric interfaces) have been extensively studied in various research fields ranging from purely physical foundations to application-oriented regimes [1, 2, 3] (and references therein). In this chapter, we have given an overview of our previous work on surface plasmon resonance in prism couplers, where the metal thin film is replaced by three- and four-level quantum coherent media [16] (e.g., atomic vapors and quantum-dot materials [15]). The advantages of three- and four-level quantum coherent media include field tunability and frequency sensitivity [21]. We also considered the issue of field quantization of localized surface plasmons and surface plasmon polaritons. The aspects of quantum surface plasmonics (including quantum-coherent surface plasmonics and quantum-field surface plasmonics) considered in this chapter would provide new working principles or mechanisms for designing some photonic and quantum optical devices.

Localized surface plasmons and its plasmonic resonance on metallic nanospheres is an important concept (or physical mechanism) in electromagnetics, optics we well as chemical and biological detections. It is not only of practical significance, but also of theoretical significance for the study of electromagnetic or chromodynamic surface plasmons on nuclei and nucleons. It may be of enlightening value in description of nuclear and nucleon energy level structures. For example, proton has its internal subparticle structure (quark-gluon plasma) and the plasma-assisted surface plasmon resonance can possible take place on proton when it is scattered by high-energy gamma photons, and if electron is an elementary particle without internal structure, no such surface plasmon resonance would occur. Therefore, it may be expected that we can use "localized surface plasmons" as a key to explore the internal structure of "elementary" particles.

Over the past decade, parity-time symmetry has attracted interest in quantum mechanics and optics [23, 24, 25, 26, 27, 28]. We have suggested some parity-time symmetric surface wave modes ("parity-time symmetric electromagnetics") in optical/electromagnetic material interface structures in this chapter (we had mentioned such parity-time symmetric surface electromagnetic waves in our previous references on "parity-time symmetric acoustics" [29, 30]). Since the acoustic (elastic) displacement field equation (Navier-Poisson-Lamé equation [132, 133]) has the same structure as the Maxwell field equation, the theory of parity-time symmetric acoustics can also be established, where the parity-time symmetric acoustic surface wave modes can be suggested [29, 30]. We expect that this would enable to broaden readers' scope of mind and to suggest more intriguing optical effects in electromagnetic and acoustic metamaterials.

We compare the field equations between the Maxwell electromagnetic field and the elastic field. It can be found that the elastic longitudinal wave plays the role of an "electric" scalar potential d^0 in the Maxwell equations and the elastic displacement vector d^i with $i = 1, 2, 3$ can be identified with an equivalent "magnetic" vector potential $\mathbf{A}$ in the Maxwell electromagnetic field equations. Thus, the elastic field equations have the same structure as that of the Maxwell equations in 4-D spacetime. Therefore, elastic waves possess all the propagation properties of free electromagnetic waves. New effects in manipulating electromagnetic waves with artificial electromagnetic media can in principle also be achieved using acoustic metamaterials. This could open the minds of researchers to come up with more effects and phenomena in acoustic metamaterials for future experimental implementations.

Acknowledgments

This work is supported in part by the Natural Science Foundations of Zhejiang Province (China) under Project No. LY16A040002, the National Natural Science Foundations of China under Grants No. 11174250, 91233119. Shen is also grateful to the Foundations of "Program of Zhejiang Leading Team of Science and Technology Innovation".

References

[1] Maier S. A., *Plasmonics: Fundamentals and Applications* (Springer Science + Business Media LLC, New York, 2007), Chaps. 2–4 and Chaps. 7–11.

[2] Cao Z. Q., *Optics of Guided Waves* (Science Press of China, Beijing, 2007), Chap. 10.

[3] Chong S. Y. and Shen J. Q., Electromagnetic spin current density of surface plasmon polaritons, *Prog. Electromag. Res.* C **77**, 57–68 (2017).

[4] Imamoğlu A., Schmidt H., Woods G. and Deutsch M., Strongly interacting photons in a nonlinear cavity, *Phys. Rev. Lett.* **79**, 1467–1470 (1997).

[5] Wang H., Goorskey D., and Xiao M., Enhanced Kerr nonlinearity via atomic coherence in a three-level atomic system, *Phys. Rev. Lett.* **87**, 073601 (2001).

[6] Marangos J. P., Electromagnetically induced transparency, *J. Mod. Opt.* **45**(3), 471–503 (1998).

[7] Fleischhauer M. and Scully M. O., Quantum sensitivity limits of an optical magnetometer based on atomic phase coherence, *Phys. Rev.* A **49**, 1973–1986 (1994).

[8] Anonymous author, A material on review of electromagnetically induced transparency offered by EIT group of St. Andrews University, U. K. (St. Andrews University EIT group website: http://star-www.st-and.ac.uk/physics/research/eit).

[9] Meystre P. and Sargent III M., *Elements of Quantum Optics*, 2nd Edition (Cambridge University Press, Cambridge, 1995).

[10] Sun C. P., Wang Y. D., Li Y. and Zhang P., Fundamental concepts and methods in microcavity quantum electrodynamics, Chapter 4, pp. 139–187, in: *Recent Progress in Quantum Mechanics*, 3rd Volume, Editors: J. Zeng, G. Long and S. Pei (Tsinghua University, Beijing, China, 2003).

[11] Sun C. P., Gao Y. B., Dong H. F., Zhao S. R., Partial factorization of wave functions for a quantum dissipative system, *Phys. Rev.* E **57**(4), 3900–3904 (1998).

[12] Klimov V. V., Spontaneous emission of an excited atom placed near left-handed media, *Opt. Commun.* **211**, 183–196 (2002).

[13] Casimir H. B. G., On the attraction between two perfectly conducting plates, *Proc. K. Ned. Akad. Wet.* **51**, 793–795 (1948).

[14] Lamoreaux S. K., Demonstration of the Casimir force in the 0.6 to 6 μm range, *Phys. Rev. Lett.* **78**, 5–8 (1997).

[15] Shen J. Q., Total reflection enhancement in an incoherent-gain prism coupler towards optical switches and photonic transistors, *J. Phys. Soc. Jpn* **83**, 044401 (2014).

[16] Shen J. Q., Electromagnetically-induced-transparency plasmonics: Quantum-interference-assisted tunable surface-plasmon-polariton resonance and excitation, *Phys. Rev.* A **90**, 023814 (2014).

[17] Du Y., Zhang J. and Gao J., Atomic coherence effects in four-level system, *Acta Sin. Quantum Opt.* (in Chinese), **10**(1), 14–19 (2004).

[18] Chen J., Liu Z.-D., Zheng J. and Fang H.-J., Effect of vacuum-induced coherence in a four-level atomic system via quantum interference, *Acta Phys. Sin.* (in Chinese), **56**(11), 6441–6445 (2007).

[19] Yao J., Wu H. and Wang H., The transient optical properties in four-level atomic medium induced by quantum interference effect, *Acta Sin. Quantum Opt.* (in Chinese) **9**(3), 121–125 (2003).

[20] Greentree A. D., Smith T. B., de Echaniz S. R., Durrant A. V., Marangos J. P., Segal D. M. and Vaccaro J. A., Resonant and off-resonant transients in electromagnetically induced transparency: Turn-on and turn-off dynamics, *Phys. Rev.* A **65**, 053802 (2002).

[21] Shen J. Q., A three-level dark state and double-control single-photon logic gates via quantum coherent control, *Ann. Phys.* (Berlin) **524**(2), 85–96 (2012).

[22] Almeida V. R., Barrios C. A., Panepucci R. R. and Lipson M., All-optical control of light on a silicon chip, *Nature* **431**(7012), 1081–1084 (2004).

[23] Bender C. M. and Boettcher S., Real spectra in non-Hermitian Hamiltonians having PT symmetry, *Phys. Rev. Lett.* **80**, 5243–5246 (1998).

[24] Bender C. M., Brody D. C. and Jones H. F., Complex extension of quantum mechanics, *Phys. Rev. Lett.* **89**, 270401 (2002).

[25] Zhang H., Huang R., Zhang S.-D., Li Y., Qiu C.-W., Nori F. and Jing H., Breaking anti-PT symmetry by spinning a resonator, *Nano Lett.* **20**, 7594–7599 (2020) and references therein.

[26] Mostafazadeh A., Exact $\mathcal{PT}$-symmetry is equivalent to Hermiticity, *J. Phys. A: Math. Gen.* **36**, 7081–7091 (2003).

[27] Rüter C. E., Makris K. G., El-Ganainy R., Christodoulides D. N., Segev M. and Kip D., Observation of parity-time symmetry in optics, *Nature Phys.* **6**, 192–195 (2010).

[28] Feng L., Xu Y.-L., Fegadolli W. S., Lu M.-H., Oliveira J. E. B., Almeida V. R., Chen Y.-F. and Scherer A., Experimental demonstration of a unidirectional reflectionless parity-time metamaterial at optical frequencies, *Nature Mater.* **12**, 108–113 (2013).

[29] Shen J. Q., Strong and weak confinement of parity-time-symmetric acoustic surface wave, *Europhys. Lett.* **105**, 17006 (2014).

[30] Shen J. Q., Canonical acoustics and its application to surface acoustic wave on acoustic metamaterials, *J. Phys. Soc. Jpn.* **85**, 084401 (2016).

[31] Guo S. H., *Electrodynamics* (Higher Education Press of China, Beijing, China, 1979), Chap. 6.

[32] Liang K. M., Yu C. and Ma G. Q., *Theoretical Mechanics (Mechanics Vol. II)*, 3rd Ed. (Higher Education Press of China, Beijing, China, 1995), Chap. 12.

[33] Zhou Z. P., *Silicon-Based Photoelectronics* (Peking University Press, Beijing, 2012), Chap. 8.

[34] Wang R. C., *Introductory General Relativity* (Press of University of Science and Technology of China, Anhui Hefei, China, 1996).

[35] Jackson J. D., *Classical Electrodynamics*, 3rd Edition (John Wiley & Sons, New York, 2001), Chap. 7.

[36] Cheng D. K., *Field and Wave Electromagnetics*, 2nd Edition (Addison-Wesley, Inc., New York, 1989), Chap. 8.

[37] Maier S. A., *Plasmonics: Fundamentals and Applications* (Springer Science+Business Media, New York, 2007), Chap. 1.

[38] Johnson P. B. and Christy R. W., Optical constants of the noble metals, *Phys. Rev.* B **6**, 4370–4379 (1972).

[39] Li D., Chen Y. H., Ma J. X. and Yang W. H., *Plasma Physics* (Higher Education Press of China, Beijing, China, 2006), Chaps. 1 and 4.

[40] Boyd T. J. M. and Sanderson J. J., *The Physics of Plasmas* (Cambridge University Press, Cambridge, U. K., 2003), Chap. 1.

[41] Jackson J. D., *Classical Electrodynamics*, 3rd Edition (John Wiley & Sons, New York, 2001), Chap. 4, pp. 159–162.

[42] Coelho R. and Aladenize B., *Les diélectriques propriétés diélectriques des matériaux isolants* [*Dielectrics dielectric properties of insulating materials.*] (Editions Hermès, Paris, 1993), Chap. 5.

[43] Cook D. M., *The Theory of the Electromagnetic Field* (Prentice-Hall, Inc., New Jersey, 1975), Chap. 11.

[44] Fox M., *Optical Properties of Solids* (Oxford University Press, U. K., 2001), Chap. 7.

[45] Fox M., *Optical Properties of Solids* (Oxford University Press, U. K., 2001), Chap. 2.

[46] Yan S. S., *Fundamentals of Solid State Physics*, 3rd Edition (Peking University Press, Beijing, China, 2011), Chap. 1.

[47] Li Z. Z., *Solid State Theory*, 2nd Edition (Higher Education Press of China, Beijing, 2002), Chap. 4.

[48] Shen J. Q., Isotropic negatively-refracting atomic-vapor medium, *Mod. Phys. Lett.* B **21** (12), 717–728 (2007).

[49] Shen J. Q., Negatively refracting atomic vapor, *J. Mod. Opt.* **53**(15), 2195–2205 (2006).

[50] Shen J. Q., The dark-baryonic matter mass relation for observational verification in Verlinde's emergent gravity, *Gen. Relativ. Gravit.* **50**, Paper Number 73 (28 pages) (2018).

[51] Huang K. and Han R. Q., *Solid State Physics* (Higher Education Press of China, Beijing, China, 1988).

[52] Shen J. Q., A gravitational constant and a cosmological constant in a spin-connection gravitational gauge field theory, *J. Phys. A: Math. Theor.* **42**, 155401 (2009).

[53] Shen J. Q., A gravitational gauge field theory based on Stephenson-Kilmister-Yang gravitation with scalar and spinor fields as gravitating matter sources, *Gen. Relat. Grav.* **48**, Paper Number 62 (48 pages) (2016).

[54] Jackson J. D., *Classical Electrodynamics*, 3rd Edition (John Wiley & Sons, New York, 2001), Chap. 4, pp. 157–159.

[55] Maier S. A., *Plasmonics: Fundamentals and Applications* (Springer Science+Business Media, New York, 2007), Chap. 5.

[56] Harris S. E., Electromagnetically induced transparency, *Phys. Today* **50**(7), 36–42 (1997).

[57] Zhu S. Y. and Scully M. O., Spectral line elimination and spontaneous emission cancellation via quantum interference, *Phys. Rev. Lett.* **76**(3), 388–391 (1996).

[58] Zhu S. Y., Scully M. O., Fearn H. and Narducci L. M., Lasing without inversion, *Z. Phys.* D: At. Mol. Clus. **22**, 483–493 (1992).

[59] Li Y. -Q. and Xiao M., Transient properties of an electromagnetically induced transparency in three-level atoms, *Opt. Lett.* **20**(13), 1489–1491 (1995).

[60] Schmidt H. and Imamoğlu A., Giant Kerr nonlinearities obtained by electromagnetically induced transparency, *Opt. Lett.* **21**(23), 1936–1938 (1996).

[61] Veselago V. G., The electrodynamics of substances with simultaneously negative values of ϵ and μ, *Sov. Phys. Usp.* **10**, 509–514 (1968).

[62] Smith D. R., Padilla W. J., Vier D. C., Nemat-Nasser S. C. and Schultz S., Composite medium with simultaneously negative permeability and permittivity, *Phys. Rev. ett.* **84**, 4184–4187 (2000).

[63] Shelby R. A., Smith D. R. and Schultz S., Experimental verification of a negative index of refraction, *Science* **292**, 77–79 (2001).

[64] Pendry J. B., Negative refraction makes a perfect lens, *Phys. Rev. Lett.* **85**, 3966–3969 (2001).

[65] Zhang Z. M. and Fu C. J., Unusual photo tunneling in the presence of a layer with a negative index, *Appl. Phys. Lett.* **80**, 1097–1099 (2002).

[66] Pendry J. B., Holden A. J., Robbins D. J., and Stewart W. J., Low frequency plasmons in thin wire structures, *J. Phys. Condens. Matter* **10**, 4785–4809 (1998).

[67] J. B. Pendry, Holden A. J., Stewart W. J., and Youngs I., Extremely low frequency plasmons in metallic mesostructures, *Phys. Rev. Lett.* **76**, 4773–4776 (1996).

[68] Chen L., He S., and Shen L., Finite-size effects of a left-handed material slab on the image quality, *Phys. Rev. Lett.* **92**, 107404 (2004).

[69] Simovski C. R. and He S. L., Frequency range and explicit expressions for negative permittivity and permeability for an isotropic medium formed by a lattice of perfectly conducting Ω particles, *Phys. Lett.* A **311**, 254–263 (2003).

[70] Oktel M. Ö. and Müstecaphoğlu Ö. E., Electromagnetically induced left-handedness in a dense gas of three-level atoms, *Phys. Rev.* A **70**, 053806 (2004).

[71] Shen J. Q., Ruan Z. C. and He S., How to realize a negative refractive index material at the atomic level in an optical frequency range? *J. Zhejiang Univ. Science (China)* **5**, 1322–1326 (2004).

[72] Shen J. Q., Photonic-resonant left-handed medium, *Phys. Lett.* A **357**, 54–60 (2006).

[73] Krowne C. M. and Shen J. Q., Dressed-state mixed-parity transitions for realizing negative refractive index, *Phys. Rev.* A **79**, 023818 (2009).

[74] Scully M. O. and Zubairy M. S., *Quantum Optics* (Cambridge Univ. Press, Cambridge, England, 1997), Chap. 7.

[75] Yan M., Rickey E. G. and Zhu Y., Observation of absorptive photon switching by quantum interference, *Phys. Rev.* A **64**, 041801(4 pages) (2001).

[76] Wang H., Goorskey D. and Xiao M., Enhanced Kerr nonlinearity via atomic coherence in a three-level atomic system, *Phys. Rev. Lett.* **87**, 073601 (2001).

[77] Imamoğlu A., Schmidt H., Woods G. and Deutsch M., Strongly interacting photons in a nonlinear cavity, *Phys. Rev. Lett.* **79**, 1467–1470 (1997).

[78] Shen J. Q., Quantum-coherent plasmonics: Atomic-phase-coherence assisted surface plasmon resonance and negative optical parameters in multilevel media, Chapt. 3, pp. 121–238, in: *Plasmonics: Advances in Research and Applications* (Series of "Physics Research and Technology"), Editor: A. Bright (Nova Science Publishers, Inc., New York, 2017).

[79] John S. and Florescu M., Photonic band gap materials: toward an all-optical micro-transistor, *J. Opt.* A: Pure Appl. Opt. **3**, S103–S120 (2001).

[80] Forsberg E., Hessmo B. and Thylén Lars, Limits to modulation rates of electroabsorption modulators, *IEEE J. Quantum Electron.* **40**(4), 400–405 (2004).

[81] Meystre P. and Sargent III M., *Elements of Quantum Optics*, 2nd Edition (Springer-Verlag, Berlin Heidelberg, 1990), Chap. 5, pp. 129–177.

[82] Otto A., Exaction of nonradiative surface plasma waves in silver by the method of frustrated total reflection, *Z. Phys.* **216**, 398–410 (1968).

[83] Kretschmann E., Determination of the optical constants of metals by excitation of surface plasmons, *Z. Physik* **248**, 313–324 (1971).

[84] Ding Y., CaoZ. Q. and Shen Q. S., Improved SPR technique for determination of the thickness and optical constants of thin metal films, *Opt. Quantum Electr.* **35**, 1091–1097 (2003).

[85] Ishimaru A., Jaruwatanadilok S. and Kuga Y., Generalized surface plasmon resonance sensors using metamaterials and negative index materials, *Prog. Electrom. Res.* **51**, 139–152 (2005).

[86] Weber W. H. and McCarthy S. L., Surface-plasmon resonance as a sensitive optical probe of metal-film properties, *Phys. Rev.* B **12**, 5643–5650 (1975).

[87] Regalado L. E., Machorro R. and Siqueiros J. M., Attenuated-total-reflection technique for the determination of optical constants, *Appl. Opt.* **30**, 3176–3180 (1991).

[88] Homola J., Yee S. S. and Gauglitz G., Surface plasmon resonance sensors, *Sensors Actuat* B: Chemical **54**, 3–15 (1999).

[89] Jiang Y., Cao Z., Chen G., Dou X. and Chen Y., Low voltage electro-optic polymer light modulator using attenuated total internal reflection, *Opt. Laser Technol.* **33**(6), 417–420 (2001).

[90] Schuck P., Use of surface plasmon resonance to probe the equilibrium and dynamic aspects of interactions between biological macromolecules, *Annu. Rev. Biophys. Biomol. Struct.* **26**, 541–566 (1997) and references therein.

[91] Barnes W. L., Dereux A. and Ebbesen T. W., Surface plasmon subwavelength optics, *Nature* **424**, 824–830 (2003).

[92] Ozbay E., Merging photonics and electronics at nanoscale dimensions, *Science* **311**, 189–193 (2006).

[93] Yoon J., Song S. H. and Kim J.-H., Extraction efficiency of highly confined surface plasmon-polaritons to far-field radiation: an upper limit, *Opt. Express* **16**(2), 1269–1279 (2008).

[94] Cao Z. Q., *Optics of Guided Waves* (Science Press of China, Beijing, 2007), Chap. 9.

[95] Petrosyan L. S., Kirakosyan A. S. and Shahbazyan T.V., Extraordinary electron transmission through a periodic array of quantum dots, *Phys. Rev. Lett.* **107**, 196802 (2011).

[96] Ziolkowski R. W., Superluminal transmission of information through an electromagnetic metamaterial, *Phys. Rev.* E **63**, 046604 (2001) and references therein.

[97] Shen J. Q., Quantum-vacuum effects and quantum coherence in artificial electromagnetic materials, pp. 157-179, in: *Advances in Condensed Matter and Materials Research*, Vol. 6, Editors: H. Geelvinck and S. Reynst (Nova Science Publishers, Inc., New York, 2010) and references therein.

[98] Zheltikov A. M., Phase coherence control and subcycle transient detection in nonlinear Raman scattering with ultrashort laser pulses, *Phys. Rev.* A **74**, 053403 (2006).

[99] Gandman A., Chuntonov L., Rybak L. and Amitay Z., Coherent phase control of resonance-mediated (2+1) three-photon absorption, *Phys. Rev.* A **75**, 031401(R) (2007).

[100] Cohen J. L. and Berman P. R., Amplification without inversion: understanding probability amplitudes, quantum interference, and Feynman rules in a strongly driven system, *Phys. Rev.* A **55**, 3900–3917 (1997).

[101] Champenois C., Morigi G., and Eschner J., Quantum coherence and population trapping in three-photon processes, *Phys. Rev.* A **74**, 053404 (2006).

[102] Shen J. Q. and Zhang P., Double-control quantum interferences in a four-level atomic system, *Opt. Express* **15**, 6484–6493 (2007).

[103] Shen J. Q., Transient evolutional behaviours of double-control electromagnetically induced transparency, *New J. Phys.* **9**, 374–388 (2007).

[104] Paspalakis E. and Knight P. L., Transparency, slow light and enhanced nonlinear optics in a four-level scheme, *J. Opt.* B: Quantum Semiclass. Opt. **4**, S372–S375 (2002).

[105] Paspalakis E. and Knight P. L., Transparency and parametric generation in a four-level system, *J. Mod. Opt.* **49**, 87–95 (2002).

[106] Yeh P., *Optical Waves in Layered Media* (John Wiley & Sons, Inc., New Jersey, 2005), Chaps. 4–6.

[107] Sun G., Khurgin J. B. and Soref R. A., Plasmonic light-emission enhancement with isolated metal nanoparticles and their coupled arrays, *J. Opt. Soc. Am.* B **25**, 1748–1755 (2008).

[108] Prade B., Vinet J. Y. and Mysyrowicz A., Guided optical waves in planar heterostructures with negative dielectric constant, *Phys. Rev.* B **44**, 13556–13572 (1991).

[109] Ni G. J. and Chen S. Q., *Advanced Quantum Mechanics* (in Chinese), 2nd Edition (Fudan University Press, Shanghai, China, 2004), Chap. 4.

[110] Zeng J. Y., *Quantum Mechanics Volume II* (in Chinese), 3rd Edition (Science Press of China, Beijing, China, 2000), Chap. 12.

[111] Feng X., Jin L. C. and Liu C. F., Solving the mystery of muon anomalous magnetic moment with supercomputer: lattice quantum chromodynamics (in Chinese), *Wechat Official Account of Modern Physics Magazine (Beijing), China, Sept. 2nd, 2021.*

[112] Koschmieder E. L., Theory of the anomalous magnetic moment of the electron, *arXiv: 1001.3371 (physics.gen-ph)*, https://arxiv.org/pdf/1001.3371.pdf (2013).

[113] Aoyama T., Asmussen N., Benayoun M., Bijnens J. and Zhevlakov A. S., The anomalous magnetic moment of the muon in the standard model, *Phys. Rept.* **887**, 1–166 (2020).

[114] Schwinger J. S., On quantum electrodynamics and the magnetic moment of the electron, *Phys. Rev.* **73**, 416–417 (1948).

[115] Levine M. J. and Wright J., Anomalous magnetic moment of the electron, *Phys. Rev.* D **8**, 3171–3179 (1973).

[116] Aoyama T., Hayakawa M., Kinoshita T. and Nio M., Quantum electrodynamics calculation of lepton anomalous magnetic moments: numerical approach to the perturbation theory of QED, *Prog. Theor. Exp. Phys.* **2012**, 01A107 (2012).

[117] Aoyama T., Kinoshita T. and Nio M., Theory of the anomalous magnetic moment of the electron, *Atoms*, **7**(1), Paper Number 28 (27 pages) (2019).

[118] Aoyama T., Hayakawa M., Kinoshita T. and Nio M., Automated calculation scheme for α^n contributions of QED to lepton $g-2$: Generating renormalized amplitudes for diagrams without lepton loops, *Nucl. Phys.* B **740**, 138–180 (2006).

[119] Aoyama T., Hayakawa M., Kinoshita T. and Nio M., Automated calculation scheme for α^n contributions of QED to lepton $g-2$: New treatment of infrared divergence for diagrams without lepton loops, *Nucl. Phys.* B **796**, 184–210 (2008).

[120] Shen J. Q., *Surface plasma waves (Surface plasmon polaritons)* (Classroom Lecture I of "Electromagnetic Wave Theory" Graduate Course at Zhejiang University, Hangzhou, China, 2012–2013).

[121] Zhou B. R., *Quantum Field Theory* (Higher Education Press of China, Beijing, China, 2007), Chaps. 1–3.

[122] Bjorken J. D. and Drell S. D., *Relativistic Quantum Fields* (Mcgraw-Hill Companies, Inc., New York, 1965), Chap. 14.

[123] Shen J. Q., Gravitational gauge theory of spinorial and vectorial gravitating matter fields, in: *Gauge Theories and Differential Geometry*, Editor: L. Bailey (Nova Science Publishers, Inc. New York, 2016), Chap. 2, pp. 45–95.

[124] Zayats A. V., Smolyaninov I. I. and Maradudin A. A., Nano-optics of surface plasmon polaritons, *Phys. Rept.* **408**(3), 131–314 (2005).

[125] Pitarke J. M., Silkin V. M., Chulkov E. V. and Echenique P. M., Theory of surface plasmons and surface-plasmon polaritons, *Rept. Prog. Phys.* **70**(1), 1–87 (2007).

[126] Guidry M., *Gauge Field Theories: An Introduction with Applications* (Wiley, New York, 1992), Appendix A.

[127] Lu H. X., Wang X. Q., Liu H. J. and Zhang Y. D., A generalized multiphoton Jaynes-Cummings model for two collectively radiating atoms solved via supersymmetric unitary transformation, *Mod. Phys. Lett.* B **15**, 479–485 (2001).

[128] Fan H. Y., Yuan H. C. and Wu H., *Invariant Eigenvalue Operator Formulation in Quantum Mechanics* (in Chinese) (Shanghai Jiaotong University Press, Shanghai, China, 2011), Chap. 8.

[129] Fan H. Y. and Hu L. Y., *Optical Transformation: From Quantum to Classical* (Series of New Progress in Quantum Physics, in Chinese) (Shanghai Jiaotong University Press, Shanghai, China, 2010), Chap. 12.

[130] Shen J.-Q. and Zhu H.-Y., Exact solutions to the time-dependent supersymmetric Jaynes-Cummings model and the Chiao-Wu model, *Ann. Phys.*(Leipzig) **12**(3), 131–145 (2003).

[131] Liang J. Q. and Wei L. F., *Progress in Quantum Physics* (Science Press of China, Beijing, China, 2011), Chap. 6.

[132] Cheng J. C., *Fundamentals of Theoretical Physics* (Science Press of China, Beijing, China, 2007), Chap. 3.

[133] Liang K. M., *Mechanics II* (Higher Education Press, Beijing, China, 1995), Chap. 10., pp. 335–357 and 365–370.

[134] Cheng Y., Xu J. Y. and Liu X. J., One-dimensional structured ultrasonic metamaterials with simultaneously negative dynamic density and modulus, *Phys. Rev.* B **77**, 045134 (2008).

[135] Zhu R., Huang G. L. and Hu G. K., Effective dynamic properties and multi-resonant design of acoustic metamaterials, *J. Vib. Acoust.* **134**, 031006 (2012).

[136] Fokin V., Ambati M., Sun C. and Zhang X., Method for retrieving effective properties of locally resonant acoustic metamaterials, *Phys. Rev.* B **76**, 144302 (2007).

[137] Christensen J., Fernandez-Dominguez A. I., De Leon-Perez F., Martin-Moreno L. and Garcia-Vidal F. J., Collimation of sound assisted by acoustic surface waves, *Nature Phys.* **3**, 851–852 (2007).

[138] Ambati M., Fang N., Sun C. and Zhang X., Surface resonant states and superlensing in acoustic metamaterials, *Phys. Rev.* B **75**, 195447 (2007).

[139] Wang S. J., *Vacuum Structure, Gravity Origin and Dark Energy Problem* (Science Press of China, Beijing, 2016), Chaps. 8–10.

[140] Kleinert H., Gravity as a theory of defects in a crystal with only second gradient elasticity, *Ann. Phys.*(Berlin) **499**(2), 117–119 (1987).

[141] Kleinert H. and Zaanen J., Nematic world crystal model of gravity explaining absence of torsion in spacetime, *Phys. Lett.* A **324**(5-6), 361–365 (2004).

[142] Danielewski M., The Planck-Kleinert crystal, *Z. Naturfoschung* A **62**(10-11), 564–568 (2007).

Bibliography

2D materials: properties and devices

LCCN 2016043383

Type of material Book

Main title 2D materials: properties and devices / edited by Phaedon Avouris (IBM T.J. Watson Research Center, New York), Tony F. Heinz (Stanford University and SLAC National Accelerator Laboratory), Tony Low (University of Minnesota).

Published/Produced Cambridge, United Kingdom; New York, NY: Cambridge University Press, 2017.
©2017

Links Contributor biographical information https://www.loc.gov/catdir/enhancements/fy1702/2016043383-b.html
Publisher description https://www.loc.gov/catdir/enhancements/fy1702/2016043383-d.html
Table of contents only https://www.loc.gov/catdir/enhancements/fy1702/2016043383-t.html

ISBN 9781107163713 (hardback; alk. paper)
1107163714 (hardback; alk. paper)

LC classification TA455.G65 A15 2017

Variant title Two dimensional materials

Related names Avouris, Phaedon, 1945- editor.
Heinz, Tony F., editor.
Low, Tony, editor.

Contents Graphene: basic properties / Mikhail I. Katsnelson and Annalisa Fasolino -- Electrical transport in graphene: carrier scattering by impurities and phonons / Jian-Hao Chen -- Optical properties of graphene / Feng Wang and Sufei Shi -- Graphene mechanical properties / C. DiMarco, R. Li, S. Rastogi, J. Hone and J.W . Kysar -- Vibrations in graphene / Ado Jorio, Luiz Gustavo Cançado and

Leandro M. Malard -- Thermal properties of graphene: from physics to applications / Alexander A. Balandin -- Graphene plasmonics / Frank Koppens, Mark B. Lundeberg, Marco Polini, Tony Low and Phaedon Avouris -- Electron optics with graphene p-n junctions / James R. Williams -- Graphene electronics / Chen Wang, Xidong Duan and Xiangfeng Duan -- Graphene: optoelectronic devices / Thomas Mueller and Phaedon Avouris -- Graphene spintronics / Aron W. Cummings, Sergio O. Valenzuela, Frank Ortmann and Stephan Roche -- Graphene-BN heterostructures / Lei Wang, James Hone and Cory. R. Dean -- Controlled growth of graphene crystals by chemical vapor deposition: from solid metals to liquid metals / Dechao Geng and Kian Ping Loh -- Electronic properties and strain engineering in semiconducting transition metal dichalcogenides / Rafael Roldán and Francisco Guinea -- Valley-spin physics in 2D semiconducting transition metal dichalcogenides / Hongyi Yu and Wang Yao -- Electrical transport in MoS2, a prototypical semiconducting TMDC / Andras Kis -- Optical properties of TMD heterostructures / Pasqual Rivera, Wang Yao and Xiaodong Xu -- TMDs - optoelectronic devices / Thomas Mueller -- Large area synthesis / Yumeng Shi and Lain-Jong Li -- Defects in two-dimensional materials / Xiaolong Zou and Boris I. Yakobson -- Theoretical overview of black phosphorus / Tony Low, Andrey Chaves, Wei Ji, Jesse Maassen and Traian Dumitrica -- Anisotropic properties of black phosphorus / Yuchen Du, Zhe Luo, Han Liu, Xianfan Xu and Peide D. Ye -- Optical properties and optoelectronic applications of black phosphorus / Andres Castellanos-Gomez and Mo Li -- Silicene, germanene and stanene / Guy Le Lay, Eric Salomon and Thierry Angot -- Predictions of single-layer honeycomb structures from first-principles / S. Ciraci and S. Cahangirov.

Subjects	Graphene.
	Nanostructured materials.
	Thin films.
Notes	Includes bibliographical references and index.

3rd International Conference on Nanotechnologies and Biomedical Engineering: ICNBME-2015, September 23-26, 2015, Chisinau, Republic of Moldova

LCCN	2019768693
Type of material	Book
Main title	3rd International Conference on Nanotechnologies and Biomedical Engineering: ICNBME-2015, September 23-26, 2015, Chisinau, Republic of Moldova / edited by Victor Sontea, Ion Tiginyanu.
Edition	1st ed. 2016.
Published/Produced	Singapore: Springer Singapore: Imprint: Springer, 2016.
Description	1 online resource (XXI, 564 pages 549 illustrations in color.)
	text file PDF
ISBN	9789812877369
Related names	Sontea, Victor, editor.
	Tiginyanu, Ion, editor.
Summary	This volume presents the proceedings of the 3rd International Conference on Nanotechnologies and Biomedical Engineering which was held on September 23-26, 2015 in Chisinau, Republic of Moldova. ICNBME-2015 continues the series of International Conferences in the field of nanotechnologies and biomedical engineering. It aims at bringing together scientists and engineers dealing with fundamental and applied research for reporting on the latest theoretical developments and applications involved in the fields. Topics include Nanotechnologies and nanomaterials Plasmonics and metamaterials Bio-micro/nano technologies Biomaterials Biosensors and sensors systems Biomedical instrumentation Biomedical signal processing Biomedical imaging and image

	processing Molecular, cellular and tissue engineering Clinical engineering, health technology management and assessment; Health informatics, e-health and telemedicine Biomedical engineering education Nuclear and radiation safety and security Innovations and technology transfer.
Subjects	Biological physics.
	Biomedical materials.
	Biomedical engineering.
	Biophysics.
	Health informatics.
	Regenerative medicine.
	Tissue engineering.
	Biomedical Engineering and Bioengineering.
	Biological and Medical Physics, Biophysics.
	Biomaterials.
	Health Informatics.
	Regenerative Medicine/Tissue Engineering.
Notes	Description based on publisher-supplied MARC data.
Additional formats	Printed edition: 9789812877352
	Printed edition: 9789812877376
Series	IFMBE Proceedings, 1680-0737; 55
	IFMBE Proceedings, 1680-0737; 55

Active Plasmonic Devices: Based on Magnetoplasmonic Nanostructures

LCCN	2019766063
Type of material	Book
Personal name	Martín Becerra, Diana, author.
Main title	Active Plasmonic Devices: Based on Magnetoplasmonic Nanostructures / by Diana Martín Becerra.
Edition	1st ed. 2017.
Published/Produced	Cham: Springer International Publishing: Imprint: Springer, 2017.
Description	1 online resource (XXIII, 113 pages 55 illustrations, 4 illustrations in color.)
	text file PDF
ISBN	9783319484112

Summary

This thesis investigates the effect of the magnetic field on propagating surface plasmon polaritons (SPPs), or surface plasmons for short. Above all, it focuses on using the magnetic field as an external agent to modify the properties of the SPPs, and therefore achieving active devices. Surface plasmons are evanescent waves that arise at metal-dielectric interfaces. They can be strongly confined (beyond the light diffraction limit), and provide a strong enhancement of the electromagnetic field at the interface. These waves have led to the development of plasmonic circuitry, which is a key candidate as an alternative to electronic circuitry and traditional optical telecommunication devices, since it is faster than the former and less bulky than the latter. Adopting both a theoretical and an experimental point of view, the book analyzes the magnetic modulation in SPPs by means of an interferometer engraved in a multilayer combining Au and Company In this interferometer, which acts like a modulator, the SPP magnetic modulation is studied in detail, as are the parameters that have a relevant impact on it, simple ways to enhance it, its spectral dependence, and the highly promising possibility of using this system for biosensing. The thesis ultimately arrives at the conclusion that this method can provide values of modulations similar to other active methods used in plasmonics.

Contents

Motivation and Objectives -- Active Plasmonics and Magnetoplasmonics -- Magnetoplasmonic Interferometry -- Magnetic Modulation of SPP in Au/Co/Au Trilayers -- Sensing Capability of the Interferometers -- Near Field Magnetoplasmonic Interferometry -- General Conclusions.

Subjects

Interfaces (Physical sciences).
Lasers.
Materials-Surfaces.
Nanoscale science.
Nanoscience.

	Nanostructures.
	Nanotechnology.
	Photonics.
	Surfaces (Physics).
	Thin films.
	Optics, Lasers, Photonics, Optical Devices.
	Nanoscale Science and Technology.
	Nanotechnology.
	Surface and Interface Science, Thin Films.
	Surfaces and Interfaces, Thin Films.
Notes	Description based on publisher-supplied MARC data.
Additional formats	Print version: Active plasmonic devices. 9783319484105 (DLC) 2016955326
	Printed edition: 9783319484105
	Printed edition: 9783319484129
	Printed edition: 9783319839363
Series	Springer Theses, Recognizing Outstanding Ph.D. Research, 2190-5053
	Springer Theses, Recognizing Outstanding Ph.D. Research, 2190-5053

An Introduction to Graphene Plasmonics

LCCN	2016003760
Type of material	Book
Personal name	Gonçalves, P. A. D., author.
Main title	An introduction to graphene plasmonics / P.A.D Gonçalves, University of Minho, Portugal, N.M.R Peres, University of Minho, Portugal.
Published/Produced	New Jersey: World Scientific, [2016]
	©2016
Description	xxix, 431 pages: illustrations; 24 cm
ISBN	9789814749978 (hardcover; alk. paper)
	9814749974 (hardcover; alk. paper)
	9789814749985 (softcover; alk. paper)
	9814749982 (softcover; alk. paper)
LC classification	QC176.8.P55 G66 2016
Related names	Peres, N. M. R., 1967- author.
Summary	"The book aims at the graduate student entering the field of graphene plasmonics. It is essentially self-

	contained and brings together a number of different topics about the field that are scattered in a vast literature. The text is composed of eleven chapters and of a set of detailed appendices. It can be read in two different ways: Reading only the chapters to get acquainted with the field of plasmonics in graphene or reading the chapters and studying the appendices to get a working knowledge of the topic. Thus, the book provides a good training for the students on the different aspects of the field"-- Provided by publisher.
Contents	Electromagnetic properties of solids in a nutshell -- Surface plasmon-polaritons at dielectric-metal interfaces -- Graphene surface plasmons -- Excitation of graphene surface plasmons -- Launching plasmons using a metallic antenna -- Plasmonics in a periodic array of graphene-based microribbons -- Plasmons in graphene nanostructures and in onedimensional defects -- Excitation of surface plasmon-polaritons using dielectric gratings -- Excitation of plasmons by an emitting dipole.
Subjects	Plasmonics. Graphene--Optical properties. Nanophotonics.
Notes	Includes bibliographical references and index.

An Introduction to Organic Lasers

LCCN	2017471052
Type of material	Book
Personal name	Boudrioua, Azzedine, author.
Main title	An introduction to organic lasers / Azzedine Boudrioua, Mahmoud Chakaroun, Alexis Fischer.
Published/Produced	London: ISTE Press Ltd; Oxford: Elsevier Ltd, 2017. ©2017
Description	xiv, 212 pages: illustrations; 24 cm.
ISBN	9781785481581 hardcover 1785481584 hardcover
LC classification	QC689.55.O7 B68 2017
Related names	Chakaroun, Mahmoud, author. Fischer, Alexis, author.

Summary	"One of the biggest challenges of organic optoelectronics is the realization of the first organic laser diode (electrically pumped) which has a very strong potential for many applications. Similar to what happened in the field of inorganic optoelectronics when transforming LEDs into LDs, the race is on to transform an OLED into an OLD. This involves the development of innovative solutions to overcome the difficulties inherent in organic materials and the electric pump. This book presents the elements of physics, materials and technologies that allow us to understand the basics of organic lasers and to capture the progress made. It also provides guidance for future developments towards the organic laser diode."-- Back cover.
Contents	Organic semiconductors -- Organic light-emitting diodes -- Organic lasers -- Organic plasmonics: toward organic nanolasers.
Subjects	Lasers. Organic electronics. Organic semiconductors. Light emitting diodes. Lasers. Light emitting diodes. Organic electronics. Organic semiconductors.
Notes	Includes bibliographical references and index.
Series	Advanced lasers set Advanced lasers set.

Anisotropic and Shape-Selective Nanomaterials: Structure-Property Relationships

LCCN	2019769810
Type of material	Book
Main title	Anisotropic and Shape-Selective Nanomaterials: Structure-Property Relationships / edited by Simona E. Hunyadi Murph, George K. Larsen, Kaitlin J. Coopersmith.
Edition	1st ed. 2017.

Published/Produced	Cham: Springer International Publishing: Imprint: Springer, 2017.
Description	1 online resource (X, 470 pages 228 illustrations, 160 illustrations in color.) text file PDF
ISBN	9783319596624
Related names	Coopersmith, Kaitlin J, editor. Hunyadi Murph, Simona E, editor. Larsen, George K, editor.
Summary	This book reviews recent advances in the synthesis, characterization, and physico-chemical properties of anisotropic nanomaterials. It highlights various emerging applications of nanomaterials, including sensing and imaging, (bio)medical applications, environmental protection, plasmonics, catalysis, and energy. It provides an excellent and comprehensive overview of the effect that morphology and nanometric dimension has on the physico-chemical properties of various materials and how this leads to novel applications.
Contents	Part I: Introduction and Fundamentals -- An Introduction to Nanotechnology -- Nanoscale Materials: Fundamentals and Emergent Properties -- Synthetic Strategies for Anisotropic and Shape-Selective Nanomaterials -- Characterization of Anisotropic and Shape-Selective Nanomaterials: Methods and Challenges -- Part 2: Effect of the morphology and the nanometric dimension of materials on their physico- chemical properties -- Anisotropic Metallic and Metallic Oxide Nanostructures - Correlation between their shape and properties -- Putting Nanoparticles to Work: Self-propelled Inorganic Micro- and Nanomotors -- Prospects for rational control of nanocrystal shape through successive ionic layer adsorption and reaction (SILAR) and related approaches -- Plasmon Drag Effect. Theory and Experiment -- Dimensional Variations in Nanohybrids: Property Alterations, Applications, and Considerations for Toxicological

	Implications -- Assemblies and Superstructures of Inorganic Colloidal Nanocrystals -- Nanostructured Catalysts for the Electrochemical Reduction of CO2 -- Strategies for the Synthesis of Anisotropic Catalytic Nanoparticles -- Biomedical Applications of Anisotropic Gold Nanoparticles -- Application of Gold Nanorods in Cardiovascular Science -- Architectured Nanomembranes -- Summary and Final Thoughts.
Subjects	Inorganic chemistry. Lasers. Nanochemistry. Nanotechnology. Photonics. Nanochemistry. Inorganic Chemistry. Nanotechnology. Optics, Lasers, Photonics, Optical Devices.
Notes	Description based on publisher-supplied MARC data.
Additional formats	Print version: Anisotropic and shape-selective nanomaterials. 9783319596617 (DLC) 2017943180 Printed edition: 9783319596617 Printed edition: 9783319596631 Printed edition: 9783319866741
Series	Nanostructure Science and Technology, 1571-5744 Nanostructure Science and Technology, 1571-5744

Anti-Reflection and Light Trapping in c-Si Solar Cells

LCCN	2019761867
Type of material	Book
Personal name	Solanki, Chetan Singh, author.
Main title	Anti-reflection and Light Trapping in c-Si Solar Cells / by Chetan Singh Solanki, Hemant Kumar Singh.
Edition	1st ed. 2017.
Published/Produced	Singapore: Springer Singapore: Imprint: Springer, 2017.
Description	1 online resource (XXIX, 186 pages 128 illustrations, 88 illustrations in color.) text file PDF

ISBN	9789811047718
Related names	Singh, Hemant Kumar, author.
Summary	This book offers essential insights into c-Si based solar cells and fundamentals of reflection, refraction, and light trapping. The basic physics and technology for light trapping in c-Si based solar cells are covered, from traditional to advanced light trapping structures. Further, the book discusses the latest developments in plasmonics for c-Si solar cell applications, along with their future scope and the requirements for further research. The book offers a valuable guide for graduate students, researchers and professionals interested in the latest trends in solar cell technologies.
Contents	Solar cells: Overview -- c-Si cell: Physics and Technology -- Principle of dielectric based anti-reflection and light trapping -- Principle of texturization for enhanced light trapping -- Texturing process of c-Si wafers -- Anti-reflection coatings with textured surface for c-Si solar cells -- Advancement in traditional lighting trapping structures -- Plasmonics based advanced anti-reflection and light trapping: Principles and Technology -- Plasmonics based light trapping for c-Si solar cell applications -- Future scope in advanced lighting trapping structure development.
Subjects	Electronic materials. Energy storage. Lasers. Optical materials. Photonics. Renewable energy sources. Energy Storage. Optical and Electronic Materials. Optics, Lasers, Photonics, Optical Devices. Renewable and Green Energy.
Notes	Description based on publisher-supplied MARC data.
Additional formats	Print version: Anti-reflection and light trapping in c-Si solar cells. 9789811047701 (DLC) 2017940816

	Printed edition: 9789811047701
	Printed edition: 9789811047725
	Printed edition: 9789811352263
Series	Green Energy and Technology, 1865-3529
	Green Energy and Technology, 1865-3529

Bimetallic Nanostructures: Shape-Controlled Synthesis for Catalysis, Plasmonics, and Sensing Applications

LCCN	2018017086
Type of material	Book
Main title	Bimetallic nanostructures: shape-controlled synthesis for catalysis, plasmonics, and sensing applications / edited by Professor Ya-Wen Zhang.
Published/Produced	Hoboken, NJ: John Wiley & Sons, 2018.
Description	1 online resource.
ISBN	9781119214625 (epub)
	9781119214632 (pdf)
LC classification	QD505
Related names	Zhang, Ya-Wen, 1965- editor.
Contents	Introduction of bimetallic nanostructures -- Theoretical models for bimetallic surfaces and nanoalloys -- In situ characterization techniques of bimetallics -- Bimetallic nanopolyhedrons and nanospheres -- Bimetallic convex and concave nanostructures -- Bimetallic nanoframes and nanoporous structures -- Bimetallic dendritic nanostructures -- Bimetallic ultrathin nanowires -- Bimetallic nanoplates and nanosheets -- Electrocatalysis -- Heterogeneous catalysis -- Plasmonics -- Sensing.
Subjects	Metal catalysts.
	Nanostructured materials.
Notes	Includes bibliographical references and index.
	Description based on print version record and CIP data provided by publisher.
Additional formats	Print version: Bimetallic nanostructures Hoboken, NJ: John Wiley & Sons, 2018 9781119214649 (DLC) 2018010417

Bimetallic Nanostructures: Shape-Controlled Synthesis for Catalysis, Plasmonics, and Sensing Applications

LCCN	2018010417
Type of material	Book
Main title	Bimetallic nanostructures: shape-controlled synthesis for catalysis, plasmonics, and sensing applications / edited by Professor Ya-Wen Zhang.
Published/Produced	Hoboken, NJ: John Wiley & Sons, 2018.
ISBN	9781119214649 (cloth)
LC classification	QD505 .B556 2018
Related names	Zhang, Ya-Wen, 1965- editor.
Contents	Introduction of bimetallic nanostructures -- Theoretical models for bimetallic surfaces and nanoalloys -- In situ characterization techniques of bimetallics -- Bimetallic nanopolyhedrons and nanospheres -- Bimetallic convex and concave nanostructures -- Bimetallic nanoframes and nanoporous structures -- Bimetallic dendritic nanostructures -- Bimetallic ultrathin nanowires -- Bimetallic nanoplates and nanosheets -- Electrocatalysis -- Heterogeneous catalysis -- Plasmonics -- Sensing.
Subjects	Metal catalysts. Nanostructured materials.
Notes	Includes bibliographical references and index.
Additional formats	Online version: Bimetallic nanostructures Hoboken, NJ: John Wiley & Sons, 2018 9781119214632 (DLC) 2018017086

Chiral Nanophotonics: Chiral Optical Properties of Plasmonic Systems

LCCN	2019766083
Type of material	Book
Personal name	Schäferling, Martin, author.
Main title	Chiral Nanophotonics: Chiral Optical Properties of Plasmonic Systems / by Martin Schäferling.
Edition	1st ed. 2017.
Published/Produced	Cham: Springer International Publishing: Imprint: Springer, 2017.

Description 1 online resource (XV, 159 pages 79 illustrations, 75 illustrations in color.)
text file PDF
ISBN 9783319422640
Summary This book describes the physics behind the optical properties of plasmonic nanostructures focusing on chiral aspects. It explains in detail how the geometry determines chiral near-fields and how to tailor their shape and strength. Electromagnetic fields with strong optical chirality interact strongly with chiral molecules and, therefore, can be used for enhancing the sensitivity of chiroptical spectroscopy techniques. Besides a short review of the latest results in the field of plasmonically enhanced enantiomer discrimination, this book introduces the concept of chiral plasmonic near-field sources for enhanced chiroptical spectroscopy. The discussion of the fundamental properties of these light sources provides the theoretical basis for further optimizations and is of interest for researchers at the intersection of nano-optics, plasmonics and stereochemistry.
Contents Introduction -- Chirality in Nature and Science -- Chiroptical Spectroscopy- Chiral Properties of Light -- Enhancement of Chiral Fields by Geometrically Chiral Structures -- Chiral Fields of Macroscopically Achiral Arrangements -- Formation of Chiral Fields Near Symmetric Structures -- Chiral Eigenmodes of Geometrically Chiral Structures -- Conclusions and Outlook.
Subjects Electrodynamics.
Lasers.
Nanoscale science.
Nanoscience.
Nanostructures.
Nanotechnology.
Optics.
Photonics.
Optics, Lasers, Photonics, Optical Devices.

	Classical Electrodynamics.
	Nanoscale Science and Technology.
	Nanotechnology and Microengineering.
Notes	Description based on publisher-supplied MARC data.
Additional formats	Print version: Chiral nanophotonics. 9783319422633 (DLC) 2016947502
	Printed edition: 9783319422633
	Printed edition: 9783319422657
	Printed edition: 9783319825410
Series	Springer Series in Optical Sciences, 0342-4111; 205
	Springer Series in Optical Sciences, 0342-4111; 205

Coaxial Lithography

LCCN	2019750004
Type of material	Book
Personal name	Ozel, Tuncay. author.
Main title	Coaxial Lithography / by Tuncay Ozel.
Edition	1st ed. 2016.
Published/Produced	Cham: Springer International Publishing: Imprint: Springer, 2016.
Description	1 online resource (XXIX, 92 pages 53 illustrations, 4 illustrations in color.)
	text file PDF
ISBN	9783319454146
Summary	This thesis focuses on the electrochemical synthesis of multi-segmented nanowires. In contrast to previous work, which was largely limited to one-dimensional modifications, Tuncay Ozel presents a technique, termed coaxial Lithography (COAL), which allows for the synthesis of coaxial nanowires in a parallel fashion with sub-10 nanometer resolution in both the axial and radial dimensions. This work has significantly expanded current synthetic capabilities with respect to materials generality and the ability to tailor two-dimensional growth in the formation of core-shell structures. These developments have enabled fundamental and applied studies which were not previously possible. The COAL technique will increase the capabilities of

many researchers who are interested in studying light-matter interactions, nanoparticle assembly, solution-dispersible nanoparticles and labels, semiconductor device physics and nanowire biomimetic probe preparation. The methodology and results presented in this thesis appeal to researchers in nanomaterial synthesis, plasmonics, biology, photovoltaics, and photocatalysis.

Contents Introduction to Plasmonics, Templated Electrochemical Synthesis, and On-Wire Lithography -- 1D Nanowire Synthesis: Extending the OWL Toolbox with Semiconductors to Explore Plasmon-Exciton Interactions in the Form of Long-Range Optical Nanoscale Rulers -- Hybrid Semiconductor Core-Shell Nanowires with Tunable Plasmonic Nanoantennas -- 2D Nanowire Synthesis: Invention of Coaxial Lithography -- Solution Dispersible Metal Nanorings: Independent Control of Architectural Parameters and Materials Generality -- Conclusions and Outlook on Templated Electrochemical Synthesis Using Coaxial Lithography.

Subjects Electrochemistry.
Lasers.
Nanochemistry.
Nanotechnology.
Photonics.
Electrochemistry.
Nanochemistry.
Nanotechnology.
Optics, Lasers, Photonics, Optical Devices.

Notes Description based on publisher-supplied MARC data.

Additional formats Print version: Coaxial lithography. 9783319454139 (DLC) 2016950377
Printed edition: 9783319454139
Printed edition: 9783319454153
Printed edition: 9783319832876

Series Springer Theses, Recognizing Outstanding Ph.D. Research, 2190-5053

Springer Theses, Recognizing Outstanding Ph.D. Research, 2190-5053

Compendium on Electromagnetic Analysis: From Electrostatics to Photonics: Fundamentals and Applications for Physicists and Engineers

LCCN	2018055910
Type of material	Book
Personal name	Tsukerman, Igor, author.
Main title	Compendium on electromagnetic analysis: from electrostatics to photonics: fundamentals and applications for physicists and engineers / editor-in-chief, Igor Tsukerman (University of Akron, USA), editor, Michael Donahue (National Institute of Standards & Technology, USA), Yilmaz Sozer (University of Akron, USA), Thomas Bauernfeind (Graz University of Technology, Austria) and Vadim A Markel (University of Pennsylvania, USA
Published/Produced	New Jersey: World Scientific, [2020]
Description	volumes: illustrations (some color); 25 cm
ISBN	9789813270169 (set; hardcover) 9789813270251 (v. 1; hardcover) 9789813270275 (v. 2; hardcover) 9789813270299 (v. 3; hardcover) 9789813270312 (v. 4; hardcover) 9789813270336 (v. 5; hardcover) (set; ebook) (set; ebook other)
LC classification	TK7872.M25 T86 2020
Related names	Donahue, Michael, author. Sozer, Yilmaz, author.
Summary	"The five-volume set may serve as a comprehensive reference on electromagnetic analysis and its applications at all frequencies, from static fields to optics and photonics. The material includes micro- and nanomagnetics, the new generation of electric machines, renewable energy, hybrid vehicles, low-noise motors; antennas and microwave devices, plasmonics, metamaterials, lasers, and more. Written at a level accessible to both graduate students and

	engineers, Electromagnetic Analysis is a comprehensive reference, covering methods and applications at all frequencies (from statics to optical). Each volume contains pedagogical/tutorial material of high archival value as well as chapters on state-of-the-art developments"-- Provided by publisher.
Contents	v. 1 Electrostatic and Magnetic Phenomena -- v. 2 The New Generation of Electric Machines -- v. 3 Antennas, Antenna Arrays and Microwave Devices -- v. 4 Optics and Photonics I -- v. 5 Optics and Photonics II.
Subjects	Electromagnetic devices. Electromagnetism.
Notes	Includes bibliographical references and index.

Computational Photonic Sensors

LCCN	2019745816
Type of material	Book
Main title	Computational Photonic Sensors / edited by Mohamed Farhat O. Hameed, Salah Obayya.
Edition	1st ed. 2019.
Published/Produced	Cham: Springer International Publishing: Imprint: Springer, 2019.
Description	1 online resource (VIII, 452 pages 273 illustrations, 219 illustrations in color.) text file PDF
ISBN	9783319765563
Related names	Hameed, Mohamed Farhat O. editor. Obayya, Salah. editor.
Summary	This book provides a comprehensive overview of the photonic sensing field by covering plasmonics, photonic crystal, and SOI techniques from theory to real sensing applications. A literature review of ultra-sensitive photonic sensors, including their design and application in industry, makes this a self-contained and comprehensive resource for different types of sensors, with high value to the biosensor sector in particular. The book is organized into four parts: Part

I covers the basic theory of wave propagation, basic principles of sensing, surface plasmon resonance, and silicon photonics; Part II details the computational modeling techniques for the analysis and prediction of photonic sensors; Part III and Part IV cover the various mechanisms and light matter interaction scenarios behind the design of photonic sensors including photonic crystal fiber sensors and SOI sensors. This book is appropriate for academics and researchers specializing in photonic sensors; graduate students in the early and intermediate stages working in the areas of photonics, sensors, biophysics, and biomedical engineering; and to biomedical, environmental, and chemical engineers. Provides high quality material for researchers and engineers to harness their knowledge in the area of photonic sensors Covers the applications of photonic sensors in bio-molecular sensing, environmental issues, temperature variations, pressure sensing, DNA sequence sensors, and multifunctional sensors Explains the fundamental role that new technologies, such as liquid crystal, photonic crystal, plasmonics, and SOI, play in optimizing photonic sensor performance.

Contents

Part 1. Fundamentals -- Introduction to Optical Waveguides -- Fundamentals of Photonic Crystals -- Basic Principles of Surface Plasmon Resonance -- Introduction to Silicon Photonics -- Basic Principles of Biosensing -- Part 2. Computational Modeling Techniques -- Finite Element Method for Sensing Applications -- FDTD in Cartesian and Spherical Grids -- Part 3. Photonic Crystal Fiber Sensors -- Temperature Sensors Based on Plasmonic Photonic Crystal Fiber -- Microstructured Optical Fiber Based Plasmonic Sensors -- Multifunctional Plasmonic Photonic Crystal Fiber Biosensors -- Photonic Crystal Fiber Pressure Sensors -- Development of Photonic Crystal Fiber Based Gas/Chemical Sensors -- Part 4. Silicon-on-Insulator Sensors -- Silicon Nanowires for

	DNA Sensing -- Compact Photonic SOI Sensors -- Silicon Ring-Resonator Based Biochips -- SOI Waveguide Based Biochemical Sensors.
Subjects	Chemical engineering.
	Crystallography.
	Lasers.
	Nanotechnology.
	Photonics.
	Nanotechnology and Microengineering.
	Crystallography and Scattering Methods.
	Industrial Chemistry/Chemical Engineering.
	Nanotechnology.
	Optics, Lasers, Photonics, Optical Devices.
Notes	Description based on publisher-supplied MARC data.
Additional formats	Print version: Computational photonic sensors. 9783319765556 (DLC) 2018933458
	Printed edition: 9783030095222
	Printed edition: 9783319765556
	Printed edition: 9783319765570

Contemporary Optoelectronics: Materials, Metamaterials and Device Applications

LCCN	2019747699
Type of material	Book
Main title	Contemporary Optoelectronics: Materials, Metamaterials and Device Applications / edited by Oleksiy Shulika, Igor Sukhoivanov.
Edition	1st ed. 2016.
Published/Produced	Dordrecht: Springer Netherlands: Imprint: Springer, 2016.
Description	1 online resource (X, 234 pages 121 illustrations, 66 illustrations in color.)
	text file PDF
ISBN	9789401773157
Related names	Shulika, Oleksiy. editor.
	Sukhoivanov, Igor. editor.
Summary	This book presents a collection of extended contributions on the physics and application of optoelectronic materials and metamaterials. The book

is divided into three parts, respectively covering materials, metamaterials and optoelectronic devices. Individual chapters cover topics including phonon-polariton interaction, semiconductor and nonlinear organic materials, metallic, dielectric and gyrotropic metamaterials, singular optics, parity-time symmetry, nonlinear plasmonics, microstructured optical fibers, passive nonlinear shaping of ultrashort pulses, and pulse-preserving supercontinuum generation. The book contains both experimental and theoretical studies, and each contribution is a self-contained exposition of a particular topic, featuring an extensive reference list. The book will be a useful resource for graduate and postgraduate students, researchers and engineers involved in optoelectronics/photonics, quantum electronics, optics, and adjacent areas of science and technology.

Contents From the Contents: Advanced Optoelectronics -- III-V Nanowires for Optoelectronic Applications -- Advances in Optoelectronic Approaches for Wideband and Programmable Processing of Ultrafast Signals -- From Order to Chaos and back: a High-Level Coupling Approach for Cryptography of Transmitted Data.

Subjects Electronic materials.
Lasers.
Microwaves.
Optical engineering.
Optical materials.
Photonics.
Optics, Lasers, Photonics, Optical Devices.
Microwaves, RF and Optical Engineering.
Optical and Electronic Materials.

Notes Description based on publisher-supplied MARC data.

Additional formats Print version: Contemporary optoelectronics. 9789401773140 (DLC) 2015945602
Printed edition: 9789401773140
Printed edition: 9789401773164
Printed edition: 9789402402940

Series	Springer Series in Optical Sciences, 0342-4111; 199 Springer Series in Optical Sciences, 0342-4111; 199

Diffractive Optics and Nanophotonics

LCCN	2017029740
Type of material	Book
Main title	Diffractive optics and nanophotonics / edited by Victor A. Soifer.
Published/Produced	Boca Raton, FL: CRC Press, Taylor & Francis Group, [2018] ©2017
ISBN	9781498754477 (hardback; alk. paper) 1498754473 (hardback; alk. paper)
LC classification	QC415 .D55 2018
Related names	Soĭfer, V. A. (Viktor Aleksandrovich), editor.
Summary	"Nanophotonics deals with the diffraction of light from objects with nanoscale features. This book is devoted to achievements in diffractive optics, focusing on the creation of new nanophotonic components and devices, as well as instrumentation and available information technology. It describes methods of calculation of diffractive optical elements to solve actual problems of nanophotonics. Coverage includes mathematical methods for calculation of diffraction gratings, calculation of modes of inhomogeneous waveguides, integral methods of calculation of electromagnetic field near the focus, and methods of calculation of diffractive optical elements generating vortex laser beams"-- Provided by publisher.
Contents	The Fourier modal method and its use in plasmonics and the theory of resonant diffraction gratings -- Components of photonic crystals -- Sharp focusing of light with microoptics components -- Optical trapping and manipulation of micro- and nano-objects.
Subjects	Diffraction. Nanophotonics.

Electromagnetic Theory and Plasmonics for Engineers

LCCN	2018952889
Type of material	Book
Main title	Electromagnetic theory and plasmonics for engineers.
Published/Produced	New York, NY: Springer Berlin Heidelberg, 2018.
ISBN	9789811323508

Electromagnetic Theory and Plasmonics for Engineers

LCCN	2019751529
Type of material	Book
Personal name	Nickelson, Liudmila, author.
Main title	Electromagnetic Theory and Plasmonics for Engineers / by Liudmila Nickelson.
Edition	1st ed. 2019.
Published/Produced	Singapore: Springer Singapore: Imprint: Springer, 2019.
Description	1 online resource (XXIII, 749 pages 353 illustrations, 185 illustrations in color.) text file PDF
ISBN	9789811323522
Summary	This book presents the theory of electromagnetic (EM) waves for upper undergraduate, graduate and PhD-level students in engineering. It focuses on physics and microwave theory based on Maxwell's equations and the boundary conditions important for studying the operation of waveguides and resonators in a wide frequency range, namely, from approximately 10**9 to 10**16 hertz. The author also highlights various current topics in EM field theory, such as plasmonic (comprising a noble metal) waveguides and analyses of attenuations by filled waveguide dielectrics or semiconductors and also by conducting waveguide walls. Featuring a wide variety of illustrations, the book presents the calculated and schematic distributions of EM fields and currents in waveguides and resonators. Further, test questions are presented at the end of each chapter.

Contents	Vector Analysis in Annex to Electromagnetics -- Static and Stationary Fields -- Fields and Materials -- Maxwell's Equations and Boundary Conditions -- Plane Electromagnetic Wave Propagation -- Reflection and Transmission of Plane Electromagnetic Waves -- Rectangular Hollow Metallic Waveguides and Resonators -- Cylindrical Hollow Metallic Waveguides and Cavity Resonators -- Plasmonics.
Subjects	Electrical engineering. Electrodynamics. Image processing. Microwaves. Optical engineering. Optics. Signal processing. Speech processing systems. Microwaves, RF and Optical Engineering. Classical Electrodynamics. Communications Engineering, Networks. Signal, Image and Speech Processing.
Notes	Description based on publisher-supplied MARC data.
Additional formats	Print version: Electromagnetic theory and plasmonics for engineers. 9789811323508 (DLC) 2018952889 Printed edition: 9789811323508 Printed edition: 9789811323515

Evanescent Waves in Optics: An Introduction to Plasmonics

LCCN	2017954456
Type of material	Book
Personal name	Bertolotti, Mario.
Main title	Evanescent waves in optics: an introduction to plasmonics / Mario Bertolotti.
Edition	1st edition.
Published/Produced	New York, NY: Springer Berlin Heidelberg, 2017.
ISBN	9783319612607 (hardback: alk. paper)

Evanescent Waves in Optics: An Introduction to Plasmonics

LCCN	2019746780

Type of material	Book
Personal name	Bertolotti, Mario. author.
Main title	Evanescent Waves in Optics: An Introduction to Plasmonics / by Mario Bertolotti, Concita Sibilia, Angela M. Guzman.
Edition	1st ed. 2017.
Published/Produced	Cham: Springer International Publishing: Imprint: Springer, 2017.
Description	1 online resource (IX, 259 pages 131 illustrations, 24 illustrations in color.) text file PDF
ISBN	9783319612614
Related names	M. Guzman, Angela. author. Sibilia, Concita. author.
Summary	This monograph provides an introductory discussion of evanescent waves and plasmons, describes their properties and uses, and shows how they are fundamental when operating with nanoscale optics. Far field optics is not suitable for the design, description, and operation of devices at this nanometre scale. Instead one must work with models based on near-field optics and surface evanescent waves. The new discipline of plasmonics has grown to encompass the generation and application of plasmons both as a travelling excitation in a nanostructure and as a stationary enhancement of the electrical field near metal nanosurfaces. The book begins with a brief review of the basic concepts of electromagnetism, then introduces evanescent waves through reflection and refraction, and shows how they appear in diffraction problems, before discussing the role that they play in optical waveguides and sensors. The application of evanescent waves in super-resolution devices is briefly presented, before plasmons are introduced. The surface plasmon polaritons (SPPs) are then treated, highlighting their potential applications also in ultra-compact circuitry. The book concludes with a discussion of the quantization of evanescent waves

and quantum information processing. The book is intended for students and researchers who wish to enter the field or to have some insight into the matter. It is not a textbook but simply an introduction to more complete and in-depth discussions. The field of plasmonics has exploded in the last ten years, and most of the material treated in this book is scattered in original or review papers. A short comprehensive treatment is missing; this book is intended to provide just that.

Contents	Preface -- Chapter I Basic electromagnetics -- Chapter II Evanescent Waves -- Chapter III Evanescent Waves in Waveguides -- Chapter IV High resolution optical microscopes (in which evanescent waves play a role) -- Chapter V Plasmons -- Chapter VI Applications of plasmons -- Chapter VII Quantization of evanescent waves, surface plasmons and surface plasmon polaritons.
Subjects	Electrodynamics. Electronic materials. Lasers. Microwaves. Optical engineering. Optical materials. Optics. Photonics. Quantum optics. Semiconductors. Optics, Lasers, Photonics, Optical Devices. Classical Electrodynamics. Microwaves, RF and Optical Engineering. Optical and Electronic Materials. Quantum Optics. Semiconductors.
Notes	Description based on publisher-supplied MARC data.
Additional formats	Print version: Evanescent waves in optics: an introduction to plasmonics 9783319612607 (DLC) 2017954456 Printed edition: 9783319612607

Printed edition: 9783319612621
Printed edition: 9783319870359
Series Springer Series in Optical Sciences, 0342-4111; 206
Springer Series in Optical Sciences, 0342-4111; 206

Fano Resonances in Optics and Microwaves: Physics and Applications
LCCN 2019742967
Type of material Book
Main title Fano Resonances in Optics and Microwaves: Physics and Applications / edited by Eugene Kamenetskii, Almas Sadreev, Andrey Miroshnichenko.
Edition 1st ed. 2018.
Published/Produced Cham: Springer International Publishing: Imprint: Springer, 2018.
Description 1 online resource (XXIII, 582 pages 259 illustrations, 206 illustrations in color.)
text file PDF
ISBN 9783319997315
Related names Kamenetskii, Eugene. editor.
Miroshnichenko, Andrey. editor.
Sadreev, Almas. editor.
Summary This book discusses the development of Fano-based techniques and reveals the characteristic properties of various wave processes by studying interference phenomena. It explains that the interaction of discrete (localized) states with a continuum of propagation modes leads to Fano interference effects in transmission, and explores novel coherent effects such as bound states in the continuum accompanied by collapse of Fano resonance. Originating in atomic physics, Fano resonances have become one of the most appealing phenomena of wave scattering in optics, microwaves, and terahertz techniques. The generation of extremely strong and confined fields at a deep subwavelength scale, far beyond the diffraction limit, plays a central role in modern plasmonics, magnonics, and in photonic and metamaterial structures. Fano resonance effects take advantage of the coupling of these bound states with

a continuum of radiative electromagnetic waves. With their unique physical properties and unusual combination of classical and quantum structures, Fano resonances have an application potential in a wide range of fields, from telecommunication to ultrasensitive biosensing, medical instrumentation and data storage. Including contributions by international experts and covering the essential aspects of Fano-resonance effects, including theory, modeling and design, proven and potential applications in practical devices, fabrication, characterization and measurement, this book enables readers to acquire the multifaceted understanding required for these multidisciplinary challenges.

Contents Theory of Fano Resonances -- Bound States in a Continuum -- Fano Resonances and Field-Matter Chirality -- Fano Resonances in Plasmonic Structures and Metamaterials -- Fano Resonances in Microwave Structures -- Fano Resonances for Biosensing.

Subjects Atomic structure.
Electrodynamics.
Measurement.
Microscopy.
Microwaves.
Molecular structure.
Optical engineering.
Optics.
Physical measurements.
Semiconductors.
Spectroscopy.
Atomic/Molecular Structure and Spectra.
Classical Electrodynamics.
Measurement Science and Instrumentation.
Microwaves, RF and Optical Engineering.
Semiconductors.
Spectroscopy and Microscopy.

Notes Description based on publisher-supplied MARC data.

Additional formats Print version: Fano resonances in optics and microwaves. 9783319997308 (DLC) 2018953587

	Printed edition: 9783319997308
	Printed edition: 9783319997322
Series	Springer Series in Optical Sciences, 0342-4111; 219
	Springer Series in Optical Sciences, 0342-4111; 219

Fiber Optic Sensors Based on Plasmonics

LCCN	2015472031
Type of material	Book
Personal name	Gupta, Banshi Dhar, author.
Main title	Fiber optic sensors based on plasmonics / Banshi Dhar Gupta, Indian Institute of Technology Delhi, India, Sachin Kumar Srivastava, Ilse Katz Institiute of Nanoscale Science and Technology, Israel, Roli Verma, Indian Institute of Technology Delhi, India.
Published/Produced	New Jersey: World Scientific, [2015]
Description	xiv, 267 pages; 24 cm
ISBN	981461954X
	9789814619547
LC classification	MLCM 2018/46527 (T)
Subjects	Optical fibers--Industrial applications.
	Optical fibers in biochemistry.
	Optical fiber detectors.
	Plasmons (Physics)
Notes	Minimal Level Cataloging Plus.

From Classical to Quantum Plasmonics in Three and Two Dimensions

LCCN	2016956178
Type of material	Book
Main title	From classical to quantum plasmonics in three and two dimensions.
Published/Produced	New York, NY: Springer Berlin Heidelberg, 2016.
Links	Publisher description https://www.loc.gov/catdir/enhancements/fy1622/2016956178-d.html
	Table of contents only https://www.loc.gov/catdir/enhancements/fy1622/2016956178-t.html
ISBN	9783319485614

From Classical to Quantum Plasmonics in Three and Two Dimensions

LCCN	2019765781
Type of material	Book
Personal name	Christensen, Thomas, author.
Main title	From Classical to Quantum Plasmonics in Three and Two Dimensions / by Thomas Christensen.
Edition	1st ed. 2017.
Published/Produced	Cham: Springer International Publishing: Imprint: Springer, 2017.
Description	1 online resource (XX, 181 pages 39 illustrations, 14 illustrations in color.) text file PDF
ISBN	9783319485621
Summary	This thesis provides a comprehensive introduction to two active research directions within the field of plasmonics: (i) nonclassical, or quantum, aspects of the plasmonic response; and (ii) two-dimensional plasmonics, a recent innovation in the field stimulated by the advent of two-dimensional materials. It discusses the fundamentals of this field in detail, and explores several current research directions. Nonclassical plasmonics has been spurred on in recent years by the tremendous technological progress in nanofabrication and optical characterization; today, it is possible to investigate the plasmonic features of nanostructures with characteristic features in the few nanometer range. The book describes and analyzes the breakdown of the classical theory under these conditions and explores several alternatives and extensions. The unique electronic and dimensional features of novel two-dimensional materials, such as graphene, lie at the core of plasmonics' most rapidly developing subfield; two-dimensional plasmonics. This thesis provides a clear and comprehensive exposition of the central features for interested researchers looking for an entry point to this riveting area..
Contents	Introduction -- Fundamentals of Plasmonics -- Nonclassical Plasmonics -- Electronic Properties of

Graphene -- Classical Graphene Plasmonics -- Nonclassical Graphene Plasmonics -- Outlook and Conclusions.

Subjects Physics.
Physics, general.

Notes Description based on publisher-supplied MARC data.

Additional formats Print version: From classical to quantum plasmonics in three and two dimensions. 9783319485614 (DLC) 2016956178
Printed edition: 9783319485614
Printed edition: 9783319485638
Printed edition: 9783319839738

Series Springer Theses, Recognizing Outstanding Ph.D. Research, 2190-5053
Springer Theses, Recognizing Outstanding Ph.D. Research, 2190-5053

Frontiers of Engineering: Reports on Leading-Edge Engineering from the 2015 Symposium

LCCN 2017385892

Type of material Book

Main title Frontiers of engineering: reports on leading-edge engineering from the 2015 Symposium / National Academy of Engineering.

Published/Produced Washington, D. C.: National Academies Press, [2016]
©2016

Description xi, 161 pages: illustrations; 23 cm

ISBN 0309379539
9780309379533

LC classification TK5105.59 .F76 2016

Related names National Academy of Engineering, issuing body.
Symposium on Frontiers of Engineering (2015: Irvine, Calif.)

Contents Cybersecurity and privacy: introduction / David Brumley and Daniela Oliveira -- Security at different layers of abstractions: application, operating systems, and hardware / Bryan D. Payne -- Computer security and privacy: where human factors meet engineering /

Franziska Roesner -- On the technical debt of medical device security / Kevin Fu -- Challenges of engineering cybersecurity: a government perspective / Tomas Vagoun -- Engineering the search for Earth-like exoplanets: introduction / Sara Seager and Mitchel Walker -- Engineering the James Webb space telescope / Amy Lo -- Starlight suppression: technologies for direct imaging of exoplanets / Dmitry Savransky -- Realizing large structures in space / Jeremy Banik -- Sensing controls for spaces-based planet finding / Jonathan T. Black -- Optical and mechanical metamaterials: introduction / Jennifer Dionne and Luke Sweatlock -- Materials by design: using architecture and nanomaterial size effects to attain unexplored properties / Julia Greer -- Mechanical metamaterials: design, fabrication, and performance / Christopher Spadaccini -- Metamaterial-based device engineering / Andrea Alù -- Catching light rays: refractory plasmonics for energy conversion, data storage, and medical applications / Alexandra Boltasseva and Urcan Guler --Forcasting natural disasters: introduction / James Done and Amir AghaKouchak -- An integrated approach to assess and manage hurricane risk in a changing climate / Ning Lin -- Moving from risk assessment to risk reduction: an economic perspective on decision making in natural disasters / Jeffrey Czajkowski.

Subjects Computer networks--Security measures--Congresses.
Data protection--Congresses.
Extrasolar planets--Congresses.
Metamaterials--Congresses.
Natural disasters--Forecasting--Congresses.
Computer networks--Security measures.
Data protection.
Extrasolar planets.
Metamaterials.
Natural disasters--Forecasting.

Form/Genre	Conference papers and proceedings. Conference papers and proceedings.
Notes	Includes bibliographical references.

Fundamentals and Applications of Nano Silicon in Plasmonics and Fullerines: Current and Future Trends

LCCN	2021275252
Type of material	Book
Personal name	Nayfeh, Munir H. (Munir Hasan), author.
Main title	Fundamentals and Applications of Nano Silicon in Plasmonics and Fullerines: Current and Future Trends / Munir Nayfeh.
Published/Produced	Amsterdam, Netherlands; Cambridge, MA, United States: Elsevier, [2018]
Description	xvi, 586 pages; 24 cm.
ISBN	0323480578 hardcover 9780323480574 hardcover
Subjects	Nanosilicon.
Series	Micro and nano technologies Micro & nano technologies.

Fundamentals of van der Waals and Casimir Interactions

LCCN	2019743874
Type of material	Book
Personal name	Sernelius, Bo E. author.
Main title	Fundamentals of van der Waals and Casimir Interactions / by Bo E. Sernelius.
Edition	1st ed. 2018.
Published/Produced	Cham: Springer International Publishing: Imprint: Springer, 2018.
Description	1 online resource (XVIII, 416 pages 109 illustrations, 78 illustrations in color.) text file PDF
ISBN	9783319998312
Summary	This book presents a self-contained derivation of van der Waals and Casimir type dispersion forces, covering the interactions between two atoms but also between microscopic, mesoscopic, and macroscopic objects of various shapes and materials. It also

presents detailed and general prescriptions for finding the normal modes and the interactions in layered systems of planar, spherical and cylindrical types, with two-dimensional sheets, such as graphene incorporated in the formalism. A detailed derivation of the van der Waals force and Casimir-Polder force between two polarizable atoms serves as the starting point for the discussion of forces: Dispersion forces, of van der Waals and Casimir type, act on bodies of all size, from atoms up to macroscopic objects. The smaller the object the more these forces dominate and as a result they play a key role in modern nanotechnology through effects such as stiction. They show up in almost all fields of science, including physics, chemistry, biology, medicine, and even cosmology. Written by a condensed matter physicist in the language of condensed matter physics, the book shows readers how to obtain the electromagnetic normal modes, which for metallic systems, is especially useful in the field of plasmonics.

Contents Introduction -- Part I - Background Material -- Electromagnetic -- Complex Analysis -- Statistical Physics -- Electromagnetic Normal Modes -- Different Approaches -- General Method to find the Normal Modes in Layered Structures -- Part II - Non-retarded Formalism: van der Waals -- Van der Waals Force -- Van der Waals Interaction in Planar Structures -- Van der Waals Interaction in Spherical Structures -- Van der Waals Interaction in Cylindrical Structures -- Part III - Fully Retarded Formalism: Casimir -- Casimir Interaction -- Dispersion Interaction in Planar Structures -- Dispersion Interaction in Spherical Structures -- Dispersion Interaction in Cylindrical Structures -- Summary and Outlook.

Subjects Atoms.
Physics.
Solid state physics.

	Solid State Physics.
	Atomic, Molecular, Optical and Plasma Physics.
	Mathematical Methods in Physics.
	Numerical and Computational Physics, Simulation.
Notes	Description based on publisher-supplied MARC data.
Additional formats	Print version: Fundamentals of Van der Waals and Casimir interactions. 9783319998305 (DLC) 2018953297
	Printed edition: 9783030076375
	Printed edition: 9783319998305
	Printed edition: 9783319998329
Series	Springer Series on Atomic, Optical, and Plasma Physics, 1615-5653; 102
	Springer Series on Atomic, Optical, and Plasma Physics, 1615-5653; 102

Graphene Science Handbook. Nanostructure and Atomic Arrangement

LCCN	2015042989
Type of material	Book
Main title	Graphene science handbook. Nanostructure and atomic arrangement / edited by Mahmood Aliofkhazraei, Nasar Ali, William I. Milne, Cengiz S. Ozkan, Stanislaw Mitura, and Juana L. Gervasoni.
Published/Produced	Boca Raton: CRC Press, Taylor & Francis Group, [2016]
	©2016
Description	xviii, 583 pages: illustrations; 29 cm
ISBN	9781466591370 (hardcover; alk. paper)
	1466591374 (hardcover; alk. paper)
LC classification	QD341.H9 G687 2016
Portion of title	Nanostructure and atomic arrangement
Related names	Aliofkhazraei, Mahmood, editor.
	Ali, Nasar, editor.
	Milne, W. I. (William I.), editor.
	Ozkan, Cengiz S., editor.
	Mitura, Stanislaw, 1951- editor.
	Gervasoni, Juana L., editor.
Summary	"Graphene is the strongest material ever studied and can be an efficient substitute for silicon. There is no

other major reference work of this scope on the topic of graphene, which is one of the most researched materials of the twenty-first century. The set includes contributions from top researchers in the field and a foreword written by two Nobel laureates in physics. This volume in the set focuses on nanostructure and atomic arrangement"-- Provided by publisher.

Contents Graphene heterostructures / Zheng Liu -- Atomic-scale defects and impurities on grapheme / R. Martinazzo -- Atomic arrangement and its effects on electronic structures of graphene from tight-binding description / Sirichok Jungthawan -- Graphene plasmonics: light-matter interactions at the atomic scale / Pai-Yen Chen -- Graphene/polymer nanocomposites: crystal structure, mechanical and thermal properties / Carlos Velasco -- Graphene-like structures as cages for doxorubicin / Vojtech Adam -- Mathematical modeling for hydrogen storage inside graphene-based materials / Yue Chan -- Morphology of cylindrical carbon nanostructures grown by catalytic chemical vapor deposition method / Subrata Ray -- sp2 to sp3 phase transformation in graphene-like nanofilms / Jinlong Yang -- Symmetry and topology of graphenes / A.R. Ashrafi -- N-doped graphene for supercapacitors / Dingsheng Yuan -- Electrical and optical properties and applications of doped graphene sheets / Kim Soo Young -- Chemical modifications of graphene via covalent bonding / Jingquan Liu -- Functionalization and vacancy effects on hydrogen binding in grapheme / A. Tapia -- Modifications of electronic properties of graphene by boron (B) and nitrogen (N) substitution / Debnarayan Jana -- Electronic structure and topological disorder in sp2 phases of carbon / David Drabold -- 3D macroscopic graphene assemblies / Marcus A. Worsley -- 3D-AFM-hyperfine imaging of graphene mono-layers deposit on YBCO-superconducting surface / Khaled Elsabawy -- Phonon spectrum and vibrational thermodynamic characteristics of graphene nanofilms / A. Feher -- Tuning atomic and electronic properties of graphene by selective doping / Cecile Malardier-Jugroot -- Scanning electron microscopy of graphene / Yoshikazu Homma -- Tunneling current of the contact of the curved graphene nanoribbon with metal and quantum dots / Natalia N. Yanyushkina -- Using few-layer graphene sheets as ultimate reference of quantitative transmission electron microscopy / Fengqi Song -- Computational modeling of graphene and carbon nanotube structures in the terahertz, near-infrared and optical regimes / Mario F. Pantoja -- Design and properties of graphene based three dimensional (3D) architectures / Ludovic Dumee -- Electronic structure of graphene based materials and their carrier transport properties / Liang Gengchiau -- Graphene-enabled heterostructures: role in future-generation carbon electronics / Bin Yu -- Recent progresses and understanding of lithium storage behavior of graphene nanosheet anode for lithium ion batteries / Xueliang A. Sun -- Study of the transmission, transport and electronic structure properties of periodic and aperiodic graphene-based structures / Isaac Rodriguez-Vargas -- Benefits of few-layer graphene structures for various applications / Irina Antonova -- Designing carbon-based thin films from graphene-like nanostructures / Cecilia Goyenola -- Graphene based hybrid composites / Antonio F. Avila -- Graphene dispersion by polymers and hybridization with nanoparticles / Jiang-Jen Lin -- Magnetocaloric effect of graphenes / Mario Reis -- Mode-locked of fiber laser employing graphene based saturable absorber / Isabella Huang.

Subjects Graphene--Handbooks, manuals, etc.
Nanostructured materials--Handbooks, manuals, etc.

Notes Includes bibliographical references and index.

High field plasmonics

LCCN 2016948806

Type of material Book
Main title High field plasmonics.
Published/Produced New York, NY: Springer Berlin Heidelberg, 2016.
Links Contributor biographical information https://www.loc.gov/catdir/enhancements/fy1619/2016948806-b.html
Publisher description https://www.loc.gov/catdir/enhancements/fy1619/2016948806-d.html
Table of contents only https://www.loc.gov/catdir/enhancements/fy1619/2016948806-t.html
ISBN 9783319442891

High Field Plasmonics
LCCN 2019765981
Type of material Book
Personal name Fedeli, Luca, author.
Main title High Field Plasmonics / by Luca Fedeli.
Edition 1st ed. 2017.
Published/Produced Cham: Springer International Publishing: Imprint: Springer, 2017.
Description 1 online resource (XIX, 180 pages 83 illustrations, 30 illustrations in color.)
text file PDF
ISBN 9783319442907
Summary This thesis describes pioneering research on the extension of plasmonics schemes to the regime of high-intensity lasers. By presenting a rich and balanced mix of experimentation, theory and simulation, it provides a comprehensive overview of the emerging field of high field plasmonics, including open issues and perspectives for future research. Combining specially designed targets and innovative materials with ultrashort, high-contrast laser pulses, the author experimentally demonstrates the effects of plasmon excitation on electron and ion emission. Lastly, the work investigates possible further developments with the help of numerical simulations, revealing the potential of plasmonics effects in the relativistic regime for advances in laser-driven

	sources of radiation, and for the manipulation of extreme light at the sub-micron scale.
Contents	High Intensity Laser-Plasma Interaction and High Field Plasmonics -- Numerical Tools -- Electron Acceleration with Grating Targets -- Foam Targets for Enhanced Ion Acceleration -- Numerical Exploration of High Field Plasmonics in Different Scenarios.
Subjects	Lasers. Particle acceleration. Photonics. Physics. Plasma (Ionized gases). Optics, Lasers, Photonics, Optical Devices. Numerical and Computational Physics, Simulation. Particle Acceleration and Detection, Beam Physics. Plasma Physics.
Notes	Description based on publisher-supplied MARC data.
Additional formats	Print version: High field plasmonics. 9783319442891 (DLC) 2016948806 Printed edition: 9783319442891 Printed edition: 9783319442914 Printed edition: 9783319830315
Series	Springer Theses, Recognizing Outstanding Ph.D. Research, 2190-5053 Springer Theses, Recognizing Outstanding Ph.D. Research, 2190-5053

Introduction to Nanophotonics

LCCN	2021945081
Type of material	Book
Personal name	Benisty, Henri, author.
Main title	Introduction to nanophotonics / Henri Benisty, Jean-Jacques Greffet, Philippe Lalanne.
Published/Produced	New York: Oxford University Press, 2022.
ISBN	9780198786139 (hardback)
Related names	Greffet, Jean-Jacques, author. Lalanne, Philippe, author.

Summary	"The aim of this book is to cover the scope of Nanophotonics, a discipline that has emerged around the turn of the millennium. It results from the merge of different communities working in different aspects of light-matter interaction at the nanoscale. These include near-field optics and super-resolution microscopy, photonic crystals, diffractive optics, plasmonics, optoelectronics, synthesis of metallic and semiconductor nanoparticles, two-dimensional materials and metamaterials. Our feeling when we started the project was that a book covering most of these aspects altogether was lacking. The field is so rapidly evolving that it is impossible to summarize all the recent breakthroughs. The goal of this book is to provide a self-contained discussion of the fundamentals of the different subfields involved in nanophotonics. The current project is a collaborative project between three researchers that have been actively involved in the field from different communities. Henri Benisty has a background in semiconductor physics and optoelectronics, Jean-Jacques Greffet has a background in near-field optics and light scattering, Philippe Lalanne has a background in diffractive optics and photonic crystals. All of them made significant contributions to the advancement of the field. The book material is based on lectures that have been given by them at the Institut d'Optique Graduate School (Palaiseau, Bordeaux and Saint-Etienne)"-- Provided by publisher.
Series	Oxford graduate texts

Introduction to Nanotechnology

LCCN	2021949921
Type of material	Book
Personal name	Bertino, Massimo, author.
Main title	Introduction to nanotechnology / Massimo Bertino.
Published/Produced	Hackensack: World Scientific, 2021.
ISBN	9789811231605 (hardcover)

	9789811233036 (paperback)
	(ebook)
	(mobi)
Summary	"This textbook is conceived for a one-semester course at the upper undergraduate or freshman graduate level. The book was written With the fact that nanotechnology is a vast field where the applications range from paint to nanomedicine, through plasmonics and catalysis. An introductory course must be a compromise between a quantitative and a qualitative treatment. For that, this textbook is more quantitative than others in the market, which often do not treat the key concepts with enough depth. This textbook focuses on the key physical and chemical principles and uses many formulas and equations within with the one-semester time constraint"-- Provided by publisher.

Metal Nanoparticles and Clusters: Advances in Synthesis, Properties and Applications

LCCN	2019759674
Type of material	Book
Main title	Metal Nanoparticles and Clusters: Advances in Synthesis, Properties and Applications / edited by Francis Leonard Deepak.
Edition	1st ed. 2018.
Published/Produced	Cham: Springer International Publishing: Imprint: Springer, 2018.
Description	1 online resource (XII, 426 pages 196 illustrations, 169 illustrations in color.) text file PDF
ISBN	9783319680538
Related names	Deepak, Francis Leonard. editor.
Summary	This book covers the continually expanding field of metal nanoparticles and clusters, in particular their size-dependent properties and quantum phenomena. The approaches to the organization of atoms that form clusters and nanoparticles have been advancing rapidly in recent times. These advancements are

	described through a combination of experimental and computational approaches and are covered in detail by the authors. Recent highlights of the various emerging properties and applications ranging from plasmonics to catalysis are showcased.
Contents	From Nano- to Angstrom- technology: Introduction to Metal Nanoparticles and Clusters -- Recent Developments in the Synthesis of Metal Nanocrystals -- Structure of Metal Nanoparticles -- Spectroscopic and computational studies of metal nanoparticles and clusters -- Surface enhanced spectroscopies: Optical Techniques and Plasmonic Nanostructures -- Surface Plasmons in Quantum-sized noble-metal clusters-Theoretical Aspects -- Nanoparticles in supported and colloidal catalysis -- Catalytic efficiency of metallic nanoparticles-Theoretical Aspects -- Advanced electron microscopy of nanoalloys -- Imaging clusters under an electron microscope -- Thermodynamics of Metal Nanoparticles -- Gold and silver fluorescent nanoclusters as an emerging probe for chemical and biological sensors -- Plasmonic and NIR light sensitive nanomaterials for photothermal cancer therapy.
Subjects	Lasers. Metals. Nanochemistry. Nanotechnology. Photonics. Metallic Materials. Nanochemistry. Nanotechnology and Microengineering. Optics, Lasers, Photonics, Optical Devices.
Notes	Description based on publisher-supplied MARC data.
Additional formats	Print version: Metal nanoparticles and clusters: advances in synthesis, properties and applications. 9783319680521 (DLC) 2017958015 Printed edition: 9783319680521 Printed edition: 9783319680545 Printed edition: 9783319885384

Metal Nanostructures for Photonics

LCCN	2020304433
Type of material	Book
Main title	Metal nanostructures for photonics / edited by Luciana Reyes Pires Kassab, Cid Bartolomeu de Araujo; series editor, David L. Andrews.
Published/Produced	Amsterdam, Netherlands; Cambridge, MA: Elsevier, [2019]
Description	xix, 338 pages: illustrations; 24 cm.
ISBN	9780081023785 (paperback) (ePub ebook)
LC classification	TA418.9.N35 M5243 2019
Related names	Kassab, Luciana Reyes Pires, editor. Araujo, Cid Bartolomeu de, editor.
Contents	Metal-Dielectric Nanocomposites Based on Germanate and Tellurite Glasses / Luciana R.P. Kassab, Cid B. de Araújo -- Effects of Spherical Metallic Nanoparticle Plasmon on 4f-4f Luminescence: A Theoretical Approach / Albano N. Carneiro Neto, Marcos A. Couto dos Santos, Oscar L. Malta, Renata Reisfeld -- Nonlinear Optics With Metal-Dielectric Nanocomposites / Raúl Rangel-Rojo, Héctor Sánchez-Esquivel, Bonifacio Can-Uc, A. Crespo-Sosa, A. Oliver -- High-Order Nonlinearities of Metal-Dielectric Nanocomposites / Albert S. Reyna, Cid B. de Araújo -- Optical Response of Nanostructures: From Pure to Alloyed Metals / Zackery A. Benson, Chen Gong, Marina S. Leite -- Multipolar Second Harmonic Generation From Metallic Nanoparticles / Jérémy Butet, Anthony Maurice, Emeric Bergmann, Guillaume Bachelier, Isabelle Russier-Antoine, Cédric Ray, Oriane Bonhomme, Christian Jonin, Emmanuel Benichou, Pierre-François Brevet -- A Numerical Study of Plasmonic Nanostructures for Linear and Nonlinear Quantum Elements / Günter Kewes, Oliver Benson -- Fabrication of Metal Dielectric Nanocomposites by Ion Implantation and

	Characterization by Nonlinear Optics Techniques / Andrey L. Stepanov -- Nanostructuring Solar Cells Using Metallic Nanoparticles / Brendan Brady, Peng Hui Wang, Volker Steenhoff, Alexandre G. Brolo -- Plasmonics for the Characterization of Metal Organic Films and Nanoparticles / Tommaso Del Rosso, Walter Margulis, Jake Fontana, Isabel C.S. Carvalho -- Biopolymer-Metal Composites / Robson R. Silva, Sidney J.L. Ribeiro, Hernane S. Barud, Helida O. Barud, Osvaldo N. Oliveira, Jr, J.R. Mejía-Salazar -- Pedestal Doped Waveguides for Infrared Light Amplification / Luciana R.P. Kassab, Davinson M. da Silva.
Subjects	Nanostructured materials. Nanophotonics. Metals.
Notes	Includes bibliographical references and index.
Series	Nanophotonics series

Molecular Electronic Control over Tunneling Charge Transfer Plasmons Modes

LCCN	2019767794
Type of material	Book
Personal name	Tan, Shu Fen, author.
Main title	Molecular Electronic Control Over Tunneling Charge Transfer Plasmons Modes / by Shu Fen Tan.
Edition	1st ed. 2018.
Published/Produced	Singapore: Springer Singapore: Imprint: Springer, 2018.
Description	1 online resource (XXXIV, 115 pages 55 illustrations, 46 illustrations in color.) text file PDF
ISBN	9789811088032
Summary	This thesis describes the controlled immobilization of molecules between two cuboidal metal nanoparticles by means of a self-assembly method to control the quantum plasmon resonances. It demonstrates that quantum-plasmonics is possible at length scales that are useful for real applications. Light can interact

with certain metals and can be captured in the form of plasmons, which are collective, ultra-fast oscillations of electrons that can be manipulated at the nano-scale. Surface plasmons are considered as a promising phenomenon for potentially bridging the gap between fast-operating-speed optics and nano-scale electronics. Quantum tunneling has been predicted to occur across two closely separated plasmonic resonators at length scales (<0.3 nm) that are not accessible using present-day nanofabrication techniques. Unlike top-down nanofabrication, the molecules between the closely-spaced metal nanoparticles could control the gap sizes down to sub-nanometer scales and act as the frequency controllers in the terahertz regime, providing a new control parameter in the fabrication of electrical circuits facilitated by quantum plasmon tunneling.

Contents General Introduction -- Plasmonics Properties, Stability and Chemical Reactivity of Metal Nanoparticles - A Literature Review -- Self-Assembly of Silver Nanoparticles with Sub-Nanometer Separation -- Quantum Plasmon Resonances Controlled by Molecular Tunnel Junctions -- Stability of Gold and Silver Nanoparticles under Electron Beam Irradiation -- Real-Time Imaging of Chemical Reaction between Gold and Silver Nanoparticles -- Real-Time Imaging of Au@Ag Core-Shell Nanoparticles Formation -- General Conclusions and Outlook.

Subjects Microwaves.
Nanochemistry.
Nanotechnology.
Optical engineering.
Quantum physics.
Nanochemistry.
Microwaves, RF and Optical Engineering.
Nanotechnology.
Quantum Physics.

Notes Description based on publisher-supplied MARC data.

Additional formats	Print version: Molecular electronic control over tunneling charge transfer plasmons modes. 9789811088025 (DLC) 2018939945 Printed edition: 9789811088025 Printed edition: 9789811088049 Printed edition: 9789811342448
Series	Springer Theses, Recognizing Outstanding Ph.D. Research, 2190-5053 Springer Theses, Recognizing Outstanding Ph.D. Research, 2190-5053

Multifunctional Gold Nanostars for Cancer Theranostics

LCCN	2019746542
Type of material	Book
Personal name	Liu, Yang. author.
Main title	Multifunctional Gold Nanostars for Cancer Theranostics / by Yang Liu.
Edition	1st ed. 2018.
Published/Produced	Cham: Springer International Publishing: Imprint: Springer, 2018.
Description	1 online resource (XX, 71 pages 28 illustrations in color.) text file PDF
ISBN	9783319749204
Summary	This thesis presents the development of theranostic gold nanostars (GNS) for multimodality cancer imaging and therapy. Furthermore, it demonstrates that a novel two-pronged treatment, combining immune-checkpoint inhibition and GNS-mediated photothermal nanotherapy, can not only eradicate primary treated tumors but also trigger immune responses to treat distant untreated tumors in a mouse animal model. Cancer has become a significant threat to human health with more than eight million deaths each year, and novel methods for cancer management to improve patients' overall survival are urgently needed. The developed multifunctional GNS nanoprobe with tip-enhanced plasmonics in the near-infrared region can be combined with (1) surface-

	enhanced Raman spectroscopy (SERS), (2) two-photon photoluminescence (TPL), (3) X-ray computed tomography (CT), (4) magnetic resonance imaging (MRI), (5) positron emission tomography (PET), and (6) photothermal therapy (PTT) for cancer imaging and treatment. The ability of the GNS nanoprobe to detect submillimeter intracranial brain tumors was demonstrated using PET scan - a superior non-invasive imaging modality - in a mouse animal model. In addition, delayed rechallenge with repeated cancer cell injection in cured mice did not lead to new tumor formation, indicating generation of a memorized immune response to cancer. The biocompatible gold nanostars with superior capabilities for cancer imaging and treatment have great potential for translational medicine applications.
Contents	Introduction -- Multifunctional GNS nanoprobe development and characterization -- In vivo evaluation of GNS nanoprobe -- GNS toxicity investigation -- Sensitive brain tumor detection using GNS nanoprobe -- Photoimmunotherapy for cancer metastasis treatment -- Conclusion and future outlook.
Subjects	Biomedical materials. Biomedical engineering. Cancer research. Medicinal chemistry. Nanochemistry. Spectroscopy. Nanochemistry. Biomaterials. Biomedical Engineering and Bioengineering. Cancer Research. Medicinal Chemistry. Spectroscopy/Spectrometry.
Notes	Description based on publisher-supplied MARC data.

Additional formats	Print version: Multifunctional gold nanostars for cancer theranostics. 9783319749198 (DLC) 2018931449 Printed edition: 9783030091040 Printed edition: 9783319749198 Printed edition: 9783319749211
Series	Springer Theses, Recognizing Outstanding Ph.D. Research, 2190-5053 Springer Theses, Recognizing Outstanding Ph.D. Research, 2190-5053

Nanoengineering: The Skills and Tools Making Technology Invisible

LCCN	2019304381
Type of material	Book
Personal name	Berger, Michael (Founder of Nanowerk), author.
Main title	Nanoengineering: the skills and tools making technology invisible / by Michael Berger.
Published/Produced	London, UK: Royal Society of Chemistry, [2020] ©2020
Description	xvi, 430 pages: illustrations (some color); 25 cm
ISBN	9781788018678 (hardback) 1788018672 (hardback)
LC classification	T174.7 .B474 2020
Summary	"While our five senses are doing a reasonably good job at representing the world around us on a macro-scale, we have no existing intuitive representation of the nanoworld, ruled by laws entirely foreign to our experience. Following in the footsteps of Nano-Society and Nanotechnology: The Future is Tiny, this title introduces a new collection of stories demonstrating recent research in the field of nanotechnology. The book gives a personal perspective on how nanotechnologies are created and developed, and will appeal to anyone who has an interest in the research and future of nanotechnology." -- Amazon.
Contents	The world of graphene -- The growing landscape of two-dimensional materials -- Not found in nature: metamaterials and metasurfaces -- Plasmonics --

	Nanobiotechnology --Nanomedicine -- Characterization -- Fabrication moves to the printer -- (Mostly wearable) electronics -- Sensors and diagnostics -- Energy harvesting and storage -- For nanoengineers, anything goes.
Subjects	Nanotechnology. Nanostructured materials. Nanostructured materials. Nanotechnology.
Notes	Includes bibliographical references and index.

Nanomaterials and Nanoarchitectures: A Complex Review of Current Hot Topics and their Applications

LCCN	2019764900
Type of material	Book
Main title	Nanomaterials and Nanoarchitectures: A Complex Review of Current Hot Topics and their Applications / edited by M. Bardosova, T. Wagner.
Edition	1st ed. 2015.
Published/Produced	Dordrecht: Springer Netherlands: Imprint: Springer, 2015.
Description	1 online resource (XIV, 343 pages 230 illustrations, 132 illustrations in color.) text file PDF
ISBN	9789401799218
Related names	Bardosova, M, editor. Wagner, T, editor.
Summary	The current work consists of nine contributions describing recent progress in the interdisciplinary of Nanoscience, which involves physics, chemistry, engineering, biology and medicine and one essay outlining some important historical and socioeconomic factors pertaining to recent developments in nanoscale science and technology. All 10 chapters have been written by eminent experts in their respective fields. The authors employ the terms 'nanomaterials' as building blocks of a range of materials, 'nanoarchitecture' represents the design and 'nanotechnology' the means to produce a

particular device or functionality. Two of the chapters are devoted to novel materials and two others focus on analyzing techniques, which can be used to enable molecular control of the film architecture. Additionally, the reader will find material devoted to photonic and hybrid plasmonic-photonic crystals as well as sections which address their applications, such as the use of plasmonic particles and nanostructures for new sensing concepts and ultrasensitive detection techniques. This work will be of interest to graduate students, researchers and practitioners alike.

Contents Colloidal photonic crystal films: fabrication and tunable structural color and applications -- Plasmonics and Ultrasensitive Detection -- Functional Nanomaterials Prepared by Nanoarchitectonics-Based Supramolecular Assembly -- A brief history of nanoscience and foresight in nanotechnology -- Halloysite clay nanotube composites with sustained release of chemicals -- Colloidal Photonic Crystal Architectures for Advanced Light Management Applications -- Crystalline and Amorphous Chalcogenides, High-Tech Materials with Structural Disorder and Many Important Applications -- Plasmonic gas and chemical sensing -- Planar Hybrid Plasmonic-Photonic Crystals -- Fundamentals and Applications of Organised Molecular Films.

Subjects Interfaces (Physical sciences).
Lasers.
Nanochemistry.
Nanotechnology.
Photonics.
Surfaces (Physics).
System safety.
Thin films.
Nanochemistry.
Nanotechnology.
Optics, Lasers, Photonics, Optical Devices.

	Security Science and Technology.
	Surface and Interface Science, Thin Films.
Notes	Description based on publisher-supplied MARC data.
Additional formats	Print version: Nanomaterials and nanoarchitectures: a complex review of current hot topics and their applications 9789401799201 (DLC) 2015946315
	Printed edition: 9789401799201
	Printed edition: 9789401799225
	Printed edition: 9789401799379
Series	NATO Science for Peace and Security Series C: Environmental Security, 1874-6519
	NATO Science for Peace and Security Series C: Environmental Security, 1874-6519

Nano-Optics: Principles Enabling Basic Research and Applications

LCCN	2019742389
Type of material	Book
Main title	Nano-Optics: Principles Enabling Basic Research and Applications / edited by Baldassare Di Bartolo, John Collins, Luciano Silvestri.
Edition	1st ed. 2017.
Published/Produced	Dordrecht: Springer Netherlands: Imprint: Springer, 2017.
Description	1 online resource (XVIII, 584 pages 222 illustrations, 175 illustrations in color.)
	text file PDF
ISBN	9789402408508
Related names	Collins, John. editor.
	Di Bartolo, Baldassare. editor.
	Silvestri, Luciano. editor.
Summary	This book provides a comprehensive overview of nano-optics, including basic theory, experiment and applications, particularly in nanofabrication and optical characterization. The contributions clearly demonstrate how advances in nano-optics and photonics have stimulated progress in nanoscience and -fabrication, and vice versa. Their expert authors address topics such as three-dimensional optical lithography and microscopy beyond the Abbe

diffraction limit, optical diagnostics and sensing, optical data- and telecommunications, energy-efficient lighting, and efficient solar energy conversion. Nano-optics emerges as a key enabling technology of the 21st century. This work will appeal to a wide readership, from physics through chemistry, to biology and engineering. The contributions that appear in this volume were presented at a NATO Advanced Study Institute held in Erice, 4-19 July, 2015.

Contents Part I: Lectures -- Light-Matter Interactions: A Coupled Oscillator Description -- Luminescence Spectroscopy of Nanophosphors -- Nanomaterials: basic concepts and quantum models -- Non-Radiative Processes In Nanocrystals -- 3d Optical Laser Lithography -- Nanostructures And Nanocrystals With Radiation Color Centers: Characteristics And Some Applications -- Colloidal Nanophotonics: State-Of-The-Art And Prospective -- Ultrafast Nano-Biophotonics -- Circuit Optomechanics With Diamond Integrated Optical Devices -- Terahertz Sensing At The Nanoscale -- How Latitude Location On A Micro-World Enables Real-Time Nanoparticle Sizing -- Nanoplasmonic And Microfluidic Devices For Biological Sensing -- High-Throughput And Ultra-Sensitive Biosensing And Spectroscopy By Plasmonics -- Photoemission From Nanomaterials In Strong Few-Cycle Laser Fields -- Luminescence Of Er 3+ Ions In Nanocrystalline Glass-Ceramics -- Localization Of Yb3+, Er3+ And Co2+ Dopants In An Optical Glass Ceramic Of Mgal2o4 Spinel Nano-Crystals Embedded In Sio2 Glass -- Nd3+, Eu3+ And Yb3+ Ions As Structural Probes In The Scheelite-Type Cadmium Molybdates With Vacancies -- Medical Applications Of Nanomaterials -- Emission Cross Section, Füchtbauer-Ladenburg Equation, And Purcell Factor -- Surface Plasmon Enhanced Fluorescence Of Glycine-Dimer-Functionalized Silver Nanoparticles -- Part II: Interdisciplinary

	Lecturers -- Andrea Pozzo: The Art of Perspective -- Part III: Short Seminars -- Part IV: Posters -- Part V: List of Participants -- Index.
Subjects	Lasers.
	Microscopy.
	Nanoscale science.
	Nanoscience.
	Nanostructures.
	Photonics.
	Spectroscopy.
	System safety.
	Nanoscale Science and Technology.
	Optics, Lasers, Photonics, Optical Devices.
	Security Science and Technology.
	Spectroscopy and Microscopy.
Notes	Description based on publisher-supplied MARC data.
Additional formats	Print version: Nano-optics: principles enabling basic research and applications 9789402408485 (DLC) 2017931968
	Printed edition: 9789402408485
	Printed edition: 9789402408492
	Printed edition: 9789402408690
Series	NATO Science for Peace and Security Series B: Physics and Biophysics, 1874-6500
	NATO Science for Peace and Security Series B: Physics and Biophysics, 1874-6500

Nanophotonics and Plasmonics: An Integrated View

LCCN	2017002486
Type of material	Book
Main title	Nanophotonics and plasmonics: an integrated view / edited by, Ching Eng Png, Yuriy Akimov.
Published/Produced	Boca Raton: CRC Press, Taylor & Francis Group, [2017]
	©2017
Description	xii, 348 pages; 24 cm.
ISBN	9781498758673 (hardback; alk. paper)
	1498758673 (hardback; alk. paper)
LC classification	TA1530 .N364 2017

Related names	Png, Ching Eng, 1974- editor. Akimov, Yuriy, 1979- editor.
Subjects	Nanophotonics. Plasmonics. Nanostructured materials--Optical properties.
Notes	Includes bibliographical references and index.
Series	Series in optics and optoelectronics Series in optics and optoelectronics (CRC Press)

Nanophotonics

LCCN	2019769281
Type of material	Book
Personal name	McGurn, Arthur, author.
Main title	Nanophotonics / by Arthur McGurn.
Edition	1st ed. 2018.
Published/Produced	Cham: Springer International Publishing: Imprint: Springer, 2018.
Description	1 online resource (XII, 558 pages 98 illustrations, 10 illustrations in color.) text file PDF
ISBN	9783319770727
Summary	This book gives a readable introduction to the important, rapidly developing, field of nanophotonics. It provides a quick understanding of the basic elements of the field, allowing students and newcomers to progress rapidly to the frontiers of their interests. Topics include: The basic mathematical techniques needed for the study of the materials of nanophotonic technology; photonic crystals and their applications as laser resonators, waveguides, and circuits of waveguides; the application of photonic crystals technology in the design of optical diodes and transistors; the basic properties needed for the design and understanding of new types of engineered materials known as metamaterials; and a consideration of how and why these engineered materials have been formulated in the lab, as well as their applications as negative refractive index materials, as perfect lens, as cloaking devices, and

their effects on Cherenkov and other types of radiation. Additionally, the book introduces the new field of plasmonics and reviews its important features. The role of plasmon-polaritons in the scattering and transmission of light by rough surfaces and the enhanced transmission of light by plasmon-polariton supporting surfaces is addressed. The important problems of subwavelength resolution are treated with discussions of applications in a number of scientific fields. The basic principles of near-field optical microscopy are presented with a number of important applications. The basics of atomic cavity physics, photonic entanglement and its relation to some of the basic properties of quantum computing, and the physics associated with the study of optical lattices are presented.

Contents General Introduction and overview of the text: Qualitative remarks about the systems and effects found in nanophotonics and their important technological applications -- Mathematical preliminaries -- Photonic crystals -- Metamaterials -- Plasmonics -- Subwavelength focusing -- Near-field scanning optical microscopy -- Optical tweezers -- Trapped atoms -- Discussions of some of the basic techniques of experimental implementation of the ideas of nanophotonics.

Subjects Electrodynamics.
Electronic materials.
Lasers.
Microwaves.
Optical engineering.
Optical materials.
Optics.
Photonics.
Semiconductors.
Optics, Lasers, Photonics, Optical Devices.
Classical Electrodynamics.
Microwaves, RF and Optical Engineering.
Optical and Electronic Materials.

	Semiconductors.
Notes	Description based on publisher-supplied MARC data.
Additional formats	Print version: Nanophotonics. 9783319770710 (DLC) 2018933512
	Printed edition: 9783030083625
	Printed edition: 9783319770710
	Printed edition: 9783319770734
Series	Springer Series in Optical Sciences, 0342-4111; 213
	Springer Series in Optical Sciences, 0342-4111; 213

Nanoscale Materials and Devices for Electronics, Photonics and Solar Energy

LCCN	2019770054
Type of material	Book
Main title	Nanoscale Materials and Devices for Electronics, Photonics and Solar Energy / edited by Anatoli Korkin, Stephen Goodnick, Robert Nemanich.
Edition	1st ed. 2015.
Published/Produced	Cham: Springer International Publishing: Imprint: Springer, 2015.
Description	1 online resource (XII, 282 pages 135 illustrations, 65 illustrations in color.)
	text file PDF
ISBN	9783319186337
Related names	Goodnick, Stephen, editor.
	Korkin, Anatoli, editor.
	Nemanich, Robert, editor.
Summary	This book presents research dedicated to solving scientific and technological problems in many areas of electronics, photonics and renewable energy. Progress in information and renewable energy technologies requires miniaturization of devices and reduction of costs, energy and material consumption. The latest generation of electronic devices is now approaching nanometer scale dimensions; new materials are being introduced into electronics manufacturing at an unprecedented rate; and alternative technologies to mainstream CMOS are evolving. The low cost of natural energy sources have

created economic barriers to the development of alternative and more efficient solar energy systems, fuel cells and batteries. Nanotechnology is widely accepted as a source of potential solutions in securing future progress for information and energy technologies. Nanoscale Materials and Devices for Electronics, Photonics and Solar Energy features chapters that cover the following areas: atomic scale materials design, bio- and molecular electronics, high frequency electronics, fabrication of nanodevices, magnetic materials and spintronics, materials and processes for integrated and subwave optoelectronics, nanoCMOS, new materials for FETs and other devices, nanoelectronics system architecture, nano optics and lasers, non-silicon materials and devices, chemical and biosensors, quantum effects in devices, nano science and technology applications in the development of novel solar energy devices, and fuel cells and batteries.

Contents Ultra-Low-Power Device Operation -- Ultra-Low-Power Pseudospintronics Devices via Exciton Condensation in Coupled Two-Dimensional Material Systems -- Graphene-Based Photonics and Plasmonics -- Materials Challenges for Concentrated Solar Power -- Atomistic Simulations of Electronic and Optical Properties of Semiconductor Nanostructures -- What Happens When Molecules Meet Nanostructures: The Convergence of Chemistry and Electronics at the Nanoscale -- Terahertz-Wave Generation Using Graphene and Compound Semiconductor Nano-Heterostructures -- Optics of Hybrid Nanomaterials in the Strong Coupling Regime.

Subjects Electronic materials.
Lasers.
Nanotechnology.
Optical materials.
Photonics.
Renewable energy sources.

Nanotechnology.
Nanotechnology and Microengineering.
Optical and Electronic Materials.
Optics, Lasers, Photonics, Optical Devices.
Renewable and Green Energy.
Renewable and Green Energy.

Notes Description based on publisher-supplied MARC data.

Additional formats Print version: Nanoscale materials and devices for electronics, photonics and solar energy 9783319186320 (DLC) 2015945943
Printed edition: 9783319186320
Printed edition: 9783319186344
Printed edition: 9783319371184

Series Nanostructure Science and Technology, 1571-5744
Nanostructure Science and Technology, 1571-5744

Nano-Structures for Optics and Photonics: Optical Strategies for Enhancing Sensing, Imaging, Communication and Energy Conversion

LCCN 2019751144

Type of material Book

Main title Nano-Structures for Optics and Photonics: Optical Strategies for Enhancing Sensing, Imaging, Communication and Energy Conversion / edited by Baldassare Di Bartolo, John Collins, Luciano Silvestri.

Edition 1st ed. 2015.

Published/Produced Dordrecht: Springer Netherlands: Imprint: Springer, 2015.

Description 1 online resource (XXXVI, 586 pages 253 illustrations, 167 illustrations in color.)
text file PDF

ISBN 9789401791335

Related names Collins, John. editor.
Di Bartolo, Baldassare. editor.
Silvestri, Luciano. editor.

Summary The contributions in this volume were presented at a NATO Advanced Study Institute held in Erice, Italy, 4-19 July 2013. Many aspects of important research into nanophotonics, plasmonics, semiconductor materials and devices, instrumentation for bio sensing to name just a few, are covered in depth in this volume. The growing connection between optics and electronics, due to the increasing important role plaid by semiconductor materials and devices, find their expression in the term photonics, which also reflects the importance of the photon aspect of light in the description of the performance of several optical systems. Nano-structures have unique capabilities that allow the enhanced performance of processes of interest in optical and photonic devices. In particular these structures permit the nanoscale manipulation of photons, electrons and atoms; they represent a very hot topic of research and are relevant to many devices and applications. The various subjects bridge over the disciplines of physics, biology and chemistry, making this volume of interest to people working in these fields. The emphasis is on the principles behind each technique and on examining the full potential of each technique.

Contents Part I Lectures -- Nanoplasmonics: Fundamentals and Applications -- Transformation Optics -- Localized Light-Matter Interactions with Optical Antennas -- Biosensing Instrumentation -- Ultrafast Photonics with Microstructures Fibers -- Prospects of Molecular Scale Logic Gates and Logic Circuits -- Tunability of Plasmonic Devices -- Novel Plasmonic Probes and Smart Superhydrophobic Devices, New Tools for Forthcoming Spectroscopies at the Nanoscale -- Narrow-linewidth Lasers on a Silicon Chip -- Integrated Optomechanics: Opportunities for Tunable Nanophotonic Devices -- Fluorescence-Based Studies of Nanoparticles -- Incandescent Lamp-Like White-Light Emission from Doped and Undoped Oxide Nanopowders -- An Approach in the Structural and Spectroscopic Analysis of Yb3+-doped YAG Nano-ceramics by Conjugation of TEM-EDX and Optical Techniques -- Taking Microcavity Label-free Single Molecule Detection deep into the

Protein Realm -- Fluorescence and Raman Scattering in Plasmonic Nano-structures: From Basic Science to Applications -- Principles and Applications of Rare Earth Ion-doped Nanoparticles -- Nanostructures in the Terahertz Range -- Nanomaterials for Lighting and Solar Energy Conversion -- Part II Interdisciplinary Lecture -- The Discovery of the Higgs Particle -- Part III Short Seminars -- Chemical and Strain Engineering of Functional Oxides -- Gold Photonic Crystals and Photonics Quasi-Crystals for Reproducible Surface-Enhanced Raman Substrates -- Mid-Infrared Surface Plasmon Polariton Sensors Resonant with the Vibrational Modes of Phospholipid Layers -- Structural and Spectroscopic Properties of Er3+:CdNb2O6 Phosphors -- Antibody Anchoring on QCM Gold Surfaces by UV Based Strategy -- Spectroscopic Properties of New Cubic Tungstates Doped with Eu3+ and Yb3+ Ions -- The Role of Surface Plasmon Resonance in Enhanced Transmission through Metallic Gratings -- Experimental Observation of Self-organized Nanostructures in Layered Crystals -- Nano-structuring for Molecular Motor Control -- Fingerprint Imaging Enhancement by Deposition of Columnar Thin Films Nanostructures -- Analytic Solution of the Rabi model -- Microscroll Invisibility Cloak -- The Identification of the Inorganic Pigments in the Cultural Heritage Objects using Surface-enhanced Raman Scattering -- Polymeric Microresonators for Biosensing Applications -- On the Possibility of Indicating Protein Conformational Changes via Energy Transfer between Intrinsic Fluorophores -- Part IV Posters -- Surface Plasmons and Strong Light-Matter Coupling in Metallic Nanoshells -- Photoluminescence and Electroluminescence Characterization of a-Si:H/c-Ci Interfaces -- Structural, Optoelectronic and Electrical Properties of GaAs Microcrystals Grown from (001) Si Nano-areas -- Plasmonic Sensors for Aromatic

Hydrocarbon Detection -- Optical Annealing of Black Silicon -- Strongly Confined Gap Plasmon Modes in Graphene Sandwiches and Graphene-on-silicon -- Investigation of Polarized Light Emitting Diode -- Linking Optical Properties and Nanostructure of NiCrOx Cermet Nanocomposite for Solar Thermal Application -- Ultrafast Active Control of Plasmonic Resonances at THz Frequencies -- An Eigenvector Expansion Method for Localized Plasmon Modes: Application to Extinction and Electron Energy Loss Spectra of Isolated and Coupled Metallic Nanoparticles -- Characterisations of New Nd3+-Doped Scheelite-Type Molybdates for Laser Materials -- Cluster Implantation and Deposition Apparatus: Design and Capabilities -- Roles of Surface Pattern Morphology and Sunlight Incoherence on Solar Cell Optimization -- Processing and Characterization of Nanowire Arrays for Photodetectors -- Structural Analysis of Ge-Ga-Se/S-CsCl Glasses and Glass-ceramics for Application in Chalcogenide and Chalcohalide Photonics -- Regularities in Holographic Formation of Periodic Structures in the System Polymer-Metal Nanoparticles -- Modeling the Effect of Different Dimensions in High Contrast Grating Mirror -- InAs/GaAs Quantum Dots Covered by Graded GaAsSb Strain Reducing Layer -- Compact Wavelength- and Pulse-Duration-Tunable Ultrafast Laser System for Coherent Raman Microscopy -- Random Lasing in ZnSe and CdSe Semiconductor Powders -- Design of an Ultra-Broadband Super Absorbing Plasmonic Metamaterials -- High-order Harmonic Generation and Plasmonics -- Low Dispersion Propagation of Broad-band THz Pulses in a Two-Wire Waveguide -- Femtosecond Lase Micromachining of a-Si:H -- Low-level Monoclinic Content Detection in Zirconia Implants using Raman Spectroscopy -- Modeling Additive Color Effect in Natural Photonic Polycrystals using the Layer

Homogenization Method: the Case of the Diamond Weevil -- Transformation of Optical Properties for gamma-Irradiated Lithium Fluoride Crystals under Transition from Bulk to Nanocrystals -- Characterization of Polyethylene Glycol Self-Assembled Monolayers by means of Sum-Frequency Generation Spectroscopy for Biosensor Applications -- InAs Quantum Dots Covered by GaAsSb Strain Reducing Layer -- Maximizing Intensity in TiO2 Waveguides for Nonlinear Optics -- An Improved Method for T-matrix Calculations of Light Scattering by Spheroidal Particles -- Automated Algorithms for Multilayer Thin Film Filter Design Using Metamaterials -- Surface Charge Effects on the Interaction Between a Solid-Supported Model Lipid Membrane and AuNPs Studied by SFG Spectroscopy -- Photoluminescence of colour centres in thermally-evaporated LiF films for radiation imaging detectors -- P3HT:PCBM Based Organic Solar Cells: Structure Optimization and Improving External Quantum Efficiency by Plasmonic Nanoparticles Incorporation -- Optical Properties Nanocomposite Composed of Ag Nanoparticles Embedded in a DLC Film -- Part V List of Participants.

Subjects Lasers.
Nanoscale science.
Nanoscience.
Nanostructures.
Nanotechnology.
Photonics.
Nanoscale Science and Technology.
Nanotechnology.
Optics, Lasers, Photonics, Optical Devices.

Notes Description based on publisher-supplied MARC data.

Additional formats Print version: Nano-structures for optics and photonics: optical strategies for enhancing sensing, imaging, communication and energy conversion 9789401791328 (DLC) 2014950450
Printed edition: 9789401791328

	Printed edition: 9789401791342
	Printed edition: 9789401791427
Series	NATO Science for Peace and Security Series B: Physics and Biophysics, 1874-6500
	NATO Science for Peace and Security Series B: Physics and Biophysics, 1874-6500

Narrow Plasmon Resonances in Hybrid Systems

LCCN	2019743845
Type of material	Book
Personal name	Thomas, Philip A. author.
Main title	Narrow Plasmon Resonances in Hybrid Systems / by Philip A. Thomas.
Edition	1st ed. 2018.
Published/Produced	Cham: Springer International Publishing: Imprint: Springer, 2018.
Description	1 online resource (XVII, 114 pages 49 illustrations, 37 illustrations in color.)
	text file PDF
ISBN	9783319975269
Summary	Advances in understanding the interactions between light and subwavelength materials have enabled the author and his collaborators to tailor unique optical responses at the nanoscale. In particular, metallic nanostructures capable of supporting surface plasmons can be designed to possess spectrally narrow plasmon resonances, which are of particular interest due to their exceptional sensitivity to their local environment. In turn, combining plasmonic nanostructures with other materials in hybrid systems allows this sensitivity to be exploited in a broad range of applications. In this book the author explores two different approaches to attaining narrow plasmon resonances: in gold nanoparticle arrays by utilising diffraction coupling, and in copper thin films covered by a protective graphene layer. The performance of these resonances is then considered in a number of applications. Nanoparticle arrays are used along with an atomic heterostructure as elements in a

nanomechanical electro-optical modulator that is capable of strong, broadband modulation. Strong coupling between diffraction-coupled plasmon resonances and a gold nanoparticle array and guided modes in a dielectric slab is used to construct a hybrid waveguide. Lastly, the extreme phase sensitivity of graphene-protected copper is used to detect trace quantities of small toxins in solution far below the detection limit of commercial surface plasmon resonance sensors.

Contents Plasmonics -- Two-dimensional Materials -- Super-narrow, Extremely High Quality Collective Plasmon Resonances at Telecommunication Wavelengths -- Nanomechanical Electro-optical Modulator Based on Atomic Heterostructures -- Strong Coupling of Diffraction Coupled Plasmons and Optical Waveguide Modes in Gold Stripe-dielectric Nanostructures at Telecom Wavelengths -- Phase-sensitive Detection of HT-2 Mycotoxin Using Graphene-protected Copper Plasmonics -- Conclusions and Future Work.

Subjects Electronic circuits.
Electronic materials.
Interfaces (Physical sciences).
Nanotechnology.
Optical materials.
Surfaces (Physics).
Thin films.
Surface and Interface Science, Thin Films.
Electronic Circuits and Devices.
Nanotechnology.
Optical and Electronic Materials.

Notes Description based on publisher-supplied MARC data.

Additional formats Print version: Narrow plasmon resonances in hybrid systems. 9783319975252 (DLC) 2018950793
Printed edition: 9783030073671
Printed edition: 9783319975252
Printed edition: 9783319975276

Series	Springer Theses, Recognizing Outstanding Ph.D. Research, 2190-5053 Springer Theses, Recognizing Outstanding Ph.D. Research, 2190-5053

Neoclassical Theory of Electromagnetic Interactions: A Single Theory for Macroscopic and Microscopic Scales

LCCN	2019737784
Type of material	Book
Personal name	Babin, Anatoli. author.
Main title	Neoclassical Theory of Electromagnetic Interactions: A Single Theory for Macroscopic and Microscopic Scales / by Anatoli Babin, Alexander Figotin.
Edition	1st ed. 2016.
Published/Produced	London: Springer London: Imprint: Springer, 2016.
Description	1 online resource (XXIII, 696 pages) text file PDF
ISBN	9781447172840
Related names	Figotin, Alexander. author.
Summary	In this monograph, the authors present their recently developed theory of electromagnetic interactions. This neoclassical approach extends the classical electromagnetic theory down to atomic scales and allows the explanation of various non-classical phenomena in the same framework. While the classical Maxwell-Lorentz electromagnetism theory succeeds in describing the physical reality at macroscopic scales, it struggles at atomic scales. Here, quantum mechanics traditionally takes over to describe non-classical phenomena such as the hydrogen spectrum and de Broglie waves. By means of modifying the classical theory, the approach presented here is able to consistently explain quantum-mechanical effects, and while similar to quantum mechanics in some respects, this neoclassical theory also differs markedly from it. In particular, the newly developed framework omits probabilistic interpretations of the wave function and features a new fundamental spatial scale which, at the

size of the free electron, is much larger than the classical electron radius and is relevant to plasmonics and emission physics. This book will appeal to researchers interested in advanced aspects of electromagnetic theory. Treating the classical approach in detail, including non-relativistic aspects and the Lagrangian framework, and comparing the neoclassical theory with quantum mechanics and the de Broglie-Bohm theory, this work is completely self-contained.

Contents Introduction -- The History of Views on Charges, Currents and the Electromagnetic Field -- The Neoclassical Field Theory of Charged Matter: A Concise Presentation -- Part I. Classical Electromagnetic Theory and Special Relativity -- Part II. Classical Field Theory -- Part III. The Neoclassical Theory of Charges -- Part IV. The Neoclassical Theory of Charges with Spin -- Part V. Mathematical Aspects of the Theory of Distributed Elementary Charges -- Part VI. Appendices.

Subjects Partial differential equations.
Mathematical physics.
Physics.
Optics.
Electrodynamics.
Elementary particles (Physics).
Quantum field theory.
Statistical physics.
Partial Differential Equations.
Mathematical Applications in the Physical Sciences.
Mathematical Methods in Physics.
Classical Electrodynamics.
Elementary Particles, Quantum Field Theory.
Applications of Nonlinear Dynamics and Chaos Theory.

Notes Description based on publisher-supplied MARC data.

Additional formats Printed edition: 9781447172826
Printed edition: 9781447172833
Printed edition: 9781447173977

Series Theoretical and Mathematical Physics, 1864-5879
Theoretical and Mathematical Physics, 1864-5879

Optical and Molecular Physics: Theoretical Principles and Experimental Methods

LCCN	2021016927
Type of material	Book
Main title	Optical and molecular physics: theoretical principles and experimental methods / edited by Miguel A. Esteso, PhD, Ana Cristina Faria Ribeiro, PhD, Soney C. George, PhD, Ann Rose Abraham, PhD, A.K. Haghi, PhD.
Edition	First edition.
Published/Produced	Palm Bay, FL: Apple Academic Press, 2021.
Description	1 online resource
ISBN	9781003150053 (ebook) (hardback) (paperback)
LC classification	QC374
Related names	Esteso, Miguel A., editor. Ribeiro, Ana Cristina Faria, editor. George, Soney C., editor. Abraham, Ann Rose, editor. Haghi, A. K., editor.
Summary	"Optical and Molecular Physics: Theoretical Principles and Experimental Methods addresses many important applications and advances in the field. The book is divided into 5 sections: Plasmonics and carbon dots physics with applications Optical films, fibers, and materials Optical properties of advanced materials Molecular physics and diffusion Macromolecular physics The volume also looks at optical materials in the development of composite materials for the functionalization of glass, ceramic, and polymeric substrates to interact with electromagnetic radiation. The book presents state-of-the-art research in the preparation methods, optical characterization, and usage of optical materials and devices in various photonic fields. It integrates

	materials science with other engineering subjects, such as physics, chemistry, and electrical engineering. The authors discuss devices and technologies used by the electronics, magnetics, and photonics industries and offer perspectives on the manufacturing technologies used in device fabrication. Weaving together science and engineering aspects, this book maintains a careful balance between fundamentals and technological aspects to cover applications for a variety of fields. The volume will be an invaluable resource that offers a broad selection of information and research on key ideas, formulae, techniques, and results in optical and molecular physics"-- Provided by publisher.
Subjects	Optical materials. Plasmonics. Carbon. Diffusion. Nanoparticles. Macromolecules.
Notes	Includes bibliographical references and index. Description based on print version record and CIP data provided by publisher; resource not viewed.
Additional formats	Print version: Optical and molecular physics First edition. Palm Bay, FL: Apple Academic Press, 2021 9781771889834 (DLC) 2021016926

Optical and Molecular Physics: Theoretical Principles and Experimental Methods

LCCN	2021016926
Type of material	Book
Main title	Optical and molecular physics: theoretical principles and experimental methods / edited by Miguel A. Esteso, PhD, Ana Cristina Faria Ribeiro, PhD, Soney C. George, PhD, Ann Rose Abraham, PhD, A.K. Haghi, PhD.
Edition	First edition.
Published/Produced	Palm Bay, FL: Apple Academic Press, 2021.
ISBN	9781771889834 (hardback)

	9781774639405 (paperback)
	(ebook)
LC classification	QC374 .O63 2021
Related names	Esteso, Miguel A., editor.
	Ribeiro, Ana Cristina Faria, editor.
	George, Soney C., editor.
	Abraham, Ann Rose, editor.
	Haghi, A. K., editor.
Summary	"Optical and Molecular Physics: Theoretical Principles and Experimental Methods addresses many important applications and advances in the field. The book is divided into 5 sections: Plasmonics and carbon dots physics with applications Optical films, fibers, and materials Optical properties of advanced materials Molecular physics and diffusion Macromolecular physics The volume also looks at optical materials in the development of composite materials for the functionalization of glass, ceramic, and polymeric substrates to interact with electromagnetic radiation. The book presents state-of-the-art research in the preparation methods, optical characterization, and usage of optical materials and devices in various photonic fields. It integrates materials science with other engineering subjects, such as physics, chemistry, and electrical engineering. The authors discuss devices and technologies used by the electronics, magnetics, and photonics industries and offer perspectives on the manufacturing technologies used in device fabrication. Weaving together science and engineering aspects, this book maintains a careful balance between fundamentals and technological aspects to cover applications for a variety of fields. The volume will be an invaluable resource that offers a broad selection of information and research on key ideas, formulae, techniques, and results in optical and molecular physics"-- Provided by publisher.
Subjects	Optical materials.
	Plasmonics.

Carbon.
Diffusion.
Nanoparticles.
Macromolecules.

Notes Includes bibliographical references and index.

Additional formats Online version: Optical and molecular physics First edition. Palm Bay, FL: Apple Academic Press, 2021 9781003150053 (DLC) 2021016927

Optical Metamaterials: Qualitative Models: Introduction to Nano-Optics and Optical Metamaterials

LCCN 2019744627

Type of material Book

Personal name Chipouline, Arkadi. author.

Main title Optical Metamaterials: Qualitative Models: Introduction to Nano-Optics and Optical Metamaterials / by Arkadi Chipouline, Franko Küppers.

Edition 1st ed. 2018.

Published/Produced Cham: Springer International Publishing: Imprint: Springer, 2018.

Description 1 online resource (XII, 318 pages 104 illustrations, 95 illustrations in color.)
text file PDF

ISBN 9783319775203

Related names Küppers, Franko. author.

Summary This textbook bridges the gap between university courses on electrodynamics and the knowledge needed to successfully address the problem of electrodynamics of metamaterials. It appeals to both experimentalists and theoreticians who are interested in the physical basics of metamaterials and plasmonics. Focusing on qualitative fundamental treatment as opposed to quantitative numerical treatment, it covers the phenomena of artificial magnetization at high frequencies, and discusses homogenization procedures and the basics of quantum dynamics in detail. By considering different phenomena it creates a self-consistent qualitative picture to explain most observable phenomena. This allows readers to develop a better understanding of the concepts, and helps to create a conceptual approach, which is especially important in educational contexts. This clearly written book includes problems and solutions for each chapter, which can be used for seminars and homework, as well as qualitative models that are helpful to students.

Contents Phenomenological Electrodynamics of materials with negative dielectric and magnetic constants -- Homogenization of Maxwell equations - macroscopic and microscopic approaches -- Phenomenological vs multipole models -- Charge dynamics and dielectric/magnetic constants elaboration -- Plasmons/Polaritons -- Transmission of light through subwavelength structures -- Multipole approach for homogenization of metamaterials (MM) -- "Quantum" MM.

Subjects Electrodynamics.
Electronic materials.
Lasers.
Microwaves.
Nanotechnology.
Optical engineering.
Optical materials.
Optics.
Photonics.
Optical and Electronic Materials.
Classical Electrodynamics.
Microwaves, RF and Optical Engineering.
Nanotechnology.
Optics, Lasers, Photonics, Optical Devices.

Notes Description based on publisher-supplied MARC data.

Additional formats Print version: Optical metamaterials: qualitative models. 9783319775180 (DLC) 2018938640
Printed edition: 9783319775180
Printed edition: 9783319775197

Series Springer Series in Optical Sciences, 0342-4111; 211
Springer Series in Optical Sciences, 0342-4111; 211

Optical Properties of Metallic Nanoparticles: Basic Principles and Simulation

LCCN	2019741984
Type of material	Book
Personal name	Trügler, Andreas. author.
Main title	Optical Properties of Metallic Nanoparticles: Basic Principles and Simulation / by Andreas Trügler.
Edition	1st ed. 2016.
Published/Produced	Cham: Springer International Publishing: Imprint: Springer, 2016.
Description	1 online resource (XIX, 217 pages 100 illustrations, 98 illustrations in color.) text file PDF
ISBN	9783319250748
Summary	This book introduces the fascinating world of plasmonics and physics at the nanoscale, with a focus on simulations and the theoretical aspects of optics and nanotechnology. A research field with numerous applications, plasmonics bridges the gap between the micrometer length scale of light and the secrets of the nanoworld. This is achieved by binding light to charge density oscillations of metallic nanostructures, so-called surface plasmons, which allow electromagnetic radiation to be focussed down to spots as small as a few nanometers. The book is a snapshot of recent and ongoing research and at the same time outlines our present understanding of the optical properties of metallic nanoparticles, ranging from the tunability of plasmonic resonances to the ultrafast dynamics of light-matter interaction. Beginning with a gentle introduction that highlights the basics of plasmonic interactions and plasmon imaging, the author then presents a suitable theoretical framework for the description of metallic nanostructures. This model based on this framework is first solved analytically for simple systems, and subsequently through numerical simulations for more general cases where, for example, surface roughness,

nonlinear and nonlocal effects or metamaterials are investigated.

Contents Part I: Introduction and Basic principles -- Prologue -- The world of plasmons -- Theory -- Part II: Simulation: Modeling the optical response of metallic nanoparticles -- Part III: Implementations and applications.- Imaging of surface plasmons -- Influence of surface roughness -- Nonlinear optical effects of plasmonic nanoparticles -- Nonlocal response -- Metamaterials.- Outlook -- Part IV: Supplement -- Appendix - Utilities -- Appendix - MATLAB script for Mie solution.

Subjects Lasers.
Photonics.
Metals.
Nanoscale science.
Nanoscience.
Nanostructures.
Nanotechnology.
Physics.
Optics, Lasers, Photonics, Optical Devices.
Metallic Materials.
Nanoscale Science and Technology.
Nanotechnology.
Numerical and Computational Physics, Simulation.

Notes Description based on publisher-supplied MARC data.

Additional formats Printed edition: 9783319250724
Printed edition: 9783319250731
Printed edition: 9783319797304

Series Springer Series in Materials Science, 0933-033X; 232
Springer Series in Materials Science, 0933-033X; 232

Optics for Materials Scientists

LCCN 2019014399

Type of material Book

Personal name Lee, Myeongkyu, author.

Main title	Optics for materials scientists / Myeongkyu Lee, PhD.
Published/Produced	Toronto; New Jersey: Apple Academic Press, 2020.
Description	xii, 371 pages: illustrations (some color); 24 cm
ISBN	9781771887571 (hardcover: alk. paper)
LC classification	QC355.3 .L44 2020
Contents	Electromagnetic waves -- Reflection and refraction -- Superposition of waves -- Interference -- Diffraction -- Light propagation in anisotropic media -- Polarization -- Thin-film optics and reflectance control -- Interference lithography -- Metal plasmonics.
Subjects	Optics. Optical materials.
Notes	Includes bibliographical references and index.
Additional formats	Online version: Lee, Myeongkyu, author. Optics for materials scientists Toronto; New Jersey: Apple Academic Press, 2020 9780429425356 (DLC) 2019017540

Optics for Materials Scientists

LCCN	2019017540
Type of material	Book
Personal name	Lee, Myeongkyu, author.
Main title	Optics for materials scientists / Myeongkyu Lee, PhD.
Published/Produced	Toronto; New Jersey: Apple Academic Press, 2020.
Description	1 online resource.
ISBN	9780429425356 ()
LC classification	QC355.3
Contents	Electromagnetic waves -- Reflection and refraction -- Superposition of waves -- Interference -- Diffraction -- Light propagation in anisotropic media -- Polarization -- Thin-film optics and reflectance control -- Interference lithography -- Metal plasmonics.
Subjects	Optics. Optical materials.
Notes	Includes bibliographical references and index.

	Description based on print version record and CIP data provided by publisher; resource not viewed.
Additional formats	Print version: Lee, Myeongkyu, author. Optics for materials scientists Toronto; New Jersey: Apple Academic Press, 2020 9781771887571 (DLC) 2019014399

Parity-Time Symmetry and Its Applications

LCCN	2019762873
Type of material	Book
Main title	Parity-time Symmetry and Its Applications / edited by Demetrios Christodoulides, Jianke Yang.
Edition	1st ed. 2018.
Published/Produced	Singapore: Springer Singapore: Imprint: Springer, 2018.
Description	1 online resource (XV, 579 pages 225 illustrations, 214 illustrations in color.) text file PDF
ISBN	9789811312472
Related names	Christodoulides, Demetrios, editor. Yang, Jianke, editor.
Summary	This book offers a comprehensive review of the state-of-the-art theoretical and experimental advances in linear and nonlinear parity-time-symmetric systems in various physical disciplines, and surveys the emerging applications of parity-time (PT) symmetry. PT symmetry originates from quantum mechanics, where if the Schrodinger operator satisfies the PT symmetry, then its spectrum can be all real. This concept was later introduced into optics, Bose-Einstein condensates, metamaterials, electric circuits, acoustics, mechanical systems and many other fields, where a judicious balancing of gain and loss constitutes a PT-symmetric system. Even though these systems are dissipative, they exhibit many signature properties of conservative systems, which make them mathematically and physically intriguing. Important PT-symmetry applications have also emerged. This book describes the latest advances of

	PT symmetry in a wide range of physical areas, with contributions from the leading experts. It is intended for researchers and graduate students to enter this research frontier, or use it as a reference book.
Contents	Linear and Nonlinear Experiments in PT-Symmetric Photonic Mesh Lattices -- PT-Symmetry on-a-Chip: Harnessing Optical Loss for Novel Integrated Photonic Functionality -- Parity-Time Symmetry in Scattering Problems -- Scattering Theory and PT-Symmetry -- Passive PT-symmetry in laser-written optical waveguide structures -- Non-Hermitian effects due to asymmetric backscattering of light in whispering-gallery microcavities -- Exact Results for a Special PT-symmetric Optical Potential -- Parity-time-symmetric optical lattices in atomic configurations -- Effects of exceptional points in PT-symmetric waveguides -- Higher order exceptional points in discrete photonics platforms -- Non-Hermitian Optical Waveguide Couplers -- Parity-Time Symmetric Plasmonics -- PT-symmetry and non-Hermitian wave transport in microwaves and RF circuits -- Coupled nonlinear Schrodinger equations with gain and loss: modeling PT-symmetry -- Making the PT symmetry unbreakable -- Krein signature in Hamiltonian and PT-symmetric systems -- Integrable nonlocal PT-symmetric and reverse space-time nonlinear Schrodinger equations -- Construction of non-PT-symmetric complex potentials with all-real spectra -- Constant-Intensity Waves in Non-Hermitian Media -- Nonlinear Beam Propagation in a Class of Complex Non-PT-Symmetric Potentials.
Subjects	Condensed materials. Electronic materials. Lasers. Optical materials. Phase transformations (Statistical physics). Photonics. Quantum physics. Quantum Physics.

Optical and Electronic Materials.
Optics, Lasers, Photonics, Optical Devices.
Quantum Gases and Condensates.
Notes Description based on publisher-supplied MARC data.
Additional formats Printed edition: 9789811312465
Printed edition: 9789811312489
Series Springer Tracts in Modern Physics, 0081-3869; 280
Springer Tracts in Modern Physics, 0081-3869; 280

Plasma and Plasmonics
LCCN 2017060065
Type of material Book
Personal name Shah, Kushal Y. (Kushal Yogeshkumar), author.
Main title Plasma and plasmonics / Kushal Shah.
Published/Produced Berlin: Walter de Gruyter GmbH, [2018]
Description x, 158 pages: illustrations; 24 cm
ISBN 9783110569940 (softcover)
LC classification TK8585 .S53 2018
Subjects Plasmonics.
Plasmons (Physics)
Plasma.
Notes Includes bibliographical references (pages 155-156) and index.

Plasmonics: Advances in Research and Applications
LCCN 2016044000
Type of material Book
Main title Plasmonics: advances in research and applications / Angelo Bright, editor.
Published/Produced Hauppauge, New York, USA: Nova Science Publisher's Inc., [2016]
©2016
Description 1 online resource.
ISBN 9781536101867 (pdf)
LC classification QC176.8.P55
Related names Bright, Angelo, editor.
Contents Plasmonic interaction in enhanced luminescent down-shifting layers for photovoltaic devices / H. Ahmed, A. Sethi, J. Doran, and S. J. McCormack

(School of Engineering, Trinity College Dublin, Dublin, Ireland, and others) -- Several quantum-optical analogous in plasmonic nanostructures / Jun-Jun Xiao, Qiang Zhang, Zhenzhen Liu and Xiangli Liu (College of Electronic and Information Engineering, Shenzhen Graduate School, Harbin Institute of Technology, Shenzhen, China, and others) -- Quantum-coherent plasmonics: atomic-phase-coherence assisted surface plasmon resonance and negative optical parameters in multilevel media / Jian Qi Shen (Centre for Optical and Electromagnetic Research, State Key Laboratory of Modern Optical Instrumentations, Zijingang Campus, Zhejiang University, Hangzhou, China, and others).

Subjects Plasmons (Physics)
Nanostructured materials--Optical properties.
Nanoparticles.

Notes Includes bibliographical references and index.
Description based on print version record and CIP data provided by publisher.

Additional formats Print version: Plasmonics Hauppauge, New York, USA: Nova Science Publisher's Inc., [2016] 9781536101744 (DLC) 2016040727

Series Physics research and technology

Plasmonics: Advances in Research and Applications

LCCN 2016040727

Type of material Book

Main title Plasmonics: advances in research and applications / Angelo Bright, editor.

Published/Produced New York: Nova Publishers, [2017]

Description ix, 246 pages; 23 cm

ISBN 9781536101744 (hardcover)
1536101745 (hardcover)
(pdf)
(pdf)

LC classification QC176.8.P55 P533 2017

Related names Bright, Angelo, editor.

Subjects Plasmonics.

	Nanostructured materials--Optical properties.
	Nanoparticles.
Notes	Includes bibliographical references and index.
Additional formats	Online version: Plasmonics Hauppauge, New York, USA: Nova Science Publisher's Inc., [2016] 9781536101867 (DLC) 2016044000
Series	Physics research and technology
	Physics research and technology.

Plasmonics: Design, Materials, Fabrication, Characterization, and Applications XVI

LCCN	2019300649
Type of material	Book
Meeting name	Plasmonics: Design, Materials, Fabrication, Characterization, and Applications (Conference) (16th: 2018: San Diego, Calif.)
Main title	Plasmonics: design, materials, fabrication, characterization, and applications XVI / Ding Ping Tsai, Takuo Tanaka, editors.
Published/Produced	Bellingham, Washington, USA: SPIE, [2018]
Description	1 volume (various pagings): illustrations (black and white); 28 cm.
ISBN	9781510620155 (paperback)
	151062015X
	(electronic)
Related names	Tsai, Din P., 1959- editor.
	Tanaka, Takuo, editor.
	SPIE (Society), publisher.
Notes	Includes bibliographical references.
Additional formats	Also published online.
Series	Proceedings of SPIE, 0277-786X; volume 10722

Plasmonics II: 12-14 October 2016, Beijing, China

LCCN	2018300593
Type of material	Book
Main title	Plasmonics II: 12-14 October 2016, Beijing, China / Hongxing Xu [und 3 andere], editors; sponsored by SPIE, COS - Chinese Optical Society; cooperating

	organizations: Tsinghua University (China) [und 24 andere]; published by SPIE.
Published/Produced	Bellingham, Washington SPIE [2016]
Description	1 Band: Illustrationen.
ISBN	9781510604759
	1510604758
Related names	Xu, Hongxing, Herausgeber. edt
	SPIE Sponsor.
	Chinese Optical Society Sponsor.
Notes	Auf dem Deckblatt: SPIE. COS, photonics, Asia.
Series	Proceedings of SPIE. 5200-; vol. 10028

Plasmonics III: 12-13 October 2018, Beijing, China

LCCN	2019300677
Type of material	Book
Meeting name	Plasmonics (Conference) (3rd: 2018: Beijing, China)
Main title	Plasmonics III: 12-13 October 2018, Beijing, China / Hongxing Xu, Satoshi Kawata, David J. Bergman, editors.
Published/Produced	Bellingham, Washington, USA: SPIE, [2018]
Description	1 volume (various pagings): illustrations (black and white); 28 cm.
Links	1850-9999 http://spie.org/
ISBN	9781510622463 paperback
	1510622462
	electronic
Related names	Xu, Hongxing, editor.
	Kawata, Satoshi, 1966- editor.
	Bergman, D. J. (David J.), editor.
	SPIE (Society), publisher.
	Zhongguo guang xue xue hui, sponsoring body.
Notes	"ISSN 1996-756X (electronic)"--Title page verso.
	"Sponsored by SPIE, COS - Chinese Optical Society."
	"SPIE. COS. Photonics Asia"--Cover.
	Includes bibliographical references and author index.
Additional formats	Also published electronically.
Series	Proceedings of SPIE, 0277-786X; Volume 10824

Plasmonics in Biology and Medicine XII

LCCN	2019303922
Type of material	Book
Main title	Plasmonics in biology and medicine XII / edited by Joseph Lakowicz.
Published/Produced	Bellingham: SPIE, 2015.
Description	1 volume; 23 cm
ISBN	9781628414301 (pbk.)
	1628414308
Related names	Lakowicz, Joseph R., editor.
Subjects	Plasmons (Physics)--Congresses.
	Biosensors--Congresses.
	Biosensors.
	Plasmons (Physics)
Form/Genre	Conference papers and proceedings.
Series	Proceedings of SPIE

Plasmonics in Biology and Medicine XIV: 30 January 2017, San Francisco, California, United States

LCCN	2017465012
Type of material	Book
Main title	Plasmonics in Biology and Medicine XIV: 30 January 2017, San Francisco, California, United States / Tuan Vo-Dinh, Joseph R. Lakowicz, editors; sponsored and published by SPIE.
Published/Produced	Bellingham, Washington SPIE [2017]
Description	1 Band: Illustrationen.
ISBN	9781510606012 (print)
	1510606017
Related names	Vo-Dinh, Tuan, 1948- Herausgeber. edt
	Lakowicz, Joseph R., Herausgeber. edt
	SPIE Sponsor.
Notes	Auf dem Deckblatt: SPIE. Photonics West.
Series	Proceedings of SPIE. 5200-; vol. 10080
	Progress in biomedical optics and imaging; vol. 18, no. 44

Plasmonics in Biology and Medicine XV: 29 January 2017, San Francisco, California, United States

LCCN	2018459079
Type of material	Book
Meeting name	Plasmonics in Biology and Medicine (Conference) (15th: 2018: San Francisco, Calif.) author.
Main title	Plasmonics in biology and medicine XV: 29 January 2017, San Francisco, California, United States / Tuan Vo-Dinh, Joseph R. Lakowicz, editors; sponsored and published by SPIE.
Published/Produced	Bellingham, Washington, USA: SPIE, [2018]
Description	1 volume (various pagings): illustrations; 28 cm.
Links	1850-9999 http://spie.org/
ISBN	9781510615038 paperback 1510615032 electronic
LC classification	QC176.8.P55 P566 2017
Variant title	Plasmonics in biology and medicine 15 Plasmonics in biology and medicine fifteen
Related names	Vo-Dinh, Tuan, editor. Lakowicz, Joseph R., editor. SPIE (Society), publisher, sponsoring body.
Subjects	Plasmons (Physics)--Congresses. Biosensors--Congresses.
Notes	Subtitle should read: 29 January 2018. Includes bibliographical references and author index.
Additional formats	Also published electronically.
Series	Proceedings of SPIE; volume 10509 Progress in biomedical optics and imaging, 1605-7422; vol. 19, no. 43 SPIE Photonics West

Plasmonics: Design, Materials, Fabrication, Characterization, and Applications XV: 6-10 August 2017, San Diego, California, United States

LCCN	2018300621
Type of material	Book

Meeting name Plasmonics: Design, Materials, Fabrication, Characterization, and Applications Veranstaltung (15.: 2017: San Diego, Calif.)
Main title Plasmonics: Design, Materials, Fabrication, Characterization, and Applications XV: 6-10 August 2017, San Diego, California, United States / Din Ping Tsai, Takuo Tanaka, editors; sponsored and published by SPIE.
Published/Produced Bellingham, Washington SPIE [2017]
Description 1 Band: Illustrationen.
ISBN 9781510611498
1510611495
Related names Tsai, Din PIng, Herausgeber. edt
SPIE Herausgebendes Organ. Sponsor.
Series Proceedings of SPIE. 5200-; vol. 10346

Progress in Nonlinear Nano-Optics
LCCN 2019740262
Type of material Book
Main title Progress in Nonlinear Nano-Optics / edited by Shuji Sakabe, Christoph Lienau, Rüdiger Grunwald.
Edition 1st ed. 2015.
Published/Produced Cham: Springer International Publishing: Imprint: Springer, 2015.
Description 1 online resource (XXI, 273 pages 137 illustrations, 44 illustrations in color.)
text file PDF
ISBN 9783319122175
Related names Sakabe, Shuji. editor.
Lienau, Christoph. editor.
Grunwald, Rüdiger. editor.
Summary This book presents the state of the art in nonlinear nanostructures for ultrafast laser applications. Most recent results in two emerging fields are presented: (i) generation of laser-induced nanostructures in materials like metals, metal oxides and semiconductors, and (ii) ultrafast excitation and energy transfer in nanoscale physical, chemical and

hybrid systems. Particular emphasis is laid on the up-to-date controversially discussed mechanisms of sub-wavelength ripple formation including models of self-organized material transport and multiphoton excitation channels, nonlinear optics of plasmonic structures (nanotips, nanowires, 3D-metamaterials), and energy localization and transport on ultrafast time scale and spatial nanoscale. High-resolution spectroscopy, simulation and characterization techniques are reported. New applications of ultrashort-pulsed lasers for materials processing and the use of nanostructured materials for characterizing laser fields and laser-matter-interactions are discussed.

Contents Laser-Induced Nanostructures -- Nanostructures Formation at Metal Surfaces -- General Aspects and Structuring in Three Dimensions -- Nanostructure Formation in Semiconductors and Dielectrics -- Nonlinear Nano-Optics -- Plasmonics, Metamaterials and Nonlinear Optics -- Spectroscopy, Gain Materials and Characterization.

Subjects Lasers.
Photonics.
Nanoscale science.
Nanoscience.
Nanostructures.
Optical materials.
Electronic materials.
Nanotechnology.
Optics, Lasers, Photonics, Optical Devices.
Nanoscale Science and Technology.
Optical and Electronic Materials.
Nanotechnology.

Notes Description based on publisher-supplied MARC data.

Additional formats Printed edition: 9783319122182
Printed edition: 9783319122168
Printed edition: 9783319363523

Series Nano-Optics and Nanophotonics, 2192-1970

Nano-Optics and Nanophotonics, 2192-1970

Quantum Nano-Photonics

LCCN	2019743200
Type of material	Book
Main title	Quantum Nano-Photonics / edited by Baldassare Di Bartolo, Luciano Silvestri, Maura Cesaria, John Collins.
Edition	1st ed. 2018.
Published/Produced	Dordrecht: Springer Netherlands: Imprint: Springer, 2018.
Description	1 online resource (XVI, 474 pages 149 illustrations, 126 illustrations in color.) text file PDF
ISBN	9789402415445
Related names	Cesaria, Maura. editor. Collins, John. editor. Di Bartolo, Baldassare. editor. Silvestri, Luciano. editor.
Summary	This book brings together more closely researchers working in the two fields of quantum optics and nano-optics and provides a general overview of the main topics of interest in applied and fundamental research. The contributions cover, for example, single-photon emitters and emitters of entangled photon pairs based on epitaxially grown semiconductor quantum dots, nitrogen vacancy centers in diamond as single-photon emitters, coupled quantum bits based on trapped ions, integrated waveguide superconducting nanowire single-photon detectors, quantum nano-plasmonics, nanosensing, quantum aspects of biophotonics and quantum metamaterials. The articles span the bridge from pedagogical introductions on the fundamental principles to the current state-of-the-art, and are authored by pioneers and leaders in the field. Numerical simulations are presented as a powerful tool to gain insight into the physical behavior of nanophotonic

systems and provide a critical complement to experimental investigations and design of devices.

Contents Part 1: Lectures -- 1.Sergey. Gaponenko - Nano-Photonics with and without Photons -- 2.Lukas Novotny - Quantum Sensing and Engineering -- 3.Maura Cesaria and Baldassare Di Bartolo - Nanophosphoros: from rare earth activated multicolo-tuning to new efficient white light sources -- 4.John Collins - Non-Radiative Processes In Crystals and Nanocrystals -- 5.Jean-Pierr Wolf - Quantum Aspects of Biophotonics -- 6. Lora Ramunno - Simulations in Nano-Photonics -- 7.John Bowen - Terahertz Nanoscale Science and Technology -- 8.Fabrizio Pinto - Casimir Forces: Fundamental Theory, Computation, and Nano-device Applications -- 9. Diana Adliene - Nano-materials and Nano-technologies for Photon Radiation Enhanced Cancer Treatment -- 10.Oliver Benson - Fundamental Experiments and Quantum Technology Applications with Defect Centres in Diamond -- 11.Hilmi Volkan Demir - Emerging Fields of Colloidal Nanophotonics for Quality Lighting to Versatile Lasing -- 12.Joshua Caldwell - Semiconductor Nano-Photonics using Surface Polaritons -- 13.Wolfram Pernice - Waveguide integrated superconducting single photon detectors -- 14.Stephen Arnold - Whispering Gallery Mode resonators as opto-mechanical probes to nano-particle-microscavity interactios and charge -- 15.Markus Pollnau - Novel Aspects of Fabry-Perot Resonator -- 16.Rolindes Balda - Performance of Nd3+ as structural probe of rare-earth distribution in transparent nanostructured glass-ceramics -- 17.Malrgorzata Guzik - Research on the Yb 3+ ion activated cubic molybdates and molybdato-tungstates for optical transparent ceramics -- Part II: Short Seminars -- 1. Fabian Beutel - Quantum Key Distribution over Free Space -- 2.Antonino Cala' Lesina - Modelling of Coloured Metal Surfaces by

Plasmonics Nanoparticles -- 3.Onur Erdem - Exciton Dynamics of Colloidal Semiconductor Quantum Well Stacks -- 4.James Godfrey - InAsP Quantum Dots in InP Nanowire Waveguides as Sources of Quantum Light -- 5. Wladislaw Hartmann - Diamond Nano-Photonic Circuits -- 6.Isabelle Palstra - Superchiral Near Fields in Photonic Crystal Waveguides -- 7.Marya Rudenko - Neodymium doped Luminescent Composites Derived From the Sols Based On Carboxylic Acids -- 8.Konstantin Shportko - Vibrational properties of Ge-Sb-Te phase-change alloys studied by temperature-dependent IR and Raman spectroscopy -- 9.Felix Stete - Size dependence of the coupling strength in plasmon-exciton Nano-Particles -- 10.Hasan Tabanli - Introduction to Shock Wave-Boundary Layer Interaction and Unstart in Supersonic Inlets -- 11.Sevcan Tabanli - Investigation of the white light emission from Er/Nd/Yb rare earth oxides at vacuum and atmospheric pressure -- 12.Nicolai Walter - Carbon Nanotubes as integrated electrically driven Light Source -- Part III: Posters -- 1.Antonino Cala' Lesina - Plasmonic Metasurfaces for Nonlinear Structured Light1) Antonino Cala' Lesina - Plasmonic Metasurfaces for Nonlinear Structured Light -- 2.Gabriel Campargue - Cell Poration of Fixed and Live Cells by Phase Shaped femtosecond pulses -- 3.Marco Clementi - On-chip nonlinear optics in silicon rich nitride photonic crystal cavities -- 4.Tobias Frenzel - Mechanical Activity: The elastic counterpart of optical activity -- 5.Wladislaw Hartmann - Highly compact and scalable waveguide-integrated single photon spectrometer based on tailored disorder -- 6.Iyer - Symmetry Breaking and Active Fano resonance Tuning in Dolmen Nano-structures -- 7.Robert Keitel - Pump-Profile Engineering for Spatial- and Spectral-Mode Control in Two-Dimensional

Colloidal-Quantum-Dot Spasers -- 8.Caroline Klusmann - Hybridizing whispering gallery modes and plasmonic resonances in a photonic meta-device for bio-sensing applications -- 9.Julian Koepfler - Three-dimensional Chiral Photonic Crystals in the Thz Regime Exhibiting Weyl Points with Topological Charges -- 10.Frederik Mayer - Three-dimensional Fluorescent Security Features Fabricated via 3D Laser Lithography -- 11.Michael Meeker - Strong coupling effects between IR-inactive zone folded LO phonon and localized surface phonon polariton modes in SiC nanopillars -- 12.Tobias Messer - Measuring the intensity profile of arbitrary shaped laser foci using confocal microscopy -- 13.Alexander Munchinger - 3D Cubic buclking mechanical metamaterials -- 14.Stefano Orlando - Light absorbing diamond for solar energy conversion -- 15.Nikolai Passler - Nonlinear Response and Strong Coupling of Surface Phonon Polaritons -- 16.Marco Passoni - Grating couplers in Silicon-on-insulator: the role of photonic guided resonances on lineshape and bandwith -- 17.Siim Pikker - Aluminum plasmonics: fabrication and characterization of broadly tunable plasmonic surfaces for plasmon molecule strong-coupling and fluorescence enhancement -- 18.Jinq Yuan Qu - 3D Metamaterials with Negative Thermal Expansion and Negative Effective Compressibility -- 19.Andrei Ramanenka - Colloidal Spherical Silver Nanoparticles Based Plasmon Enhanced Fluorescence for Rapid Quantitative Point of Care Testing Fluorescent Immunoassay Development -- 20.Maryia Rudenko - Synthesis And Photoluminescence Of Strontium Titanate Xerogels Doped With Terbium, Ytterbium And Europium -- 21.Philip Schrinner - Integration of single photon sources with nano-photonic circuits -- 22.Konstantin Shportko - Vibrational properties

	of Ge-Sb-Te phase-change alloys studied by IR and Raman spectroscopy at different temperatures -- 23.Vladislav Stefanov - Quantum state engineering with single atom laser -- 24.Felix Stete - Signatures of Strong Coupling on Nanoparticles: Revealing Absorption Anticrossing by Tuning the Dielectric Environment -- 25.Sevcan Tabanli - Solid State Synthesis, structural and Up-Conversion Properties of Yb3+/Er3+ and Yb3+/Tm3+/Er3+ doped La2Ti2O7 Phosphors -- 26.Silvia Tofani- Static and tunable devices for terahertz focusing and beam steering -- 7.List of Participants.
Subjects	Electronic materials. Lasers. Nanoscale science. Nanoscience. Nanostructures. Optical materials. Photonics. Quantum optics. Nanoscale Science and Technology. Optical and Electronic Materials. Optics, Lasers, Photonics, Optical Devices. Quantum Optics.
Notes	Description based on publisher-supplied MARC data.
Additional formats	Print version: Quantum nano-photonics 9789402415438 (DLC) 2018950399 Printed edition: 9789402415438 Printed edition: 9789402415452 Printed edition: 9789402415469
Series	NATO Science for Peace and Security Series B: Physics and Biophysics, 1874-6500 NATO Science for Peace and Security Series B: Physics and Biophysics, 1874-6500

Quantum Nano-Plasmonics

LCCN	2020020654
Type of material	Book

Personal name	Jacak, Witold A., author.
Main title	Quantum nano-plasmonics / Witold A. Jacak, Politechnika Wroclawska, Poland.
Published/Produced	Cambridge, UK; New York, NY: Cambridge University Press, 2020.
ISBN	9781108478397 (hardback) (epub)
LC classification	QC176.8.P55 J33 2020
Summary	"With examples and clear explanation throughout, this step-by-step approach makes the quantum theory of plasmons accessible to readers without a specialized training in theory. A fully analytical formulation offers an opportunity for further development of the research and applications. The theory is focused on the random phase approximation description of plasmons in metallic nanostructures, previously defined for the bulk metal only. Particular attention is paid to the large damping of plasmons in nanostructures, including electron scattering and Lorentz friction losses; a quantum description of the plasmon photovoltaic effect is presented and there is an in-depth analysis of plasmon-polariton kinetics in metallic nano-chains. Suitable for students in the fields of plasmonics, optoelectronics, and photonics and for researchers active in the fields of photovoltaics, optoelectronics, nanoplasmonics, and nano-photonics, the book should also be useful for researchers in soft plasmonics since it includes applications to electro-signalling in neurons"--Provided by publisher.
Subjects	Plasmons (Physics) Nanostructures. Surface plasmon resonance. Quantum theory. Plasmonics.
Notes	Includes bibliographical references and index.

Additional formats	Online version: Jacak, Witold A., Quantum nano-plasmonics 1. New York: Cambridge University Press, 2020. 9781108777698 (DLC) 2020020655

Quantum Nano-Plasmonics

LCCN	2020020655
Type of material	Book
Personal name	Jacak, Witold A., author.
Main title	Quantum nano-plasmonics / Witold A. Jacak, Politechnika Wroclawska, Poland.
Published/Produced	Cambridge, UK; New York, NY: Cambridge University Press, 2020.
Description	1 online resource
ISBN	9781108777698 (epub) (hardback)
LC classification	QC176.8.P55
Summary	"With examples and clear explanation throughout, this step-by-step approach makes the quantum theory of plasmons accessible to readers without a specialized training in theory. A fully analytical formulation offers an opportunity for further development of the research and applications. The theory is focused on the random phase approximation description of plasmons in metallic nanostructures, previously defined for the bulk metal only. Particular attention is paid to the large damping of plasmons in nanostructures, including electron scattering and Lorentz friction losses; a quantum description of the plasmon photovoltaic effect is presented and there is an in-depth analysis of plasmon-polariton kinetics in metallic nano-chains. Suitable for students in the fields of plasmonics, optoelectronics, and photonics and for researchers active in the fields of photovoltaics, optoelectronics, nanoplasmonics, and nano-photonics, the book should also be useful for researchers in soft plasmonics since it includes applications to electro-signalling in neurons"--Provided by publisher.

Subjects	Plasmons (Physics) Nanostructures. Surface plasmon resonance. Quantum theory. Plasmonics.
Notes	Includes bibliographical references and index. Description based on print version record and CIP data provided by publisher; resource not viewed.
Additional formats	Print version: Jacak, Witold A.. Quantum nanoplasmonics Cambridge, UK; New York, NY: Cambridge University Press, 2020. 9781108478397 (DLC) 2020020654

Quantum Optics and Nanophotonics: Ecole d'été de Physique des Houches, Session CI, 5-30 August 2013

LCCN	2016949071
Type of material	Book
Meeting name	Ecole d'été de physique théorique (Les Houches, Haute-Savoie, France) (101st: 2013)
Main title	Quantum optics and nanophotonics: Ecole d'été de Physique des Houches, Session CI, 5-30 August 2013 / edited by Claude Fabre, Vahid Sandoghdar, Nicolas Treps, Leticia F. Cugliandolo.
Edition	First edition.
Published/Produced	Oxford, United Kingdom: Oxford University Press, 2017.
Description	xxi, 455 pages: illustrations; 26 cm
ISBN	9780198768609 (hbk.) 0198768605 (hbk.)
LC classification	QC446.2 .E26 2013
Related names	Fabre, Claude, 1951- editor. Sandoghdar, Vahid, 1966- editor. Treps, Nicolas, editor. Cugliandolo, L. F. (Leticia F.), editor.
Contents	Introduction to quantum information / Stephen M. Barnett -- Introduction to near-field optics and plasmonics / Jean-Jacques Greffet -- Transformation optics / Ulf Leonhardt -- Temporal and spectral properties of quantum light / Birgit

	Stiller, Ulrich Seyfarth, Gerd Leuchs -- Quantum optics with nitrogen-vacancy centres in diamond / Yiwen Chu, Mikhail D. Lukin -- Single-molecule spectroscopy / Michel Orrit -- Quantum measurements: a modern view for quantum optics experimentalists / Aephraim M. Steinberg -- Quantum optics and cavity QED with quantum dots in photonic crystals / Jelena Vučkovič -- Casimir forces and vacuum energy / Serge Reynaud, Astrid Lambrecht.
Subjects	Quantum optics. Nanophotonics. Nanophotonics. Quantum optics.
Notes	Includes bibliographical references.

Quantum Photonics: Pioneering Advances and Emerging Applications

LCCN	2019763120
Type of material	Book
Main title	Quantum Photonics: Pioneering Advances and Emerging Applications / edited by Robert W. Boyd, Svetlana G. Lukishova, Victor N. Zadkov.
Edition	1st ed. 2019.
Published/Produced	Cham: Springer International Publishing: Imprint: Springer, 2019.
Description	1 online resource (XXXI, 627 pages 268 illustrations, 134 illustrations in color.) text file PDF
ISBN	9783319984025
Related names	Boyd, Robert W, editor. Lukishova, Svetlana G, editor. Zadkov, Victor N, editor.
Summary	This book brings together reviews by internationally renowned experts on quantum optics and photonics. It describes novel experiments at the limit of single photons, and presents advances in this emerging research area. It also includes reprints and historical descriptions of some of the first pioneering experiments at a single-

photon level and nonlinear optics, performed before the inception of lasers and modern light detectors, often with the human eye serving as a single-photon detector. The book comprises 19 chapters, 10 of which describe modern quantum photonics results, including single-photon sources, direct measurement of the photon's spatial wave function, nonlinear interactions and non-classical light, nanophotonics for room-temperature single-photon sources, time-multiplexed methods for optical quantum information processing, the role of photon statistics in visual perception, light-by-light coherent control using metamaterials, nonlinear nanoplasmonics, nonlinear polarization optics, and ultrafast nonlinear optics in the mid-infrared.

Contents From the contents: Single-photon and nonlinear optical experiments in a pre-laser era -- Modern single-photon and quantum nonlinear optical experiments -- Experiments on quantum imaging and information -- Preparation of non-classical light sources -- Room-temperature single-photon sources -- Single-photon spectroscopy of a single molecule -- Quantum nonlinear optics with single photons enabled by strongly interacting atoms -- Atom-light interactions in photonic crystals -- Quantum plasmonics -- Nonlinear nanoplasmonics -- Nano-antennas for light -- Light-by-light coherent control using metamaterials -- Review of modern visual science on sensitivity of human eye to the low-light level -- Determining the lower limit of human vision using single-photon source -- Measurements of photon statistics with live photoreceptor cells -- Nonlinear optics with photonic crystals/in mid- and near-IR -- Nonlinear polarization optics -- Single-cycle pulses and nonlinear optical effects in THz waves.

Subjects Electrodynamics.
Electronic materials.
Lasers.

Optical materials.
Optics.
Photonics.
Quantum optics.
Quantum Optics.
Classical Electrodynamics.
Optical and Electronic Materials.
Optics, Lasers, Photonics, Optical Devices.

Notes Description based on publisher-supplied MARC data.

Additional formats Print version: Quantum photonics. 9783319984001 (DLC) 2018960218
Printed edition: 9783319984001
Printed edition: 9783319984018

Series Springer Series in Optical Sciences, 0342-4111; 217
Springer Series in Optical Sciences, 0342-4111; 217

Quantum Plasmonics

LCCN 2016951692
Type of material Book
Main title Quantum plasmonics.
Published/Produced New York, NY: Springer Berlin Heidelberg, 2016.
Links Contributor biographical information https://www.loc.gov/catdir/enhancements/fy1620/2016951692-b.html
Publisher description https://www.loc.gov/catdir/enhancements/fy1620/2016951692-d.html
Table of contents only https://www.loc.gov/catdir/enhancements/fy1620/2016951692-t.html
ISBN 9783319458199

Quantum Plasmonics

LCCN 2019747948
Type of material Book
Main title Quantum Plasmonics / edited by Sergey I. Bozhevolnyi, Luis Martin-Moreno, Francisco Garcia-Vidal.

Edition	1st ed. 2017.
Published/Produced	Cham: Springer International Publishing: Imprint: Springer, 2017.
Description	1 online resource (XVII, 327 pages 139 illustrations, 74 illustrations in color.) text file PDF
ISBN	9783319458205
Related names	Bozhevolnyi, Sergey I. editor. Garcia-Vidal, Francisco. editor. Martin-Moreno, Luis. editor.
Summary	This book presents the latest results of quantum properties of light in the nanostructured environment supporting surface plasmons, including waveguide quantum electrodynamics, quantum emitters, strong-coupling phenomena and lasing in plasmonic structures. Different approaches are described for controlling the emission and propagation of light with extreme light confinement and field enhancement provided by surface plasmons. Recent progress is reviewed in both experimental and theoretical investigations within quantum plasmonics, elucidating the fundamental physical phenomena involved and discussing the realization of quantum-controlled devices, including single-photon sources, transistors and ultra-compact circuitry at the nanoscale.
Contents	Waveguide Quantum Electrodynamics -- Quantum Emitters for Plasmonics -- New Materials for Quantum Plasmonics -- Collective Strong Coupling -- Lasing in Plasmonic Structures -- Quantum Effects in Nanogaps.
Subjects	Electronic materials. Lasers. Microwaves. Nanotechnology. Optical engineering. Optical materials. Photonics.

	Quantum optics. Semiconductors. Optics, Lasers, Photonics, Optical Devices. Microwaves, RF and Optical Engineering. Nanotechnology and Microengineering. Optical and Electronic Materials. Quantum Optics. Semiconductors.
Notes	Description based on publisher-supplied MARC data.
Additional formats	Print version: Quantum plasmonics. 9783319458199 (DLC) 2016951692 Printed edition: 9783319458199 Printed edition: 9783319458212 Printed edition: 9783319833798
Series	Springer Series in Solid-State Sciences, 0171-1873; 185 Springer Series in Solid-State Sciences, 0171-1873; 185

Recent Trends in Computational Photonics

LCCN	2019742969
Type of material	Book
Main title	Recent Trends in Computational Photonics / edited by Arti Agrawal, Trevor Benson, Richard M. De La Rue, Gregory A. Wurtz.
Edition	1st ed. 2017.
Published/Produced	Cham: Springer International Publishing: Imprint: Springer, 2017.
Description	1 online resource (XIX, 395 pages 170 illustrations, 134 illustrations in color.) text file PDF
ISBN	9783319554389
Related names	Agrawal, Arti. editor. Benson, Trevor. editor. De La Rue, Richard M. editor. Wurtz, Gregory A. editor.
Summary	This book brings together the recent cutting-edge work on computational methods in photonics and

their applications. The latest advances in techniques such as the Discontinuous Galerkin Time Domain method, Finite Element Time Domain method, Finite Difference Time Domain method as well as their applications are presented. Key aspects such as modelling of non-linear effects (Second Harmonic Generation, lasing in fibers, including gain nonlinearity in metamaterials), the acousto-optic effect, and the hydrodynamic model to explain electron response in nanoplasmonic structures are included. The application areas covered include plasmonics, metamaterials, photonic crystals, dielectric waveguides, fiber lasers. The chapters give a representative survey of the corresponding area.

Contents Introduction to Mid Infrared Fibre Lasers -- Laser Rate and Propagation Equations -- Obtaining Spectroscopic Parameters as Inputs to the Numerical Model -- Models for Fibre Lasers -- Hydrodynamic Model for Nonlinear Plasmonics -- Discussion and Conclusions.

Subjects Electrodynamics.
Electronic materials.
Lasers.
Nanotechnology.
Optical materials.
Optics.
Photonics.
Physics.
Optics, Lasers, Photonics, Optical Devices.
Classical Electrodynamics.
Nanotechnology.
Numerical and Computational Physics, Simulation.
Optical and Electronic Materials.

Notes Description based on publisher-supplied MARC data.

Additional formats Print version: Recent trends in computational photonics. 9783319554372 (DLC) 2017947773
Printed edition: 9783319554372

Printed edition: 9783319554396
Printed edition: 9783319856612
Series Springer Series in Optical Sciences, 0342-4111; 204
Springer Series in Optical Sciences, 0342-4111; 204

Reviews in Plasmonics 2015
LCCN 2015961024
Type of material Book
Personal name D. Geddes, Chris.
Main title Reviews in plasmonics 2015 / Chris D. Geddes.
Published/Produced New York, NY: Springer Berlin Heidelberg, 2016.
Links Publisher description https://www.loc.gov/catdir/enhancements/fy1611/2015961024-d.html
Table of contents only https://www.loc.gov/catdir/enhancements/fy1611/2015961024-t.html
ISBN 9783319246048

Reviews in Plasmonics 2015
LCCN 2019739476
Type of material Book
Main title Reviews in Plasmonics 2015 / edited by Chris D. Geddes.
Edition 1st ed. 2016.
Published/Produced Cham: Springer International Publishing: Imprint: Springer, 2016.
Description 1 online resource (VIII, 453 pages 237 illustrations, 62 illustrations in color.)
text file PDF
ISBN 9783319246062
Related names Geddes, Chris D. editor.
Summary Reviews in Plasmonics 2015, the second volume of the new book series from Springer, serves as a comprehensive collection of current trends and emerging hot topics in the field of Plasmonics and closely related disciplines. It summarizes the year's progress in surface plasmon phenomena and its applications, with authoritative analytical reviews

in sufficient detail to be attractive to professional researchers, yet also appealing to the wider audience of scientists in related disciplines of Plasmonics. Reviews in Plasmonics offers an essential source of reference material for any lab working in the Plasmonics field and related areas. All academics, bench scientists, and industry professionals wishing to take advantage of the latest and greatest in the continuously emerging field of Plasmonics will find it an invaluable resource.

Contents

	Muhammad Saleem, and Ahmad Shuaib.- Plasmon assisted luminescence in rare earth doped glasses, M. Reza Dousti, Raja J. Amjad -- Surface enhanced fluorescence by plasmonic nanostructures, Jun Dong, Hairong Zheng, Zhenglong Zhang, Wei Gao, Jihong Liu, Enjie He -- Remote spectroscopy below the diffraction limit, James A. Hutchison, Hiroshi Uji-i.- Index.
Subjects	Medicine. Nanotechnology. Spectroscopy. Nanoscale science. Nanoscience. Nanostructures. Surfaces (Physics). Interfaces (Physical sciences). Thin films. Metals. Biomedicine, general. Nanotechnology. Spectroscopy/Spectrometry. Nanoscale Science and Technology. Surface and Interface Science, Thin Films. Metallic Materials.
Notes	Description based on publisher-supplied MARC data.
Additional formats	Printed edition: 9783319246048 Printed edition: 9783319246055 Printed edition: 9783319796468
Series	Reviews in Plasmonics, 1555-7731; 2015 Reviews in Plasmonics, 1555-7731; 2015

Reviews in Plasmonics 2016

LCCN	2017930597
Type of material	Book
Personal name	D. Geddes, Chris.
Main title	Reviews in plasmonics 2016 / Chris D. Geddes.
Published/Produced	New York, NY: Springer Berlin Heidelberg, 2017.
ISBN	9783319480800

Reviews in Plasmonics 2016

LCCN	2019748824
Type of material	Book
Main title	Reviews in Plasmonics 2016 / edited by Chris D. Geddes.
Edition	1st ed. 2017.
Published/Produced	Cham: Springer International Publishing: Imprint: Springer, 2017.
Description	1 online resource (VIII, 240 pages 109 illustrations, 92 illustrations in color.) text file PDF
ISBN	9783319480817
Related names	Geddes, Chris D. editor.
Summary	Reviews in Plasmonics 2016, the third volume of the new book series from Springer, serves as a comprehensive collection of current trends and emerging hot topics in the field of Plasmonics and closely related disciplines. It summarizes the year's progress in surface plasmon phenomena and its applications, with authoritative analytical reviews in sufficient detail to be attractive to professional researchers, yet also appealing to the wider audience of scientists in related disciplines of Plasmonics. Reviews in Plasmonics offers an essential source of reference material for any lab working in the Plasmonics field and related areas. All academics, bench scientists, and industry professionals wishing to take advantage of the latest and greatest in the continuously emerging field of Plasmonics will find it an invaluable resource.
Contents	Preface -- Plasmonic Nanowire Waveguide for Deep Subwavelength Confinement, Huizhong Xu -- Plasmon-enhanced fluorescence of rare earth nanocrystals, Ya-Fang Zhang et al. -- Sensing through Surface Plasmon Resonance Technique, R. K. Gupta -- Fractal Plasmonic Nanoantennae, Ravi Sadananda Hegde -- Compact Slow-light Enhanced Plasmonic Waveguide Refractive Index Sensors, Yin Huang, Changjun Min, and Georgios Veronis -

- Fabrication, Properties and Applications of Plasmene Nanosheet, Qianqian Shi, Dashen Dong, Kae Jye Si and Wenlong Cheng -- Experimental observation of melting of the effective Minkowski spacetime in cobalt-based ferrofluids, Igor I. Smolyaninov, Vera N. Smolyaninova -- Surface plasmon resonance based fiber optic sensors utilizing zinc oxide thin films and nanostructures, Banshi D. Gupta and Rana Tabassum -- Frontiers of Light Dynamics in Photonic Crystals, K. Senthilnathan -- LSPR Biosensing: Recent Advances and Approaches, Pranveer Singh -- Index.

Subjects Interfaces (Physical sciences).
Medicine.
Metals.
Nanoscale science.
Nanoscience.
Nanostructures.
Nanotechnology.
Spectroscopy.
Surfaces (Physics).
Thin films.
Biomedicine, general.
Metallic Materials.
Nanoscale Science and Technology.
Nanotechnology.
Spectroscopy/Spectrometry.
Surface and Interface Science, Thin Films.

Notes Description based on publisher-supplied MARC data.

Additional formats Print version: Reviews in plasmonics 2016 9783319480800 (DLC) 2017930597
Printed edition: 9783319480800
Printed edition: 9783319480824
Printed edition: 9783319838946

Series Reviews in Plasmonics, 1555-7731; 2016
Reviews in Plasmonics, 1555-7731; 2016

Silver Nanoparticles: from Silver Halide Photography to Plasmonics

LCCN	2015936925
Type of material	Book
Personal name	Tani, Tadaaki, author.
Main title	Silver nanoparticles: from silver halide photography to plasmonics / Tadaaki Tani.
Edition	First edition.
Published/Produced	Oxford: Oxford University Press, 2015.
Description	ix, 207 pages: illustrations; 26 cm
ISBN	9780198714606 (hbk.) 0198714602 (hbk.)
LC classification	TA480.S5 T36 2015
Summary	This title provides systematic knowledge and ideas on nanoparticles of Ag and related materials. While Ag and metal nanoparticles are essential for plasmonics, silver halide (AgX) photography relies to a great extent on nanoparticles of Ag and AgX which have the same crystal structure and have been studied extensively for many years. This book has been written to combine the knowledge of nanoparticles of Ag and related materials in plasmonics and AgX photography in order to provide new ideas for metal nanoparticles in plasmonics.
Subjects	Silver. Nanoparticles. Nanoparticles. Silver.
Notes	Includes bibliographical references and index.

Solid-State Physics: Introduction to the Theory

LCCN	2019758580
Type of material	Book
Personal name	Patterson, James D, author.
Main title	Solid-State Physics: Introduction to the Theory / by James D. Patterson, Bernard C. Bailey.
Edition	3rd ed. 2018.
Published/Produced	Cham: Springer International Publishing: Imprint: Springer, 2018.

Description 1 online resource (XXV, 954 pages 285 illustrations)
text file PDF
ISBN 9783319753225
Related names Bailey, Bernard C, author.
Summary This book teaches solid state physics in a comprehensive way, covering all areas. It begins with three broad topics: how and why atoms bind together to form solids, lattice vibrations and phonons, and electrons in solids. It then applies this knowledge to interactions, especially those between electrons and phonons, metals, the Fermi surface and alloys, semiconductors, magnetism, superconductivity, dielectrics and ferroelectrics, optical properties, defects, layered materials, quantum Hall effect, mesoscopics, nanophysics and soft condensed matter. Further important topics of the book are the evolution of BEC to BCS phenomena, conducting polymers, graphene, iron pnictide superconductors, light emitting diodes, N-V centers, nanomagnetism, negative index of refraction, optical lattices, phase transitions, phononics, photonics, plasmonics, quantum computing, solar cells, spin Hall effect and spintronics. In this 3rd edition, topics such as topological insulators, quantum computing, Bose-Einstein transitions, highly correlated electron systems and several others have been added. New material on magnetism in solids, as well as a discussion of semiconductors and a changed set of problems with solutions, are also included. The book also discusses "folk theorems" to remind readers of the essence of the physics without mathematics, and includes 90 mini-biographies of prominent solid state physicists of the past and present to put a human face on the subject. An extensive solutions manual rounds out the book.
Contents Crystal Binding and Structure -- Lattice Vibrations and Thermal Properties -- Electrons in Periodic

	Potentials -- The Interaction of Electrons and Lattice Vibrations -- Metals, Alloys, and the Fermi Surface -- Semiconductors -- Magnetism, Magnons, and Magnetic Resonance -- Superconductivity -- Dielectrics and Ferroelectrics -- Optical Properties of Solids -- Defects in Solids -- Current Topics in Solid Condensed-Matter Physics.
Subjects	Electronic materials. Microwaves. Optical engineering. Optical materials. Physics. Semiconductors. Solid state physics. Solid State Physics. Mathematical Methods in Physics. Microwaves, RF and Optical Engineering. Optical and Electronic Materials. Semiconductors.
Notes	Description based on publisher-supplied MARC data.
Additional formats	Print version: Introduction to the theory. 9783319753218 (DLC) 2018932169 Printed edition: 9783319753218 Printed edition: 9783319753232

Technological Advances in Tellurite Glasses: Properties, Processing, and Applications

LCCN	2019769743
Type of material	Book
Main title	Technological Advances in Tellurite Glasses: Properties, Processing, and Applications / edited by V.A.G. Rivera, Danilo Manzani.
Edition	1st ed. 2017.
Published/Produced	Cham: Springer International Publishing: Imprint: Springer, 2017.
Description	1 online resource (IX, 335 pages 169 illustrations, 117 illustrations in color.)

Springer Series in Materials Science, 0933-033X; 254

World Scientific Handbook of Metamaterials and Plasmonics

LCCN	2017029450
Type of material	Book
Main title	World Scientific handbook of metamaterials and plasmonics / editor, Stefan A. Maier, Imperial College London, UK.
Published/Produced	Hackensack, New Jersey: World Scientific, [2018]
Description	4 volumes: illustrations; 24 cm.
ISBN	9789813227613 (set) 9813227613 (set) 9789813227620 (v. 1; hardcover; alk. paper) 9813227621 (v. 1; hardcover; alk. paper) 9789813227637 (v. 2; hardcover; alk. paper) 981322763X (v. 2; hardcover; alk. paper) 9789813227644 (v. 3; hardcover; alk. paper) 9813227648 (v. 3; hardcover; alk. paper) 9789813227651 (v. 4; hardcover; alk. paper) 9813227656 (v. 4; hardcover; alk. paper)
LC classification	TK7871.15.M48 W67 2018
Related names	Maier, Stefan A., editor.
Contents	volume 1. Fundamentals of electromagnetic metamaterials / volume editor, Ekaterina Shamonina, University of Oxford, UK -- volume 2. Elastic, acoustic, and seismic metamaterials / volume editors, Richard Craster, Imperial College London, UK, Sébastien Guenneau, Institut Fresnel, France and Aix-Marseille Université, France -- volume 3. Active nanoplasmonics and metamaterials / volume editor, Ortwin Hess, Imperial College London, UK -- volume 4. Recent progress in the field of nanoplasmonics / edited volume J. Aizpurua, Spanish Council for Scientific Research (CSIC).
Subjects	Metamaterials. Plasmons (Physics)
Notes	Includes bibliographical references and indexes.

Additional formats	Online version: World Scientific handbook of metamaterials properties Singapore; Hackensack, NJ: World Scientific, [2017] 9789813228696 (DLC) 2017035684
Series	World Scientific series in nanoscience and nanotechnology, 2301-301X; volume 16 World Scientific series in nanoscience and nanotechnology; v. 16.

Index

F

G

H

I

K

L

M

N

O

T

W